W9-BBP-267

Selected Titles in This Series

(Continued in the back of this publication)

Applied Analysis

CONTEMPORARY MATHEMATICS

221

Applied Analysis

Proceedings of a Conference on
Applied Analysis
April 19–21, 1996
Baton Rouge, Louisiana

J. Robert Dorroh
Gisèle Ruiz Goldstein
Jerome A. Goldstein
Michael Mudi Tom
Editors

American Mathematical Society
Providence, Rhode Island

1991 *Mathematics Subject Classification.* Primary 34–XX, 35–XX, 47–XX, 35B40, 35Q55, 35Q53, 35Q35, 76B15, 76B25.

Library of Congress Cataloging-in-Publication Data
Applied analysis : proceedings of a conference on applied analysis, April 19–21, 1996, Baton Rouge, Louisiana / J. Robert Dorroh... [et al.], editors.
    p. cm. — (Contemporary mathematics, ISSN 0271-4132 ; 221)
    Papers presented at sessions of the Southeastern Regional Meeting of the American Mathematical Society.
    Includes bibliographical references.
    ISBN 0-8218-0673-4 (acid-free paper)
    1. Mathematical analysis—Congresses. I. Dorroh, J. Robert (James Robert), 1937– .
II. American Mathematical Society. Southeastern Regional Meeting (1996 : Baton Rouge, La.)
III. Series: Contemporary mathematics (American Mathematical Society) ; v. 221.
QA299.6.A66   1999
515—dc21

98-34483
CIP

Contents

Preface

Much of the Southeastern Regional Meeting of the American Mathematical Society (April 19-21, 1996) was devoted to "Applied Analysis", that part of mathematical analysis devoted to problems arising in science. Among the Special Sessions of that meeting were four devoted to Applied Analysis, namely "Control Theory" (organized by Guillermo Ferreyra and Peter Wolenski), "Fluid Dynamics" (organized by Jerry Goldstein and Michael Tom), "Nonlinear Partial Differential Equations" (organized by Bob Dorroh and Gisèle Goldstein), "Transform Theory and Evolution Equations" (organized by Frank Neubrander and Lutz Weis). The present refereed proceedings volume consists of papers based on some of the invited talks in these sessions plus the paper based on Fritz Gesztesy's invited hour address.

Three of the papers (Bona-Luo, Gesztesy-Weikard, and Semenov) are state-of-the-art views of fields of current interest; these papers are simultaneously surveys and research papers. The scope of this thin volume is rather wide, encompassing a lot of new work in partial differential equations and incorporating topics in modelling, control theory, fluid dynamics, dispersive waves, and sharp regularity results, while making use of techniques from functional analysis, geometry, harmonic analysis, and probability. The conference was excellent and contained many inspiring lectures. We hope that this volume captures some of the excitement of the rapid developments taking place in Applied Analysis.

Some of the speakers received some support from an LEQSF Distinguished Speaker Series grant, which is gratefully acknowledged. We are also deeply appreciative of the American Mathematical Society staff (Ed Dunne, Chris Thivierge, et al.) for all of the help they provided with this volume.

J. Robert Dorroh (Baton Rouge, LA)
Gisèle Ruiz Goldstein (Memphis, TN)
Jerome A. Goldstein (Memphis, TN)
Michael Mudi Tom (Baton Rouge, LA)

Contemporary Mathematics
Volume **221**, 1999

Concentration Compactness and the Stability of Solitary-Wave Solutions to Nonlocal Equations

John P. Albert

ABSTRACT. In their proof of the stability of standing-wave solutions of nonlinear Schrödinger equations, Cazenave and Lions used the principle of concentration compactness to characterize the standing waves as solutions of a certain variational problem. In this article we first review the techniques introduced by Cazenave and Lions, and then discuss their application to solitary-wave solutions of nonlocal nonlinear wave equations. As an example of such an application, we include a new result on the stability of solitary-wave solutions of the Kubota-Ko-Dobbs equation for internal waves in a stratified fluid.

1. Introduction

The first mathematical treatment of the problem of stability of solitary waves was published in 1871 by Joseph Boussinesq [**Bou**], who at the time was 29 years old and just beginning a long and distinguished career in mathematical physics. The solitary waves he was concerned with are water waves with readily recognizable hump-like profiles, which are often produced by disturbances in a shallow channel and which can undergo strong interactions and travel long distances without evident change in form.

Boussinesq showed in [**Bou**] that if a water wave propagates along a flat-bottomed channel of undisturbed depth H, and has large wavelength and small amplitude relative to H, then the elevation h of the water surface considered as a function of the coordinate x along the channel and the time t will approximately satisfy the equation

$$h_{tt} - gHh_{xx} - gH(\frac{3}{2H}h^2 + \frac{H^2}{3}h_{xx})_{xx} = 0,$$

where g is the gravitational acceleration. Using this equation he obtained an explicit representation of solitary waves in terms of elementary functions (reproduced below as equation (1.2)). He then proposed to show that solitary waves are stable in the sense that a slight perturbation of a solitary wave will continue to resemble a solitary wave for all time, rather than evolving into some other wave form. Such a result

---

1991 *Mathematics Subject Classification*. Primary 35Q53, 35B35: Secondary 35S99, 35A15.

would go a long way towards explaining why solitary waves are so easily produced
and observed in experiments.

Boussinesq's proof of stability centered upon the quantity

$$\int_{-\infty}^{\infty} \left\{ \left(\frac{dh}{dx}\right)^2 - \frac{3h^3}{H^3} \right\} \, dx,$$

which he called the *moment of instability*. Like a wave's volume $\int_{-\infty}^{\infty} h \, dx$ or its
energy $\int_{-\infty}^{\infty} h^2 \, dx$, its moment of instability does not change from one instant to
the next as the wave evolves. Boussinesq asserted that, within the class of wave
profiles whose energy has a given value, those profiles which correspond to the
greatest moments of instability will differ the most from solitary-wave profiles,
while the minimum value of the moment of instability within this class is attained
at a solitary-wave profile. It follows that solitary waves must be stable, for if a
wave closely resembles a solitary wave at some time, then its moment of instability
must be close to that of a solitary wave; but since the moment of instability does
not change, then the wave must remain close to a solitary wave for all time ([**Bou**],
p. 62).

In support of his assertions, Boussinesq gave an ingenious proof, which is still
worthy of study today, that if energy is held constant then the moment of instability is minimized at a solitary wave. By modern standards, however, his overall
argument for the stability of solitary waves contains some gaps: for example he
does not explain why two functions whose moments of stability are nearly the same
must resemble each other in form.

The first rigorous proof of stability of solitary waves appeared a century later, in
Benjamin's article [**B2**] on solitary-wave solutions of the Korteweg-de Vries equation

$$(1.1) \qquad\qquad u_t + u u_x + u_{xxx} = 0.$$

(Actually the argument in [**B2**] still contained some flaws which were mended by
Bona [**Bo**] a few years later.) Equation (1.1), a model equation for water waves derived in [**KdV**] some twenty years after Boussinesq published his work, has solutions
$u(x, t) = \phi_C(x - Ct)$ which correspond to the solitary waves studied by Boussinesq:
here the wavespeed C can be any positive number, and the wave profile ϕ_C is given
as a function of its argument $\xi = x - Ct$ by

$$(1.2) \qquad\qquad \phi_C(\xi) = \frac{3C}{\cosh^2(\frac{1}{2}\sqrt{C}\xi)}.$$

The quantities referred to by Boussinesq as the "energy" and the "moment of
instability" of a water wave correspond to the functionals

$$(1.3) \qquad\qquad Q(u) = \frac{1}{2} \int_{-\infty}^{\infty} u^2 \, dx$$

and

$$(1.4) \qquad\qquad E(u) = \frac{1}{2} \int_{-\infty}^{\infty} \left[(u_x)^2 - \frac{1}{3}u^3 \right] \, dx,$$

respectively. It is an easy exercise to see that if $u = u(x, t)$ satisfies (1.1) then both
of these functionals are independent of time.

In the setting of equation (1.1), Boussinesq's assertion would be that ϕ_C is a minimizer of the functional E over the set of all admissible functions ψ satisfying $Q(\psi) = Q(\phi_C)$. Benjamin did not prove this in its entirety, but did show that ϕ_C is a local minimizer. More precisely, he showed that if $\psi \in H^1$ is sufficiently close to ϕ_C in H^1 norm, and $Q(\psi) = Q(\phi_C)$, then

$$E(\psi) - E(\phi_C) \geq A \inf_{y \in \mathbf{R}} \|\psi - \phi_C(\cdot + y)\|_{H^1}$$

where A denotes a positive constant which is independent of ψ. From this estimate, together with an elaboration of Boussinesq's original argument, one can deduce the following stability result:

THEOREM 1.1 [**B2,Bo**]. *For every $\epsilon > 0$, there exists $\delta > 0$ such that if*

$$\|u_0 - \phi_C\|_{H^1} < \delta,$$

then the solution $u(x,t)$ of (1.1) with $u(x,0) = u_0$ satisfies

$$\inf_{y \in \mathbf{R}} \|u(x,t) - \phi_C(x + y)\|_{H^1(dx)} < \epsilon$$

for all $t \in \mathbf{R}$.

(In stating this theorem we have tacitly assumed that the initial data u_0 belong to a class for which unique solutions of the initial-value problem for (1.1) exist for all time. We will continue to make this assumption without comment in what follows; for more information on well-posedness of the initial value problem for (1.1) and related equations the reader may consult [**ABFS**].)

Note that in Theorem 1.1 it cannot be concluded that

$$\|u(x,t) - \phi_C(x - Ct)\|_{H^1(dx)} < \epsilon$$

for all time: to see this, it suffices to consider the solution $u(x,t) = \phi_{C_1}(x - C_1t)$, where C_1 is close to, but not equal to C. However, Bona and Soyeur [**BS**] have recently shown that if $\|u_0 - \phi_C\|_1$ is sufficiently small then there exists a function $\gamma(t)$ such that

$$\|u(x,t) - \phi_C(x - \gamma(t))\|_{H^1(dx)} < \epsilon$$

and

$$|\gamma'(t) - C| < \theta\epsilon$$

for all t, where the constant θ is independent of ϵ and u_0.

In [**B2**] Benjamin pointed out that Boussinesq's idea could also be applied to solitary wave solutions of other equations of Hamiltonian form. Others soon took up this suggestion and extended the theory to more general settings: see, e.g., the far-reaching treatments in [**GSS1,GSS2,MS**] (some of whose results were anticipated in the physics literature [**KRZ**]). Recently, Pego and Weinstein introduced a particularly interesting variation of the theory [**PW**], showing that solitary-wave solutions of the KdV equation (1.1) are asymptotically stable: i.e., a small initial perturbation of a solitary wave will give rise to a solution $u(x,t)$ which not only resembles a solitary wave for all time, but in fact tends to the solitary wave as a limiting form as $t \to \infty$. (Boussinesq, by the way, had already noted that this phenomenon appears in experiments, but had attributed it to frictional effects—which are not modeled by the inviscid equation (1.1).) In all these papers, the solitary wave is proved stable by showing that it is a local (as opposed to global) constrained minimizer of a Hamiltonian functional E; this is done by analyzing the functional derivatives of E and the constraint functional Q at the solitary wave.

An alternate approach to proving stability of solitary waves, which does not rely on local analysis, was developed by Cazenave and P. Lions [**C,CL**] using Lions' method of *concentration compactness*. In this approach, instead of starting with a given solitary wave and attempting to prove that it realizes a local minimum of a constrained variational problem, one starts with the constrained variational problem and looks for global minimizers. When the method works, it shows not only that global minimizers exist, but also that every minimizing sequence is relatively compact up to translations (cf. Theorem 2.9 below). An easy corollary is that the set of global minimizers is a stable set for the associated initial value problem, in the sense that a solution which starts near the set will remain near it for all time.

Although the concentration-compactness method for proving stability of solitary waves has the advantage of requiring less detailed analysis than the local methods, it also produces a weaker result in that it only demonstrates stability of a set of minimizing solutions without providing information on the structure of that set, or distinguishing among its different members. Thus, when de Bouard and Saut [**dBS**] used the concentration-compactness method to prove stability of a set of traveling-wave solutions of the Kadomtsev-Petviashvili equation, they noted that their result did not establish the stability of the explicit "lump" solitary-wave solutions of this equation, since it is not known whether the lump solutions are in the stable set. Further, even if the lump solutions were known to be in this set, the possibility remains open that there are other elements in the stable set which do not resemble lump solutions, and that a perturbed lump solution may wander towards these other minimizers.

In this paper we use concentration compactness to prove stability of solitary-wave solutions of equations of the form

$$(1.5) \qquad u_t + (f(u))_x - (Lu)_x = 0,$$

where $f(u)$ is a real-valued function of u, and Lu is a Fourier multiplier operator defined by

$$\widehat{Lu}(k) = m(k)\widehat{u}(k)$$

(here the hats denote Fourier transforms). If $m(k)$, the *symbol* of L, is a polynomial function of k, then L is a differential operator; and in particular is a *local* operator in the sense that if $u = 0$ outside an open subset S of $\mathbf{R}$ then also $Lu = 0$ outside S. On the other hand, in many situations in fluid dynamics and mathematical physics, equations of the above type arise in which $m(k)$ is not a polynomial and hence the operator L is *nonlocal* (see Sections 3 and 4 below for examples).

The functionals

$$(1.6) \qquad Q(u) = \frac{1}{2} \int_{-\infty}^{\infty} u^2 \, dx$$

and

$$(1.7) \qquad E(u) = \int_{-\infty}^{\infty} \left[\frac{1}{2} u L u - F(u) \right] \, dx,$$

where $F'(u) = f(u)$ for $u \in \mathbf{R}$ and $F(0) = 0$, are integrals of motion for (1.5); i.e., if u solves (1.5) then $Q(u)$ and $E(u)$ are independent of t. The concentration-compactness method for proving stability of solitary-wave solutions of (1.5) aims to show that they are minimizers of E subject to the constraint that Q be held constant. In the original version of the method, as it appears for example in the article [**CL**] on nonlinear Schrödinger equations, important use is made of the fact that the operator L appearing in E is a local operator. Nevertheless, we show below (using an idea which has already appeared in [**ABS**]) that if L is not too far from being local then the results of [**CL**] still obtain. Of course, as explained above, the stability result in view is that the set of solitary waves which are solutions of the minimization problem is stable as a set; and to the extent that the structure of this set remains unknown, such a stability result leaves something to be desired.

The remainder of this paper is organized as follows. In Section 2, we review the concentration-compactness method by showing in detail how it is used to prove stability of solitary-wave solutions of the Korteweg-de Vries equation. The proof is broken into a series of short lemmas to make it easy to identify the parts which need to be modified in the nonlocal case. In Section 3, we illustrate how the method may be applied to nonlocal equations by using it to prove the stability of solitary-wave solutions of the Kubota-Ko-Dobbs equation for internal waves in stratified fluids. Generalizations of the result obtained in Section 3 are discussed in the concluding Section 4.

NOTATION. The set of natural numbers $\{1, 2, 3, \dots\}$ and the set of all integers are written $\mathbf{N}$ and $\mathbf{Z}$, respectively. The set of all real numbers is denoted by $\mathbf{R}$, and all integrals will be taken over $\mathbf{R}$ unless otherwise specified. The Fourier transform $\widehat{f}$ of a tempered distribution $f(x)$ on R is defined as

$$\widehat{f}(k) = \int e^{-ikx} f(x) \, dx.$$

For any tempered distribution f on $\mathbf{R}$ and any $s \in \mathbf{R}$, we define

$$\|f\|_s = \left(\int (1 + |k|^2)^{s/2} |\widehat{f}(k)|^2 \, dk \right)^{1/2},$$

and H^s denotes the Sobolev space of all f for which $\|f\|_s$ is finite. The notation $l_2(H^s)$ will be used for the Hilbert space of all sequences $\{g_j\}_{j \in \mathbf{Z}}$ such that $g_j \in H^s$ for each j and $\sum_{j \in \mathbf{Z}} \|g_j\|_s^2 < \infty$. For any measurable function f on $\mathbf{R}$ and any $p \in [1, \infty)$, we define

$$|f|_p = \left(\int |f(x)|^p \, dx \right)^{1/p},$$

and L^p denotes the space of all f for which $|f|_p$ is finite. The space L^∞ is defined as the space of all measurable functions f on $\mathbf{R}$ such that

$$|f|_\infty = \operatorname*{ess\,sup}_{x \in \mathbf{R}} |f(x)|$$

is finite. Finally, if E is a subset of $\mathbf{R}$ then $C_0^\infty(E)$ denotes the space of infinitely differentiable functions with compact support in E.

2. The concentration-compactness method

In this section we illustrate the concentration-compactness method by using it to prove Benjamin and Bona's result (Theorem 1.1 above) on the stability of solitary-wave solutions of the KdV equation (1.1). The ideas in this section are not new, but are rather a selection of arguments adapted from [**CL**] and [**L**] (see also [**C**]).

Let E and Q be as defined in (1.3) and (1.4); it is easy to see using the Sobolev embedding theorem that E and Q define continuous maps from H^1 to $\mathbf{R}$. Fix a positive number C and let ϕ_C be as in (1.2). Let $q = Q(\phi_C)$ and define the real number I_q by

$$I_q = \inf \left\{ \, E(\psi) : \psi \in H^1 \text{ and } Q(\psi) = q \, \right\}.$$

The *set of minimizers* for I_q is

$$G_q = \left\{ \, \psi \in H^1 : E(\psi) = I_q \text{ and } Q(\psi) = q \, \right\},$$

and a *minimizing sequence* for I_q is any sequence $\{f_n\}$ of functions in H^1 satisfying

$$Q(f_n) = q \quad \text{for all } n$$

and

$$\lim_{n \to \infty} E(f_n) = I_q.$$

To each minimizing sequence $\{f_n\}$, we associate a sequence of nondecreasing functions $M_n : [0, \infty) \to [0, q]$ defined by

$$M_n(r) = \sup_{y \in \mathbf{R}} \frac{1}{2} \int_{y-r}^{y+r} |f_n|^2 \, dx.$$

An elementary argument shows that any uniformly bounded sequence of nondecreasing functions on $[0, \infty)$ must have a subsequence which converges pointwise (in fact, uniformly on compact sets) to a nondecreasing limit function on $[0, \infty)$. Hence $\{M_n\}$ has such a subsequence, which we denote again by $\{M_n\}$. Let $M : [0, \infty) \to [0, q]$ be the nondecreasing function to which M_n converges, and define

$$\alpha = \lim_{r \to \infty} M(r);$$

then $0 \leq \alpha \leq q$.

The method of concentration compactness [**L1,L2**], as applied to this situation, consists of two observations. The first is that if $\alpha = q$, then the minimizing sequence $\{f_n\}$ has a subsequence which, when its terms are suitably translated, converges strongly in H^1 to an element of G_q. The second is that certain simple properties of the variational problem imply that α must equal q for *every* minimizing sequence $\{f_n\}$. It follows that not only do minimizers exist in H^1, but every minimizing sequence converges in H^1 norm to the set G_q.

We will now give the details of the method. We begin with a special case of a general result found in [**L1**] and [**Li**].

LEMMA 2.1. *Suppose $B > 0$ and $\delta > 0$ are given. Then there exists $\eta = \eta(B, \delta) > 0$ such that if $f \in H^1$ with $\|f\|_1 \leq B$ and $|f|_3 \geq \delta$, then*

$$\sup_{y \in \mathbf{R}} \int_{y-1/2}^{y+1/2} |f(x)|^3 \, dx \geq \eta.$$

PROOF. We have

$$\sum_{j\in\mathbf{Z}}\int_{j-1/2}^{j+1/2}[f^2+(f')^2]\,dx=\|f\|_1^2\le B^2=\frac{B^2}{|f|_3^3}|f|_3^3=\sum_{j\in\mathbf{Z}}\frac{B^2}{|f|_3^3}\int_{j-1/2}^{j+1/2}|f|^3\,dx.$$

Hence there exists some $j_0\in\mathbf{Z}$ for which

$$\int_{j_0-1/2}^{j_0+1/2}[f^2+(f')^2]\,dx\le\frac{B^2}{|f|_3^3}\int_{j_0-1/2}^{j_0+1/2}|f|^3\,dx.$$

Now by the Sobolev embedding theorem there exists a constant A (independent of f) such that

$$\left(\int_{j_0-1/2}^{j_0+1/2}|f|^3\,dx\right)^{1/3}\le A\left(\int_{j_0-1/2}^{j_0+1/2}[f^2+(f')^2]\,dx\right)^{1/2}.$$

Hence

$$\left(\int_{j_0-1/2}^{j_0+1/2}|f|^3\,dx\right)^{1/3}\le\left(\frac{AB}{|f|_3^{3/2}}\right)\left(\int_{j_0-1/2}^{j_0+1/2}|f|^3\,dx\right)^{1/2},$$

so

$$\int_{j_0-1/2}^{j_0+1/2}|f|^3\,dx\ge\frac{\delta^9}{A^6B^6}.$$

The proof is now concluded by taking η to be the constant on the right-hand side of the last inequality. $\qquad\square$

Next we establish some properties of the variational problem and its minimizing sequences which are independent of the value of α.

LEMMA 2.2. *For all $q_1>0$, one has*

$$-\infty<I_{q_1}<0.$$

PROOF. Choose any function $\psi\in H^1$ such that $Q(\psi)=q_1$ and $\int\psi^3\,dx>0$. For each $\theta>0$, define the function ψ_θ by $\psi_\theta(x)=\sqrt{\theta}\psi(\theta x)$. Then for all θ one has $Q(\psi_\theta)=Q(\psi)=q_1$ and

$$E(\psi_\theta)=\frac{\theta^2}{2}\int(\psi')^2\,dx-\frac{\theta^{1/2}}{6}\int\psi^3\,dx.$$

Hence by taking $\theta=\theta_0$ sufficiently small we get $E(\psi_{\theta_0})<0$, and since $I_{q_1}\le E(\psi_{\theta_0})$ it follows that $I_{q_1}<0$.

Now let ψ denote an arbitrary function in H^1 satisfying $Q(\psi)=q_1$. To prove that $I_{q_1}>-\infty$, it suffices to bound $E(\psi)$ from below by a number which is independent of ψ. Note first that from standard Sobolev embedding and interpolation theorems it follows that

$$\left|\int\psi^3\,dx\right|\le|\psi|_3^3\le A\|\psi\|_{\frac16}^3\le A\|\psi\|_0^{5/2}\|\psi\|_1^{1/2},$$

where A denotes various constants which are independent of ψ. Then Young's inequality gives

$$\left|\int\psi^3\,dx\right|\le\epsilon\|\psi\|_1^2+A_\epsilon\|\psi\|_0^{10/3}$$

8   J. P. ALBERT

where $\epsilon > 0$ is arbitrary and A_ϵ depends on ϵ but not on ψ. Therefore, since $\|\psi\|_0 = |\psi|_2 = (2Q(\psi))^{1/2}$, one has

$$\left| \int \psi^3 \, dx \right| \leq \epsilon \|\psi\|_1^2 + A_{\epsilon,q_1}$$

where again A_{ϵ,q_1} is independent of ψ. Hence

$$E(\psi) = E(\psi) + Q(\psi) - Q(\psi)$$
$$= \frac{1}{2} \int \left[(\psi')^2 + \psi^2 \right] \, dx - \frac{1}{6} \int \psi^3 \, dx - Q(\psi)$$
$$\geq \frac{1}{2}\|\psi\|_1^2 - \frac{\epsilon}{6}\|\psi\|_1^2 - \frac{1}{6} A_{\epsilon,q_1} - q_1.$$

Choosing $\epsilon = \epsilon_0 \leq 3$ then gives the lower bound

$$E(\psi) \geq -\frac{1}{6} A_{\epsilon_0,q_1} - q_1,$$

and so the proof is complete. □

LEMMA 2.3. *If $\{f_n\}$ is a minimizing sequence for I_q, then there exist constants $B > 0$ and $\delta > 0$ such that*
  1. $\|f_n\|_1 \leq B$ *for all n and*
  2. $|f_n|_3 \geq \delta$ *for all sufficiently large n.*

PROOF. To prove statement 1, write

$$\frac{1}{2}\|f_n\|_1^2 = E(f_n) + Q(f_n) + \frac{1}{6} \int f_n^3 \, dx$$
$$\leq \sup_n E(f_n) + q + \frac{A}{6} \|f_n\|_{1/6}^3$$
$$\leq A \left(1 + \|f_n\|_0^{5/2} \|f_n\|_1^{1/2} \right)$$
$$\leq A \left(1 + \|f_n\|_1^{1/2} \right),$$

where again Sobolev embedding and interpolation theorems have been used, and A denotes various constants which are independent of n. Since the square of $\|f_n\|_1$ has now been bounded by a smaller power, the existence of the desired bound B follows.

To prove statement 2, we argue by contradiction: if no such constant δ exists, then

$$\liminf_{n \to \infty} \int (f_n)^3 \, dx \leq 0,$$

so

$$I_q = \lim_{n \to \infty} \left(\frac{1}{2} \int (f_n')^2 \, dx - \frac{1}{6} \int f_n^3 \, dx \right) \geq \liminf_{n \to \infty} \left(-\frac{1}{6} \int f_n^3 \, dx \right) \geq 0,$$

contradicting Lemma 2.2. □

LEMMA 2.4. *For all $q_1, q_2 > 0$, one has*

$$I_{(q_1+q_2)} < I_{q_1} + I_{q_2}.$$

PROOF. First we claim that for all $\theta > 0$ and $q > 0$,

$$I_{\theta q} = \theta^{5/3} I_q.$$

To see this, associate to each function $\psi \in H^1$ the function ψ_θ defined by

$$\psi_\theta(x) = \theta^{2/3} \psi(\theta^{1/3} x).$$

Then

$$Q(\psi_\theta) = \theta Q(\psi),$$

while

$$E(\psi_\theta) = \theta^{5/3} E(\psi).$$

Hence

$$
\begin{aligned}
I_{\theta q} &= \inf \left\{ E(\psi_\theta) : Q(\psi_\theta) = \theta q \right\} \\
&= \inf \left\{ \theta^{5/3} E(\psi) : Q(\psi) = q \right\} \\
&= \theta^{5/3} I_q,
\end{aligned}
$$

as claimed.

Now from the claim and Lemma 2.2 it follows that for all $q_1, q_2 > 0$

$$I_{(q_1+q_2)} = (q_1 + q_2)^{5/3} I_1 < \left(q_1^{5/3} + q_2^{5/3} \right) I_1 = I_{q_1} + I_{q_2}.$$

$\square$

Now let $\{f_n\}$ be any minimizing sequence for I_q, and let $\{M_n(r)\}$, $M(r)$, and α be as defined earlier. (In what follows, we assume for convenience that $\{f_n\}$ has been replaced by a suitable subsequence, so that the corresponding functions $\{M_n\}$ converge pointwise to M.) We proceed to consider the implications for $\{f_n\}$ of each of the three possibilities $\alpha = q$, $0 < \alpha < q$, and $\alpha = 0$. The first of these is called the case of *compactness* by Lions because of the following lemma.

LEMMA 2.5 [**L1**]. *Suppose $\alpha = q$. Then there exists a sequence of real numbers $\{y_1, y_2, y_3, \dots\}$ such that*
1. *for every $z < q$ there exists $r = r(z)$ such that*

$$\frac{1}{2} \int_{y_n-r}^{y_n+r} |f_n|^2 \, dx > z$$

for all sufficiently large n.
2. *the sequence $\{\tilde{f}_n\}$ defined by*

$$\tilde{f}_n(x) = f_n(x + y_n) \quad \text{for } x \in \mathbf{R}$$

has a subsequence which converges in H^1 norm to a function $g \in G_q$. In particular, G_q is nonempty.

PROOF. Since $\alpha = q$, then there exists r_0 such that for all sufficiently large values of n we have

$$M_n(r_0) = \sup_{y \in \mathbf{R}} \frac{1}{2} \int_{y-r_0}^{y+r_0} |f_n|^2 \, dx > q/2.$$

Hence for each sufficiently large n we can find y_n such that

$$\frac{1}{2} \int_{y_n-r_0}^{y_n+r_0} |f_n|^2 \, dx > q/2.$$

Now let $z < q$ be given; clearly we may assume $z > q/2$. Again, since $\alpha = q$ then we can find $r_0(z)$ and $N(z)$ such that if $n \geq N(z)$ then

$$\frac{1}{2} \int_{y_n(z)-r_0(z)}^{y_n(z)+r_0(z)} |f_n|^2 \, dx > z$$

for some $y_n(z) \in \mathbf{R}$. Since $\frac{1}{2} \int_{\mathbf{R}} |f_n|^2 \, dx = q$, it follows that for large n the intervals $[y_n - r_0, y_n + r_0]$ and $[y_n(z) - r_0(z), y_n(z) + r_0(z)]$ must overlap. Therefore, defining $r = r(z) = 2r_0(z) + r_0$, we have that $[y_n - r, y_n + r]$ contains $[y_n(z) - r_0(z), y_n(z) + r_0(z)]$, and statement 1 then follows.

Now statement 1 implies that for every $k \in \mathbf{N}$, there exists $r_k \in \mathbf{R}$ such that for all sufficiently large n,

$$\frac{1}{2} \int_{-r_k}^{r_k} |\tilde{f}_n|^2 \, dx > q - \frac{1}{k}.$$

By Lemma 2.3.1, the sequence $\{\tilde{f}_n\}$ is uniformly bounded in H^1, and hence from the compactness of the embedding of $H^1(\Omega)$ into $L^2(\Omega)$ on bounded intervals Ω it follows that some subsequence of $\{\tilde{f}_n\}$ converges in $L^2[-r_k, r_k]$ norm to a limit function $g \in L^2[-r_k, r_k]$ satisfying

$$\frac{1}{2} \int_{-r_k}^{r_k} |g|^2 \, dx \geq q - \frac{1}{k}.$$

A Cantor diagonalization argument, together with the fact that $\frac{1}{2} \int_{\mathbf{R}} |\tilde{f}_n|^2 \, dx = q$ for all n, then shows that some subsequence of $\{\tilde{f}_n\}$ converges in $L^2(\mathbf{R})$ norm to a function $g \in L^2(\mathbf{R})$ satisfying $\frac{1}{2} \int_{\mathbf{R}} |g|^2 \, dx = q$. Again using Lemma 2.3.1, we have

$$|\tilde{f}_n - g|_3 \leq A\|\tilde{f}_n - g\|_{1/6} \leq A\|\tilde{f}_n - g\|_1^{1/6} \|\tilde{f}_n - g\|_0^{5/6} \leq A|\tilde{f}_n - g|_2^{5/6},$$

where A is independent of n; so $\tilde{f}_n \to g$ in L^3 norm also. Furthermore, by the weak compactness of the unit sphere and the weak lower semicontinuity of the norm in Hilbert space, we know that $\tilde{f}_n$ converges weakly to g in H^1, and that

$$\|g\|_1 \leq \liminf_{n \to \infty} \|\tilde{f}_n\|_1.$$

It follows that

$$E(g) \leq \lim_{n \to \infty} E(\tilde{f}_n) = I_q,$$

whence $E(g) = I_q$ and $g \in G_q$. Finally, $E(g) = \lim_{n \to \infty} E(\tilde{f}_n)$, $|g|_3 = \lim_{n \to \infty} |\tilde{f}_n|_3$, and $|g|_2 = \lim_{n \to \infty} |\tilde{f}_n|_2$ together imply that $\|g\|_1 = \lim_{n \to \infty} \|\tilde{f}_n\|_1$, and from an elementary exercise in Hilbert space theory it then follows that $\tilde{f}_n$ converges to g in H^1 norm. □

The next lemma is used to describe the behavior of $\{f_n\}$ in the case $0 < \alpha < q$.

LEMMA 2.6. *For every $\epsilon > 0$, there exist a number $N \in \mathbf{N}$ and sequences $\{g_N, g_{N+1}, \ldots\}$ and $\{h_N, h_{N+1}, \ldots\}$ of H^1 functions such that for every $n \geq N$,*
1. $|Q(g_n) - \alpha| < \epsilon$.
2. $|Q(h_n) - (q - \alpha)| < \epsilon$.
3. $E(f_n) \geq E(g_n) + E(h_n) - \epsilon$.

PROOF. Choose $\phi \in C_0^\infty[-2, 2]$ such that $\phi \equiv 1$ on $[-1, 1]$, and let $\psi \in C^\infty(\mathbf{R})$ be such that $\phi^2 + \psi^2 \equiv 1$ on $\mathbf{R}$. For each $r \in \mathbf{R}$, define $\phi_r(x) = \phi(x/r)$ and $\psi_r(x) = \psi(x/r)$.

For all sufficiently large values of r we have

$$\alpha - \epsilon < M(r) \leq M(2r) \leq \alpha.$$

Assume for the moment that such a value of r has been chosen. Then we can choose N so large that

$$\alpha - \epsilon < M_n(r) \leq M_n(2r) < \alpha + \epsilon$$

for all $n \geq N$. Hence for each $n \geq N$ we can find y_n such that

(2.1)
$$\frac{1}{2} \int_{y_n - r}^{y_n + r} |f_n|^2 \, dx > \alpha - \epsilon$$

and

(2.2)
$$\frac{1}{2} \int_{y_n - 2r}^{y_n + 2r} |f_n|^2 \, dx < \alpha + \epsilon.$$

Define $g_n(x) = \phi_r(x - y_n) f_n(x)$ and $h_n(x) = \psi_r(x - y_n) f_n(x)$. Then clearly statements 1 and 2 are satisfied by g_n and h_n.

To prove statement 3, note that

$$
\begin{aligned}
E(g_n) + E(h_n) &= \\
&= \frac{1}{2} \left[\int \phi_r^2 (f_n')^2 \, dx + 2 \int \phi_r \phi_r' f_n f_n' \, dx + \int (\phi_r')^2 f_n^2 \, dx \right] \\
&\quad + \frac{1}{2} \left[\int \psi_r^2 (f_n')^2 \, dx + 2 \int \psi_r \psi_r' f_n f_n' \, dx + \int (\psi_r')^2 f_n^2 \, dx \right] \\
&\quad - \frac{1}{6} \int \phi_r^2 f_n^3 \, dx - \frac{1}{6} \int \psi_r^2 f_n^3 \, dx \\
&\quad + \frac{1}{6} \int (\phi_r^2 - \phi_r^3) f_n^3 \, dx + \frac{1}{6} \int (\psi_r^2 - \psi_r^3) f_n^3 \, dx,
\end{aligned}
$$

where for brevity we have written simply ϕ_r and ψ_r for the functions $\phi_r(x - y_n)$ and $\psi_r(x - y_n)$. Now $\phi_r^2 + \psi_r^2 \equiv 1$, $|(\phi_r)'|_\infty = |\phi'|_\infty / r$, and $|(\psi_r)'|_\infty = |\psi'|_\infty / r$. Therefore, making use of Hölder's Inequality and Lemma 2.3.1, one can rewrite the preceding equation in the form

$$E(g_n) + E(h_n) = E(f_n) + O(1/r) + \frac{1}{6} \int \left[(\phi_r^2 - \phi_r^3) + (\psi_r^2 - \psi_r^3) \right] f_n^3 \, dx,$$

where $O(1/r)$ signifies a term bounded in absolute value by A_1/r with A_1 independent of r and n. But using (2.1) and (2.2) we obtain

$$\left| \int \left[(\phi_r^2 - \phi_r^3) + (\psi_r^2 - \psi_r^3) \right] f_n^3 \, dx \right| \leq \left(\int_{r \leq |x - y_n| \leq 2r} 2|f_n|^2 \, dx \right) \cdot |f_n|_\infty \leq A_2 \epsilon,$$

where again A_2 is independent of r and n.

It is now time to choose r, and we make the choice so large that the $O(1/r)$ term in the preceding paragraph is less than ϵ in absolute value. For the corresponding choices of sequences $\{g_n\}$ and $\{h_n\}$, statements 1 and 2 hold together with

(2.3) $E(f_n) \geq E(g_n) + E(h_n) - (A_2 + 1)\epsilon$

for all $n \geq N(r)$. Finally, we may return to the beginning of the proof and there replace ϵ by $\min(\epsilon, \epsilon/(A_2 + 1))$, thus transforming (2.3) into statement 3 (without affecting statements 1 and 2).   □

COROLLARY 2.7. *If* $0 < \alpha < q$ *then*

$$I_q \geq I_\alpha + I_{q-\alpha}.$$

PROOF. First observe that if g is a function such that $|Q(g) - \alpha| < \epsilon$, then $Q(\beta g) = \alpha$, where $\beta = \sqrt{\alpha/Q(g)}$ satisfies $|\beta - 1| < A_1 \epsilon$ with A_1 independent of g and ϵ. Hence

$$I_\alpha \leq E(\beta g) \leq E(g) + A_2 \epsilon,$$

where A_2 depends only on A_1 and $\|g\|_1$. A similar result holds for functions h such that $|Q(h) - (q - \alpha)| < \epsilon$.

From these observations and Lemma 2.6 it follows easily that there exists a subsequence $\{f_{n_k}\}$ of $\{f_n\}$ and corresponding functions g_{n_k} and h_{n_k} such that for all k,

$$E(g_{n_k}) \geq I_\alpha - \frac{1}{k},$$

$$E(h_{n_k}) \geq I_{q-\alpha} - \frac{1}{k}, \quad \text{and}$$

$$E(f_{n_k}) \geq E(g_{n_k}) + E(h_{n_k}) - \frac{1}{k}.$$

Hence

$$E(f_{n_k}) \geq I_\alpha + I_{q-\alpha} - \frac{3}{k}.$$

The desired result is now obtained by taking the limit of both sides as $k \to \infty$. □

REMARK. Because of Lemma 2.6 and its corollary, Lions calls $0 < \alpha < q$ the case of *dichotomy*: each of the minimizing functions f_n can be split into two summands which carry fixed proportions of the constraint functional Q, and which are sufficiently separated spatially that the sum of the values of E at each summand does not exceed $E(f_n)$. In fact, as explained in the remark following Theorem 2.9 below, for a general class of variational problems of the type considered here, the inequality $I_q \leq I_\beta + I_{q-\beta}$ holds for all $\beta \in (0, q)$, so that in the case of dichotomy one would have from Corollary 2.7 that

$$I_q = I_\alpha + I_{q-\alpha}.$$

Thus the two sequences of summands will themselves be minimizing sequences for I_α and $I_{q-\alpha}$ respectively.

Our final lemma shows that the possibility $\alpha = 0$ (called the case of *vanishing* by Lions) does not occur here.

LEMMA 2.8. *For every minimizing sequence, $\alpha > 0$.*

PROOF. From Lemmas 2.1 and 2.3 we conclude that there exists $\eta > 0$ and a sequence $\{y_n\}$ of real numbers such that

$$\int_{y_n-1/2}^{y_n+1/2} |f_n|^3 \, dx \geq \eta$$

for all n. Hence

$$\eta \leq |f_n|_\infty \left(\int_{y_n-1/2}^{y_n+1/2} |f_n|^2 \, dx \right) \leq AB \left(\int_{y_n-1/2}^{y_n+1/2} |f_n|^2 \, dx \right),$$

where A is the Sobolev constant in the embedding of L^∞ into H^1. It follows that

$$\alpha = \lim_{r\to\infty} M(r) \geq M(1/2) = \lim_{n\to\infty} M_n(1/2) \geq \frac{2\eta}{AB} > 0. \qquad \square$$

THEOREM 2.9. *The set G_q is not empty. Moreover, if $\{f_n\}$ is any minimizing sequence for I_q, then*
   1. *there exists a sequence $\{y_1, y_2, \ldots\}$ and an element $g \in G_q$ such that $f_n(\cdot + y_n)$ has a subsequence converging strongly in H^1 to g.*
   2.
$$\lim_{n\to\infty} \inf_{\substack{g\in G_q \\ y\in\mathbf{R}}} \|f_n(\cdot + y) - g\|_1 = 0.$$
   3.
$$\lim_{n\to\infty} \inf_{g\in G_q} \|f_n - g\|_1 = 0.$$

PROOF. From Lemmas 2.4, 2.7 and 2.8 it follows that $\alpha = q$. Hence by Lemma 2.5 the set G_q is nonempty and statement 1 of the present lemma holds.

Now suppose that statement 2 does not hold; then there exists a subsequence $\{f_{n_k}\}$ of $\{f_n\}$ and a number $\epsilon > 0$ such that

$$\inf_{\substack{g\in G_q \\ y\in\mathbf{R}}} \|f_{n_k}(\cdot + y) - g\|_1 \geq \epsilon$$

for all $k \in \mathbf{N}$. But since $\{f_{n_k}\}$ is itself a minimizing sequence for I_q, from statement 1 it follows that there exist a sequence $\{y_k\}$ and $g_0 \in G_q$ such that

$$\liminf_{k\to\infty} \|f_{n_k}(\cdot + y_k) - g_0\|_1 = 0.$$

This contradiction proves statement 2.

Finally, since the functionals E and Q are invariant under translations, then G_q clearly contains any translate of g if it contains g, and hence statement 3 follows immediately from statement 2. $\qquad \square$

REMARK. For arbitrary functionals E and Q defined on a function space X, and I_q and G_q defined as above, the preceding arguments show that under general conditions the subadditivity property $I_q < I_\beta + I_{q-\beta}$ (for all $\beta \in (0, q)$) is sufficient to imply the relative compactness (modulo translations) of all minimizing sequences for I_q. Lions [**L1**] has also given a heuristic argument, which we now paraphrase here, for the necessity of this property. Let $\beta \in (0, q)$ be given, and let $\{g_n\}$ and $\{h_n\}$ be minimizing sequences for I_β, $I_{q-\beta}$ respectively. Define $\tilde{h}_n = h_n(\cdot + y_n)$ where y_n is chosen so that the distance between the supports of g_n and $\tilde{h}_n$ tends to infinity with n. If E and Q are local operators, or at least not too nonlocal, then

$$E(g_n + \tilde{h}_n) \sim E(g_n) + E(\tilde{h}_n) \to I_\beta + I_{q-\beta}$$

and

$$Q(g_n + \tilde{h}_n) \sim Q(g_n) + Q(\tilde{h}_n) \to q,$$

as $n \to \infty$, so that

$$I_\beta + I_{q-\beta} = \lim_{n \to \infty} E(g_n + \tilde{h}_n) \geq I_q.$$

Now if $I_q = I_\beta + I_{q-\beta}$ then $g_n + \tilde{h}_n$ is a minimizing sequence for I_q; but $g_n + \tilde{h}_n$ cannot have a strongly convergent subsequence (even after translations) in any of the usual function spaces. (For example, it is an easy exercise to show that if $\frac{1}{2} \int g_n^2 = \beta > 0$ and $\frac{1}{2} \int \tilde{h}_n^2 = q - \beta > 0$ for all n, and the distance between the supports of g_n and $\tilde{h}_n$ tends to infinity with n, then there does not exist any sequence $\{y_n\}$ of real numbers such that $\{g_n(\cdot + y_n) + \tilde{h}_n(\cdot + y_n)\}$ has a convergent subsequence in L^2.) Hence if all minimizing sequences for I_q are relatively compact (modulo translations), then we must have $I_q < I_\beta + I_{q-\beta}$ for all $\beta \in (0, q)$.

An immediate consequence of Theorem 2.9 is that G_q forms a stable set for the initial-value problem for (1.1).

COROLLARY 2.10. *For every $\epsilon > 0$, there exists $\delta > 0$ such that if*

$$\inf_{g \in G_q} \|u_0 - g\|_1 < \delta,$$

then the solution $u(x, t)$ of (1.1) with $u(x, 0) = u_0$ satisfies

$$\inf_{g \in G_q} \|u(\cdot, t) - g\|_1 < \epsilon$$

for all $t \in \mathbf{R}$.

PROOF. Suppose the theorem to be false; then there exist a number $\epsilon > 0$, a sequence $\{\psi_n\}$ of functions in H^1, and a sequence of times $\{t_n\}$ such that

$$\inf_{g \in G_q} \|\psi_n - g\|_1 < \frac{1}{n}$$

and

$$\inf_{g \in G_q} \|u_n(\cdot, t_n) - g\|_1 \geq \epsilon$$

for all n, where $u_n(x, t)$ solves (1.1) with $u_n(x, 0) = \psi_n$. Then since $\psi_n \to G_q$ in H^1, and $E(g) = I_q$ and $Q(g) = q$ for $g \in G_q$, we have $E(\psi_n) \to I_q$ and $Q(\psi_n) \to q$. Choose $\{\alpha_n\}$ such that $Q(\alpha_n \psi_n) = q$ for all n; thus $\alpha_n \to 1$. Hence the sequence $f_n = \alpha_n u_n(\cdot, t_n)$ satisfies $Q(f_n) = q$ and

$$\lim_{n \to \infty} E(f_n) = \lim_{n \to \infty} E(u_n(\cdot, t_n)) = \lim_{n \to \infty} E(\psi_n) = I_q,$$

and is therefore a minimizing sequence for I_q. From Theorem 2.9.3 it follows that for all n sufficiently large there exists $g_n \in G_q$ such that $\|f_n - g_n\|_1 < \frac{\epsilon}{2}$. But then

$$\epsilon \le \|u_n(\cdot, t_n) - g_n\|_1 \le \|u_n(\cdot, t_n) - f_n\|_1 + \|f_n - g_n\|_1 \le |1 - \alpha_n| \cdot \|u_n(\cdot, t_n)\|_1 + \frac{\epsilon}{2},$$

and taking $n \to \infty$ gives $\epsilon \le \frac{\epsilon}{2}$, a contradiction. $\qquad\square$

Note that the stability result in Corollary 2.10 is weaker than that of Theorem 1.1; this is an instance of the general fact, discussed above in Section 1, that the concentration-compactness method by itself only proves stability with respect to an unspecified set of minimizers. If for example the set G_q contained two functions g_1 or g_2 which were not translates of each other, then not all minimizing sequences for I_q would converge modulo translations; since a minimizing sequence could contain two subsequences tending to g_1 and g_2 respectively. Of course if the elements of G_q were isolated points (modulo translations) in function space, then the argument used to prove Corollary 2.10 would show that each such point constitutes a stable set in itself; but nonisolated minimizers contained in G_q could fail to be individually stable.

In the present case, however, it is easy to see that G_q contains but a single function (modulo translations), and that this function is indeed a solitary-wave solution of (1.1).

PROPOSITION 2.11. *If G_q is not empty then*

$$G_q = \{\phi_C(\cdot + x_0) : x_0 \in \mathbf{R}\}.$$

PROOF. If $g(x) \in G_q$ then by the Lagrange multiplier principle (see, e.g., Theorem 7.7.2 of [**Lu**]), there exists $\lambda \in \mathbf{R}$ such that

(2.4) $$\delta E(g) + \lambda\, \delta Q(g) = 0,$$

where $\delta E(g)$ and $\delta Q(g)$ are the Fréchet derivatives of E and Q at g. Now δE and δQ are given (as distributions in H^{-1}) by

$$\delta E(g) = -g'' - \frac{1}{2}g^2,$$
$$\delta Q(g) = g;$$

and therefore (2.4) is an ordinary differential equation in g. A bootstrap argument shows that any L^2 distribution solution of (2.4) must be smooth, and an elementary argument shows that the only smooth solutions of (2.4) in L^2 are the functions $\phi_\lambda(x + x_0)$, where x_0 is arbitrary and ϕ_λ is defined by (1.2) with C replaced by λ. But it is easily seen that $Q(\phi_\lambda) = q = Q(\phi_C)$ if and only if $\lambda = C$. Hence $g(x) = \phi_C(x + x_0)$. This proves that $G_q \subset \{\phi_C(\cdot + x_0) : x_0 \in \mathbf{R}\}$, and the reverse inclusion follows (if G_q is not empty) from the translation invariance of E and Q. $\quad\square$

Combining Theorem 2.9 and its corollary with Proposition 2.11, we see that Theorem 1.1 has now been completely proved.

3. An application to a nonlocal equation

In this section we illustrate the use of the concentration-compactness method for nonlocal equations by proving the stability of solitary-wave solutions of the equation

$$(3.1) \qquad u_t + uu_x - (Lu)_x = 0,$$

derived by Kubota, Ko and Dobbs [KKD] as a model for long internal waves in a stratified fluid. Here L is the Fourier multiplier operator defined by

$$\widehat{Lw}(k) = m(k)\hat{w}(k),$$

where

$$m(k) = \beta_1 \left(k \coth(kH_1) - \frac{1}{H_1} \right) + \beta_2 \left(k \coth(kH_2) - \frac{1}{H_2} \right)$$

and β_1, β_2, H_1, H_2 are positive constants. By a *solitary-wave solution* of (3.1) we mean a solution of the form $u_S(x,t) = g(x - Ct)$, where $C \in \mathbf{R}$; but often we will abuse terminology slightly and use the term solitary wave to refer to the profile function g corresponding to such a solution. Here "solution" means "classical solution"; it is no gain of generality to consider distribution solutions, since it turns out that if $g \in L^2$ and u_S satisfies (3.1) in the sense of distributions, then g must in fact be infinitely differentiable (see Lemma 3.3 of [ABS]).

When $H_1 = H_2$ (or when one of β_1 or β_2 is zero), equation (3.1) has the structure of a completely integrable Hamiltonian system [KSA,LR,R] and explicit multisoliton solutions are known [J,JE]. In particular, the solitary-wave solutions of (3.1) in this case are given by $u(x,t) = \phi_{C,H}(x + x_0 - Ct)$, where $C > 0$ and $x_0 \in \mathbf{R}$ are arbitrary, H denotes the common value of H_1 and H_2, and

$$\phi_{C,H}(\xi) = \left[\frac{2a(\beta_1 + \beta_2)\sin aH}{\cosh a\xi + \cos aH} \right],$$

with $a \in (0, \pi/H)$ determined by the equation

$$aH \cot aH = 1 - \frac{CH}{\beta_1 + \beta_2}.$$

The stability of these solitary waves was studied in [AB], where it was shown that a result similar to Theorem 1.1 holds for $\phi_{C\,H}$ for all positive values of C and H.

In the general case when H_1 is not equal to H_2, equation (3.1) does not appear to be completely integrable (cf. [BD]), nor is there any known explicit formula for solitary waves. However, in [ABS] it is shown that solitary-wave solutions do exist for all positive values of β_1, β_2, H_1 and H_2 and all positive wavespeeds C. Moreover, when H_1 is near H_2 one has the following stability result:

THEOREM 3.1 [ABS]. *Let β_1, β_2, H, C be arbitrary positive numbers. Then there exists $\eta > 0$ such that if $H_1 = H$ and $|H_2 - H_1| < \eta$ then (3.1) has a solution $u_S(x,t) = \phi(x - Ct)$ which is stable in the following sense: for every $\epsilon > 0$, there exists $\delta > 0$ such that if*

$$\inf_{y \in \mathbf{R}} \|u_0 - \phi(\cdot + y)\|_{\frac{1}{2}} < \delta,$$

then the solution $u(x,t)$ of (3.1) with $u(x,0) = u_0$ satisfies

$$\inf_{y \in \mathbf{R}} \|u(\cdot,t) - \phi(\cdot + y)\|_{\frac{1}{2}} < \epsilon$$

for all $t \in \mathbf{R}$.

This result leaves open the question of whether stable solitary-wave solutions of (3.1) exist when H_1 is not close to H_2. The following theorem answers this question in the affirmative, albeit with possibly a broader interpretation of stability than that given in Theorem 3.1.

THEOREM 3.2. *Let β_1, β_2, H_1, H_2 be arbitrary positive numbers. For each $q > 0$ there exists a nonempty set G_q, consisting of solitary-wave solutions g of (3.1) having positive wavespeeds and satisfying $Q(g) = q$, which is stable in the following sense: for every $\epsilon > 0$ there exists $\delta > 0$ such that if*

$$\inf_{g \in G_q} \|u_0 - g\|_{\frac{1}{2}} < \delta,$$

then the solution $u(x,t)$ of (3.1) with $u(x,0) = u_0$ satisfies

$$\inf_{g \in G_q} \|u(\cdot,t) - g\|_{\frac{1}{2}} < \epsilon$$

for all $t \in \mathbf{R}$.

We will prove Theorem 3.2 following the same steps as used in the preceding section to prove Theorem 2.9 and its corollary. First define functionals Q and E on $H^{1/2}$ by

$$Q(u) = \frac{1}{2} \int_{-\infty}^{\infty} u^2 \, dx$$

and

$$E(u) = \frac{1}{2} \int_{-\infty}^{\infty} \left[uLu - \frac{1}{3}u^3 \right] dx,$$

and give I_q and G_q the same definitions as in Section 2. To each minimizing sequence for I_q we associate a number $\alpha \in [0, q]$ using the same procedure as in Section 2.

Analogues of Lemmas 2.1 through 2.4 are as follows.

LEMMA 3.3. *Suppose $B > 0$ and $\delta > 0$ are given. Then there exists $\eta = \eta(B, \delta) > 0$ such that if $f \in H^{1/2}$ with $\|f\|_{1/2} \leq B$ and $|f|_3 \geq \delta$, then*

$$\sup_{y \in \mathbf{R}} \int_{y-2}^{y+2} |f(x)|^3 \, dx \geq \eta.$$

PROOF. The proof of this lemma is contained in the proofs of Lemmas 3.7, 3.8, and 3.9 of [**ABS**]; but for completeness we recall the proof here. Choose a smooth function $\zeta : \mathbf{R} \to [0,1]$ with support in $[-2,2]$ and satisfying $\sum_{j \in \mathbf{Z}} \zeta(x - j) = 1$ for all $x \in \mathbf{R}$; and define $\zeta_j(x) = \zeta(x - j)$ for $j \in \mathbf{Z}$. The map $T : H^s \to l_2(H^s)$ defined by

$$Tf = \{\zeta_j f\}_{j \in \mathbf{Z}}$$

is clearly bounded for $s = 0$ and $s = 1$, and hence by interpolation ([**BL**], Section 5.6) is also bounded for $s = 1/2$; that is, there exists $A_0 > 0$ such that for all $f \in H^{1/2}$,

$$\sum_{j \in \mathbf{Z}} \|\zeta_j f\|_{\frac{1}{2}}^2 \leq A_0 \|f\|_{\frac{1}{2}}^2.$$

Now let A_1 be a positive number such that $\sum_{j\in\mathbf{Z}}|\zeta(x-j)|^3 \geq A_1$ for all $x \in \mathbf{R}$. We claim that for every function $f \in H^{1/2}$ which is not identically zero, there exists an integer j_0 such that

$$\|\zeta_{j_0}f\|_{\frac{1}{2}}^2 \leq \left(1 + A_2|f|_3^{-3}\right)|\zeta_{j_0}f|_3^3,$$

where $A_2 = A_0 B^2/A_1$. To see this, assume to the contrary that

$$\|\zeta_j f\|_{\frac{1}{2}}^2 > \left(1 + A_2|f|_3^{-3}\right)|\zeta_j f|_3^3$$

holds for every $j \in \mathbf{Z}$. After summing over j we obtain

$$A_0\|f\|_{\frac{1}{2}}^2 > \left(1 + A_2|f|_3^{-3}\right)\sum_{j\in\mathbf{Z}}|\zeta_j f|_3^3,$$

and hence

$$A_0 B^2 > \left(1 + A_2|f|_3^{-3}\right)A_1|f|_3^3 = A_1|f|_3^3 + A_0 B^2,$$

which is a contradiction.

Finally, observe that from the claim just proved and the assumptions of the lemma it follows that

$$\|\zeta_{j_0}f\|_{\frac{1}{2}}^2 \leq \left(1 + A_2/\delta^3\right)|\zeta_{j_0}f|_3^3,$$

whereas by the Sobolev embedding theorem one has

$$|\zeta_{j_0}f|_3 \leq A_3\|\zeta_{j_0}f\|_{\frac{1}{2}},$$

with a constant A_3 that is independent of f. Hence

$$|\zeta_{j_0}f|_3 \geq \left[A_3^2(1 + A_2/\delta^3)\right]^{-1},$$

and since

$$\int_{j_0-2}^{j_0+2} |f_j|^3\, dx \geq |\zeta_{j_0}f|_3^3,$$

the statement of the lemma follows immediately with $\eta = \left[A_3^2(1 + A_2/\delta^3)\right]^{-3}$. $\square$

LEMMA 3.4. *For all $q_1 > 0$, one has*

$$-\infty < I_{q_1} < 0.$$

PROOF. Choose $\psi \in H^{1/2}$ such that $Q(\psi) = q_1$ and $\int \psi^3\, dx > 0$, and for each $\theta > 0$, define ψ_θ by $\psi_\theta(x) = \sqrt{\theta}\psi(\theta x)$. Observe that $0 \leq m(k) \leq (\beta_1 + \beta_2)|k|$ for all $k \in \mathbf{R}$. Hence, using Parseval's identity and taking into account the action of dilation on Fourier transforms, we have

$$\int \psi_\theta L(\psi_\theta)\, dx = \frac{1}{\theta}\int m(k)|\widehat{\psi}(k/\theta)|^2\, dk$$

$$= \int m(\theta k)|\widehat{\psi}(k)|^2\, dk \leq (\beta_1 + \beta_2)\theta\|\psi\|_{\frac{1}{2}}^2.$$

Therefore

$$E(\psi_\theta) \leq \frac{(\beta_1 + \beta_2)\theta}{2}\|\psi\|_{\frac{1}{2}}^2 - \frac{\theta^{1/2}}{6}\int \psi^3\, dx,$$

and so for $\theta = \theta_0$ sufficiently small one has $E(\psi_{\theta_0}) < 0$. Since $Q(\psi_{\theta_0}) = q_1$ it follows that $I_{q_1} < 0$.

To show that $I_{q_1} > -\infty$, we proceed as in the proof of Lemma 2.2, except that here we use the estimates

$$\left| \int \psi^3 \, dx \right| \leq A \|\psi\|_{\frac{1}{6}}^3 \leq A \|\psi\|_0^2 \|\psi\|_{\frac{1}{2}} \leq \epsilon \|\psi\|_{\frac{1}{2}}^2 + A_\epsilon \|\psi\|_0^4,$$

valid for any $\epsilon > 0$ with A_ϵ depending only on ϵ, and

$$E(\psi) = E(\psi) + Q(\psi) - Q(\psi)$$

$$= \frac{1}{2} \int \left[(\psi L \psi + \psi^2] \, dx - \frac{1}{6} \int \psi^3 \, dx - Q(\psi) \right.$$

$$\geq \frac{A}{2} \|\psi\|_{\frac{1}{2}}^2 - \frac{\epsilon}{6} \|\psi\|_1^2 - \frac{1}{6} A_{\epsilon,q_1} - q_1,$$

where A_{ϵ,q_1} depends only on ϵ and q_1, and A is chosen so that $1 + m(k) \geq A|k|$ for all $k \in \mathbf{R}$. □

LEMMA 3.5. *If $\{f_n\}$ is a minimizing sequence for I_q, then there exist constants $B > 0$ and $\delta > 0$ such that*
1. *$\|f_n\|_{\frac{1}{2}} \leq B$ for all n and*
2. *$|f_n|_3 \geq \delta$ for all sufficiently large n.*

PROOF. Choosing A such that $1 + m(k) \geq A|k|$ for all $k \in \mathbf{R}$, we have by Parseval's inequality and Sobolev embedding and interpolation theorems that

$$\frac{A}{2} \|f_n\|_{\frac{1}{2}}^2 \leq E(f_n) + Q(f_n) + \frac{1}{6} \int f_n^3 \, dx$$

$$\leq \sup_n E(f_n) + q + \frac{A}{6} \|f_n\|_{1/6}^3$$

$$\leq A \left(1 + \|f_n\|_0^2 \|f_n\|_{\frac{1}{2}} \right)$$

$$\leq A \left(1 + \|f_n\|_{\frac{1}{2}} \right).$$

From here the proof is the same as the proof of Lemma 2.3. □

LEMMA 3.6. *For all $q_1, q_2 > 0$, one has*

$$I_{(q_1+q_2)} < I_{q_1} + I_{q_2}.$$

PROOF. Since $m(k)$ is not a homogeneous function of k, we cannot use the argument in the proof of Lemma 2.4. Instead, we use an argument from [**L2**], pp. 228-229. First we claim that for $\theta > 1$ and $q > 0$,

$$I_{\theta q} < \theta I_q.$$

To see this, let $\{f_n\}$ be a minimizing sequence for I_q, and define $\tilde{f}_n = \sqrt{\theta} f_n$ for all n, so that $Q(\tilde{f}_n) = \theta q$ and hence $E(\tilde{f}_n) \geq I_{\theta q}$ for all n. Then for all n we have

$$I_{\theta q} \leq E(\tilde{f}_n) = \frac{1}{2} \int \left[\tilde{f}_n L \tilde{f}_n - \frac{1}{3} \tilde{f}_n^3 \right] dx = \theta E(f_n) + \frac{1}{6} (\theta - \theta^{3/2}) \int f_n^3 \, dx.$$

Now taking $n \to \infty$ and using Lemma 3.5.2, we obtain

$$I_{\theta q} \leq \theta I_q + \frac{1}{6} (\theta - \theta^{3/2}) \delta < \theta I_q,$$

and so the claim is proved.

Now suppose one of q_1 and q_2 is greater than the other, say $q_1 > q_2$. Then from the claim just proved, it follows that

$$I_{(q_1+q_2)} = I_{q_1(1+q_2/q_1)} < (1 + \frac{q_2}{q_1})I_{q_1}$$

$$< I_{q_1} + \frac{q_2}{q_1}\left(\frac{q_1}{q_2}I_{q_2}\right) = I_{q_1} + I_{q_2},$$

as desired. Also, in the remaining case when $q_1 = q_2$, we have

$$I_{(q_1+q_2)} = I_{2q_1} < 2I_{q_1} = I_{q_1} + I_{q_2},$$

and so the proof is complete. □

The statement and proof of Lemma 2.5 go through unchanged in the present situation (except that H^1 is replaced by $H^{1/2}$) and so will not be repeated here. Before proceeding to the analogue of Lemma 2.6, we need the following result (cf. Lemma 3.10 of [**ABS**]).

LEMMA 3.7. *There exists a constant $A > 0$ such that if θ is any continuously differentiable function with θ and θ' in L^∞, and f is any L^2 function, then*

$$|[L,\theta]f|_2 \le A|\theta'|_\infty|f|_2,$$

where $[L,\theta]f$ denotes the commutator $L(\theta f) - \theta(Lf)$.

PROOF. By a standard density argument, it suffices to prove the result for arbitrary functions θ and f in $C_0^\infty(\mathbf{R})$.

Write $L = \frac{d}{dx}T$, where T is the Fourier multiplier operator defined by $\widehat{Tf}(k) = \sigma(k)\hat{f}(k)$ with $\sigma(k) = m(k)/ik$. Since $\sigma(k)$ is bounded on $\mathbf{R}$, then T is a bounded operator on L_2. Moreover, it is easily verified that

$$\sup_{k\in\mathbf{R}} |k|^n \left|\left(\frac{d}{dk}\right)^n \sigma(k)\right| < \infty$$

for all $n \in \mathbf{N}$, and hence by Theorem 35 of [**CM**] there exists $A_1 > 0$ such that

$$|[T,\theta](f')|_2 \le A_1|\theta'|_\infty|f|_2$$

for all functions θ and f in $C_0^\infty(\mathbf{R})$. Therefore

$$|[L,\theta]f|_2 = \left|T\frac{d}{dx}(\theta f) - \theta T\left(\frac{df}{dx}\right)\right|_2$$

$$\le |T(\theta'f)|_2 + |[T,\theta](f')|_2$$

$$\le \|T\| \, |\theta'|_\infty|f|_2 + A_1|\theta'|_\infty|f|_2,$$

where $\|T\|$ denotes the norm of T as an operator on L_2. Thus the lemma has been proved with $A = \|T\| + A_1$. □

LEMMA 3.8. *For every $\epsilon > 0$, there exist a number $N \in \mathbf{N}$ and sequences $\{g_N, g_{N+1}, \dots\}$ and $\{h_N, h_{N+1}, \dots\}$ of functions in $H^{1/2}$ such that for every $n \geq N$,*

1. $|Q(g_n) - \alpha| < \epsilon.$
2. $|Q(h_n) - (q - \alpha)| < \epsilon.$
3. $E(f_n) \geq E(g_n) + E(h_n) - \epsilon.$

PROOF. As in the proof of Lemma 2.6, we choose $r \in \mathbf{R}$ and $N \in \mathbf{N}$ so large that
$$\alpha - \epsilon < M_n(r) \leq M_n(2r) < \alpha + \epsilon$$
for all $n \geq N$, and define sequences $\{g_n\}$ and $\{h_n\}$ by $g_n(x) = \phi_r(x - y_n) f_n(x)$ and $h_n(x) = \psi_r(x - y_n) f_n(x)$ where ϕ and ψ are as before and y_n is chosen so that (2.1) and (2.2) hold. Then statements 1 and 2 follow, and it remains to prove the third statement. We write

$$E(g_n) + E(h_n) =$$
$$= \frac{1}{2}\left[\int \phi_r^2 f_n L f_n \, dx + \int \phi_r f_n [L, \phi_r] f_n \, dx \right]$$
$$+ \frac{1}{2}\left[\int \psi_r^2 f_n L f_n \, dx + \int \psi_r f_n [L, \psi_r] f_n \, dx \right]$$
$$- \frac{1}{6} \int \phi_r^2 f_n^3 \, dx - \frac{1}{6} \int \psi_r^2 f_n^3 \, dx$$
$$+ \frac{1}{6} \int (\phi_r^2 - \phi_r^3) f_n^3 \, dx + \frac{1}{6} \int (\psi_r^2 - \psi_r^3) f_n^3 \, dx,$$

where again for brevity the arguments of the functions $\phi_r(x - y_n)$ and $\psi_r(x - y_n)$ have been omitted. Now using Hölder's inequality and Lemmas 3.5.1 and 3.7, and arguing as in the proof of Lemma 2.6, we obtain

$$E(g_n) + E(h_n) = E(f_n) + O(1/r) + O(\epsilon),$$

where $O(1/r)$ and $O(\epsilon)$ denote terms bounded by A/r and $A\epsilon$, with constants A independent of r and n. The proof now concludes as before. $\square$

Corollary 2.7 holds in the present context without change of statement or proof, and the same is true of Lemma 2.8 except that here the last display in its proof should be modified to read

$$\eta \leq \left(\int_{y_n - 2}^{y_n + 2} |f_n|^2 \, dx \right)^{1/2} \left(\int_{y_n - 2}^{y_n + 2} |f_n|^4 \, dx \right)^{1/2}$$
$$\leq \left(\int_{y_n - 2}^{y_n + 2} |f_n|^2 \, dx \right)^{1/2} \left(\int_{-\infty}^{\infty} |f_n|^4 \, dx \right)^{1/2}$$
$$\leq AB^2 \left(\int_{y_n - 2}^{y_n + 2} |f_n|^2 \, dx \right)^{1/2},$$

where A is the Sobolev constant in the embedding of L^4 into $H^{1/2}$.

Thus all the preliminaries for the proofs of Theorem 2.9 and Corollary 2.10 have been established, and these proofs now apply in the present context without modification (except for the replacement of the H^1 norm by the $H^{1/2}$ norm). To complete the demonstration of Theorem 3.2 it remains only to justify the statement made there that the set G_q consists of solitary-wave solutions of (3.1) having positive wavespeeds. If $g \in G_q$ then by the Lagrange multiplier principle (cf. the proof of Proposition 2.11) there exists $\lambda \in \mathbf{R}$ such that

$$\delta E(g) + \lambda\, \delta Q(g) = 0,$$

where the Fréchet derivatives δE and δQ are given by

$$\delta E(g) = Lg - \frac{1}{2}g^2,$$

$$\delta Q(g) = g.$$

Hence $u(x,t) = g(x - \lambda t)$ solves (3.1), or in other words g is a solitary-wave solution of (3.1) with wavespeed λ. To see that $\lambda > 0$, note first that

$$\frac{d}{d\theta}\left[E(\theta g)\right]_{\theta=1} = \frac{d}{d\theta}\left[\frac{\theta^2}{2}\int gLg\,dx - \frac{\theta^3}{3}\int g^3\,dx\right]_{\theta=1}$$

$$= \int gLg\,dx - \int g^3\,dx = 2E(g) - \frac{1}{3}\int g^3\,dx.$$

But from Lemmas 3.4 and 3.5.2 we have $E(g) = I_q < 0$ and $\int g^3\,dx > 0$, so that

$$\frac{d}{d\theta}\left[E(\theta g)\right]_{\theta=1} < 0.$$

Now, using the definition of the Fréchet derivative, we have

$$\frac{d}{d\theta}\left[E(\theta g)\right]_{\theta=1} = \int \delta E(g) \cdot \frac{d}{d\theta}\left[\theta g\right]_{\theta=1}\,dx = -\lambda\int \delta Q(g) \cdot g\,dx = -\lambda\int g^2\,dx;$$

and since $\int g^2\,dx > 0$ it follows that $\lambda > 0$ as claimed.

REMARK. There remains the question of whether a stability result such as Theorem 3.1 holds for solitary-wave solutions of (3.1) for arbitrary values of H_1 and H_2. As noted in Section 2, such a result could be established if it could be shown that the set G_q consists of the translates of a single function. For this, in turn, it would suffice to show that solitary-wave solutions of (3.1) are unique up to translations. (Note that in the absence of a uniqueness result, we do not even know whether the solitary waves discussed in [**ABS**] are the same as those in the sets G_q.) Such a uniqueness result has indeed been proved for the case $H_1 = H_2$ [**A2,AT**], but the existing proofs rely heavily on an algebraic property of equation (3.1) which does not hold in the case $H_1 \neq H_2$.

Alternatively, a result like Theorem 3.1 may follow from a local analysis such as that appearing in [**A1**] and [**AB**]; a major obstacle to this approach, however, is the lack of an explicit formula for solitary-wave solutions when $H_1 \neq H_2$.

4. Further results

Clearly the proof of Theorem 3.2 did not rely heavily on special properties of the operator L appearing in equation (3.1), and similar results may be obtained in more general settings. For an equation of type (1.5), for example, one has the following theorem on stability of solitary waves.

THEOREM 4.1. *Suppose that the functions $f(u)$ and $m(k)$ satisfy the following assumptions:*

1. *Either*
$$f(u) = |u|^{p+1} \quad \text{for some } p > 1$$
or
$$f(u) = u^{p+1} \quad \text{for some } p \in \mathbf{N}.$$

2. *There exist positive constants A_1 and A_2 and a number $s > p/2$ such that $A_1 |k|^s \le m(k) \le A_2 |k|^s$ holds for all sufficiently large values of $|k|$.*

3. *For all $k \in \mathbf{R}$, $m(k) \ge 0$.*

4. *The function $m(k)$ is infinitely differentiable at all nonzero values of k, and for each $j \in \{0, 1, 2, \dots\}$ there exist constants B_1, B_2, k_0, and k_1 such that*

(4.1)
$$\left| \left(\frac{d}{dk} \right)^j \left(\frac{m(k)}{k} \right) \right| \le B_1 |k|^{-j} \quad \text{for } 0 < |k| < k_0,$$

and

(4.2)
$$\left| \left(\frac{d}{dk} \right)^j \left(\frac{\sqrt{m(k)}}{k^{s/2}} \right) \right| \le B_2 |k|^{-j} \quad \text{for } |k| > k_1.$$

Then for each $q > 0$ there exists a nonempty set G_q, consisting of solitary-wave solutions g of (1.5) having positive wavespeeds and satisfying $Q(g) = q$, which is stable in the following sense: for every $\epsilon > 0$ there exists $\delta > 0$ such that if

$$\inf_{g \in G_q} \|u_0 - g\|_{s/2} < \delta,$$

then the solution $u(x, t)$ of (1.5) with $u(x, 0) = u_0$ satisfies

$$\inf_{g \in G_q} \|u(\cdot, t) - g\|_{s/2} < \epsilon$$

for all $t \in \mathbf{R}$.

PROOF. The proof proceeds along the same lines as the proofs given above in Sections 2 and 3 for Theorems 1.1 and Theorem 3.2. First, we define invariant functionals Q and E as in (1.6) and (1.7), and the problem of minimizing E subject to constant Q is considered. It is straightforward to generalize Lemmas 3.3 through 3.6 to the present situation; we therefore omit the details, except to mention that assumptions 1 and 2 of Theorem 4.1 guarantee that the Sobolev embedding theorem can be used in the same way as in Section 3, and that assumption 3 guarantees that the proof of Lemma 3.5.2 also applies to prove the analogous result here.

The next two lemmas generalize Lemmas 3.7 and 3.8.

LEMMA 4.2. *Suppose (4.1) and (4.2) hold for the symbol $m(k)$. Let*

$$P = \left[\frac{s}{2} \right] + 1,$$

where the brackets denote the greatest integer function, and let M be the operator defined by

$$\widehat{Mu}(k) = \sqrt{m(k)}\, \widehat{u}(k).$$

Then M can be written as $M = M_1 + M_2$, where the operators M_1 and M_2 have the following properties:

1. *There exists a constant $A > 0$ such that for every function θ which is in L^∞ and has derivative θ' in L^∞, and every function f in L^2,*

$$|[(M_1)^2, \theta]f|_2 \le A|\theta'|_\infty |f|_2$$

   and

$$|[M_1 M_2, \theta]f|_2 \le A|\theta'|_\infty |f|_2.$$

2. *There exists a constant $A > 0$ such that for every function θ which is in L^∞ and has derivatives up to order P in L^∞, and every function f in L^2,*

$$|[M_2, \theta]f|_2 \le A \left(\sum_{i=1}^{P} \left| \frac{d^i \theta}{dx^i} \right|_\infty \right) \|f\|_{s/2}.$$

PROOF. As in the proof of Lemma 3.7, we may assume that θ and f are functions in $C_0^\infty(\mathbf{R})$. Choose $\chi(k) \in C_0^\infty(\mathbf{R})$ such that $\chi(k) = 1$ for $|k| < 1$ and $\chi(k) = 0$ for $|k| > 2$. Define $m_1(k) = \chi(k)\sqrt{m(k)}$ and $m_2(k) = (1 - \chi(k))\sqrt{m(k)}$, and let M_1 and M_2 be Fourier multiplier operators with symbols m_1 and m_2, respectively.

We can write $(M_1)^2 = \frac{d}{dx}T_1$ and $M_1 M_2 = \frac{d}{dx}T_2$, where T_1 is the Fourier multiplier operator with symbol $\sigma_1(k)$ given by

$$\sigma_1(k) = \frac{(m_1(k))^2}{ik},$$

and T_2 is the Fourier multiplier operator with symbol $\sigma_2(k)$ given by

$$\sigma_2(k) = \frac{m_1(k)m_2(k)}{ik}.$$

From (4.1) it follows that for $j = 0, 1, 2, \ldots$ there exist constants $\tilde{B}_j^1$ such that

$$\left| \left(\frac{d\sigma_1}{dk} \right)^j \right| \le \tilde{B}_j^1 \, |k|^{-j} \quad \text{for all } k \in \mathbf{R}.$$

Also, since $m_1(k)m_2(k)$ has compact support in k and is supported outside a neighborhood of $k = 0$, one has trivially that

$$\left| \left(\frac{d\sigma_2}{dk} \right)^j \right| \le \tilde{B}_j^1 \, |k|^{-j} \quad \text{for all } k \in \mathbf{R}$$

(with perhaps different values of the constants $\tilde{B}_j^1$). The desired estimates on $|[(M_1)^2, \theta]f|_2$ and $|[M_1 M_2, \theta]f|_2$ now follow from the same argument used to prove Lemma 3.7.

To obtain the desired estimate for $|[M_2, \theta]f|_2$, write $M_2 = \left(\frac{d}{dx} \right)^P T_3$, where T_3 is the Fourier multiplier operator with symbol $\sigma_3(k)$ given by

$$\sigma_3(k) = \frac{m_2(k)}{(ik)^P}.$$

From (4.2) it follows that for $j = 0, 1, 2, \ldots$ there exist constants $\tilde{B}_j^2$ such that

$$\left| \frac{d^j \sigma_3}{dk^j} \right| \leq \tilde{B}_j^2 \, |k|^{-j} \quad \text{for all } k \in \mathbf{R},$$

and hence Theorem 35 of [**CM**] implies that

$$(4.3) \qquad \left| [T_3, \theta] \frac{df}{dx} \right|_2 \leq A \left| \frac{d\theta}{dx} \right|_\infty |f|_2,$$

where A is independent of θ and f. Note also that since σ_3 is a bounded function of k, then T_3 is a bounded operator on L^2.

Now observe that

$$[M_2, \theta] f = \left[(d/dx)^P T_3, \theta \right] f$$

$$= T_3 \left(\frac{d^P (\theta f)}{dx^P} \right) - \theta T_3 \left(\frac{d^P f}{dx^P} \right)$$

$$= [T_3, \theta] \left(\frac{d^P f}{dx^P} \right) + T_3 \left(\sum_{i=1}^{P} a_i^P \frac{d^i \theta}{dx^i} \frac{d^{P-i} f}{dx^{P-i}} \right),$$

where a_i^P are constants which come from Liebniz' rule. From (4.3) and the boundedness of T_3 as an operator on L^2, it then follows that

$$\|[M_2, \theta] f|_2 \leq A \left(\sum_{i=1}^{P} \left| \frac{d^i \theta}{dx^i} \right|_\infty \right) \left(\sup_{0 \leq i \leq P-1} \left| \frac{d^i f}{dx^i} \right|_2 \right).$$

But since $P - 1 = [s/2] \leq s/2$, the supremum on the right-hand side of the last expression is controlled by $\|f\|_{s/2}$. This completes the proof of the Lemma. □

As in Sections 2 and 3 above, we denote an arbitrary minimizing sequence for the variational problem by $\{f_n\}$, and associate to it functions $\{M_n\}$ and a number $\alpha \in [0, q]$. From $\{f_n\}$ we may extract a subsequence (again denoted by $\{f_n\}$), whose indices are those of a pointwise convergent subsequence of $\{M_n\}$.

LEMMA 4.3. *For every $\epsilon > 0$, there exist a number $N \in \mathbf{N}$ and sequences $\{g_N, g_{N+1}, \ldots\}$ and $\{h_N, h_{N+1}, \ldots\}$ of functions in $H^{s/2}$ such that for every $n \geq N$,*

1. $|Q(g_n) - \alpha| < \epsilon.$
2. $|Q(h_n) - (q - \alpha)| < \epsilon.$
3. $E(f_n) \geq E(g_n) + E(h_n) - \epsilon.$

PROOF. Define functions ϕ_r, ψ_r, $g_n = \phi_r f_n$, and $h_n = \psi_r f_n$ as in Lemma 3.8. We must show that

$$E(g_n) + E(h_n) = E(f_n) + O(1/r) + O(\epsilon),$$

where $O(1/r)$ and $O(\epsilon)$ denote terms majorized by A/r and $A\epsilon$, with constants A depending only on $\|f_n\|_{s/2}$.

Since $M^2 = L$, and M is self-adjoint on $H^{s/2}$, we can rewrite (1.7) as

$$E(u) = \int \left[\frac{1}{2} (Mu)^2 - F(u) \right] dx.$$

Therefore (dropping the subscripts from f_n and g_n for clarity of notation),

$$(4.4) \qquad E(g) = \int \left(\frac{1}{2}(M_1(g))^2 + M_1(g)M_2(g) + \frac{1}{2}(M_2(g))^2 - F(g) \right) \, dx.$$

Since M_1 and M_2 are self-adjoint on $H^{s/2}$, the integrals of the first three terms on the right-hand side of (4.4) can be expanded as

$$\int (M_1(\phi_r f))^2 \, dx = \int \left(\phi_r^2 f(M_1)^2 f + \phi_r f[(M_1)^2, \phi_r]f \right) \, dx,$$

$$\int M_1(\phi_r f)M_2(\phi_r f) \, dx = \int \left(\phi_r^2 f M_1 M_2 f + \phi_r f[M_1 M_2, \phi_r]f \right) \, dx,$$

and

$$\int (M_2(\phi_r f))^2 \, dx = \int \left(\phi_r^2(M_2 f)^2 + 2\phi_r(M_2 f)[M_2, \phi_r]f + ([M_2, \phi_r]f)^2 \right) \, dx.$$

In these expansions, the integrals of all terms containing commutators are $O(1/r)$: this follows easily from Lemma 4.2 and Hölder's inequality, together with the estimates $\left| d^i \phi_r/dx^i \right|_\infty = O(1/r)$ for $i \geq 1$ and the fact that M_1 and M_2 are bounded operators from $H^{s/2}$ to L^2. Also, from the definitions of ϕ_r and F, it follows as in the proof of Lemma 2.6 that

$$\int F(\phi_r f) \, dx = \int \phi_r^2 F(f) \, dx + \int (\phi_r^{p+2} - \phi_r^2) F(f) \, dx$$

$$= \int \phi_r^2 F(f) \, dx + O(\epsilon).$$

Substitution of these estimates into (4.4) yields the result

$$E(g) = \int \phi_r^2 \left(\frac{1}{2} f M_1^2 f + f M_1 M_2 f + \frac{1}{2}(M_2 f)^2 - F(f) \right) \, dx + O(1/r) + O(\epsilon).$$

Exactly the same argument shows that for $h = \psi_r f$ one has

$$E(h) = \int \psi_r^2 \left(\frac{1}{2} f M_1^2 f + f M_1 M_2 f + \frac{1}{2}(M_2 f)^2 - F(f) \right) \, dx + O(1/r) + O(\epsilon).$$

Adding these two results and recalling that $\phi_r^2 + \psi_r^2 = 1$, one obtains that

$$E(g) + E(h) = E(f) + O(1/r) + O(\epsilon).$$

From here the proof of the Lemma concludes in the same way as the proof of Lemma 2.6. $\square$

The obvious analogues for the present situation of Corollary 2.7, Lemma 2.8, Theorem 2.9, and Corollary 2.10, and the observation that the wavespeeds of the solitary waves in G_q are positive, may now be established by exactly the same arguments used in Sections 2 and 3. Thus the proof of Theorem 4.1 is completed. $\square$

Examples of equations to which the assumptions of Theorem 4.1 apply are the Benjamin-Ono equation [**B1**], the Smith equation [**S**], and the Pritchard equation [**P**], which correspond to the dispersion operators with symbols $m(k) = |k|$, $m(k) = \sqrt{1 + k^2} - 1$, and $m(k) = k^2(1 + K_0(k))$ respectively. (Here $K_0(k)$ denotes a modified

Bessel function. This form of the symbol for Pritchard's equation, which differs slightly from that found in [**P**], has been suggested to the author by J. Bona as an appropriate model for the dispersive properties of the waves treated in [**P**].) For the Benjamin-Ono equation in the case $f(u) = u^2$, there is a unique solitary-wave solution (up to translation) [**AmT1**], and hence in this case the stability result of Theorem 4.1 implies a stronger result like that of Theorem 1.1. Such a result is already known (cf. [**BBSSS**]), but for the Benjamin-Ono equation with other nonlinearities $f(u)$ and for the Smith and Pritchard equations, Theorem 4.1 appears to be the only stability result available to date.

For the fifth-order KdV-type equation

$$(4.5) \qquad u_t + u^p u_x + u_{xxx} - \delta u_{xxxxx} = 0,$$

the assumptions of Theorem 4.1 hold with $s = 4$ if $p < 8$ and $\delta > 0$, and so in this case Theorem 4.1 implies existence of non-empty stable sets of solitary-wave solutions G_q for each $q > 0$. Each element g of G_q, being a solitary-wave solution of (4.5) with positive wavespeed, will satisfy the equation

$$(4.6) \qquad Cg - g'' + \delta g'''' = \frac{g^{p+1}}{p+1},$$

for some $C > 0$. In their study [**AmT2**] of equation (4.6) in the case $p = 1$, Amick and Toland showed that for all $C > 0$ and $\delta > 0$, even solutions $g(x)$ exist which satisfy $g(0) > 0$ and which decay to zero as $|x| \to \infty$; the decay being monotonic in $|x|$ if $C \le 1/(4\delta)$ and oscillatory otherwise. We do not know which (if any) of these solutions of (4.6) are also contained in the stable sets G_q.

An interesting question is whether a result such as Theorem 4.1 holds for equations in which the symbol $m(k)$ is not everywhere positive, i.e., when assumption 3 does not hold. An equation of possible physical interest in which this situation obtains is the Benjamin equation

$$u_t + uu_x + \mathcal{H}u_{xx} + \delta u_{xxx} = 0,$$

derived in [**B3**] as a model for interfacial waves in a stratified fluid in the presence of strong surface tension. Here $\mathcal{H}$ denotes the Hilbert transform, $\delta > 0$, and the equation fits the form (1.5) with symbol $m(k) = -|k| + \delta k^2$. It was shown in [**ABR**] that for large values of C, there exist oscillatory solitary-wave solutions $u(x, t) = \phi(x - Ct)$ which have a strong stability property like that expressed in Theorem 1.1; but it remained an open question whether stable solitary-wave solutions exist with the smaller wavespeeds C for which the assumptions underlying the model are valid. Recently, J. Angulo [**An**] has settled this question in the affirmative, with a proof based in part on an idea from [**K**]. Subsequently, in [**AL**], a modification of Angulo's proof was used to show that the assumption that $m(k) \ge 0$ can be dropped from Theorem 4.1.

Finally, another interesting direction for generalization of the above arguments would be to systems of nonlocal equations, such as those derived by Liu, Kubota, and Ko for modeling interactions between waves on neighboring pycnoclines within a stratified fluid (cf. [**ABS,LKK**]).

Acknowledgements

The author is grateful to IMPA and to the Universidade Estadual de Campinas for their hospitality during the writing of this article. He is also indebted to Orlando Lopes for an important conversation.

References

[ABFS] L. Abdelouhab, J. Bona, M. Felland, and J.-C. Saut, *Nonlocal models for nonlinear dispersive waves*, Pays. D **40** (1989), 360–392.

[A1]   J. Albert, *Positivity properties and stability of solitary wave solutions of model equations for long waves*, Comm. Partial Differential Equations **17** (1992), 1–22.

[A2]   J. Albert, *Positivity properties and uniqueness of solitary-wave solutions of the intermediate long-wave equation*, Evolution equations (G. Ferreyra, G. Goldstein, and F. Neubrander, eds.), Marcel Dekker, New York, 1994, pp. 11–20.

[AB]   J. Albert and J. Bona, *Total positivity and the stability of internal waves in stratified fluids of finite depth*, IMA J. Appl. Math. **46** (1991), 1–19.

[ABR]   J. Albert, J. Bona, and J. Restrepo, *Solitary-wave solutions of the Benjamin equation*, to appear.

[ABS]   J. Albert, J. Bona, and J.-C. Saut, *Model equations for waves in stratified fluids*, Proc. Roy. Soc. London Ser. A **453** (1997), 1233–1260.

[AL]   J. Albert and F. Linares, *Stability of solitary-wave solutions to long-wave equations with general dispersion*, Mat. Contemp., to appear.

[AT]   J. Albert and J. Toland, *On the exact solutions of the intermediate long-wave equation*, Differential Integral Equations **7** (1994), 601–612.

[AmT1] C. Amick and J. Toland, *Uniqueness of Benjamin's solitary-wave solution of the Benjamin-Ono equation*, IMA J. Appl. Math. **46** (1991), 21–28.

[AmT2] C. Amick and J. Toland, *Homoclinic orbits in the dynamic phase-space analogy of an elastic strut*, European J. Appl. Math. **3** (1992), 97–114.

[An]   J. Angulo Pava, *Existence and stability of solitary-wave solutions of the Benjamin equation*, to appear.

[BD]   R. Benguria and M. C. Depassier, *Equations of the Korteweg-de Vries type with non-trivial conserved quantities*, J. Phys. A **22** (1989), 4135–4142.

[B1]   T. B. Benjamin, *Internal waves of permanent form in fluids of great depth*, J. Fluid Mech. **29** (1967), 559–592.

[B2]   T. B. Benjamin, *The stability of solitary waves*, Proc. Roy. Soc. London Ser. A **328** (1972), 153–183.

[B3]   T. B. Benjamin, *A new kind of solitary wave*, J. Fluid Mech. **245** (1992), 401–411.

[BBBSS] D. Bennett, J. Bona, R. Brown, S. Stansfield, and J. Stroughair, *The stability of internal waves*, Math. Proc. Cambridge Philos. Soc. **94** (1983), 351–379.

[BL]   J. Bergh and J. Löfstrom, *Interpolation spaces: an introduction*, Springer-Verlag, New York, 1976.

[Bo]   J. Bona, *On the stability theory of solitary waves*, Proc. Roy. Soc. London Ser. A **344** (1975), 363–374.

[BS]   J. Bona and A. Soyeur, *On the stability of solitary-wave solutions of model equations for long waves*, J. Nonlinear Sci. **4** (1994), 449–470.

[Bou]   J. Boussinesq, *Théorie des ondes et des remous qui se propagent le long d'un canal rectangulaire horizontal, en communiquant au liquide contenu dans ce canal des vitesses sensiblement pareilles de la surface au fond*, J. Math. Pures Appl. (2) **17** (1872), 55–108.

[C]   T. Cazenave, *An introduction to nonlinear Schrödinger equations*, Textos de métodos matemáticos v. 22, Universidade Federal do Rio de Janeiro, Rio de Janeiro, 1989.

[CL]   T. Cazenave and P.-L. Lions, *Orbital stability of standing waves for some nonlinear Schrödinger equations*, Comm. Math. Phys. **85** (1982), 549–561.

[CM]   R. Coifman and Y. Meyer, *Au delà des opérateurs pseudo-différentiels*, Asterisque no. 57, Société Mathématique de France, Paris, 1978.

[dBS]   A. de Bouard and J.-C. Saut, *Remarks on the stability of generalized KP solitary waves*, to appear.

[GSS1] M. Grillakis, J. Shatah, and W. Strauss, *Stability theory of solitary waves in the presence of symmetry* I, J. Funct. Anal. **74** (1987), 160–197.

[GSS2]   M. Grillakis, J. Shatah, and W. Strauss, *Stability theory of solitary waves in the presence of symmetry* II, J. Funct. Anal. **94** (1990), 308–348.

[J]   R. Joseph, *Solitary waves in a finite depth fluid*, J. Phys. A **10** (1977), L225–L227.

[JE]   R. Joseph and R. Egri, *Multi-soliton solutions in a finite-depth fluid*, J. Phys. A **11** (1978), L97–L102.

[K]   S. Kichenassamy, *Existence of solitary waves for water-wave models*, Nonlinearity **10** (1997), 133-151.

[KSA]   Y. Kodama, J. Satsuma, and M. Ablowitz, *Nonlinear intermediate long-wave equation: analysis and method of solution*, Phys. Rev. Lett. **46** (1981), 687–690.

[KdV]   D.J. Korteweg and G. de Vries, *On the change of form of long waves advancing in a rectangular canal, and on a new type of long stationary waves*, Phil. Mag. **39** (1895), 422–443.

[KKD]   T. Kubota, D. Ko, and L. Dobbs, *Weakly nonlinear interval gravity waves in stratified fluids of finite depth*, J. Hydrodynamics **12** (1978), 157–165.

[KRZ]   E. Kuznetsov, A. Rubenchik, and V. Zakharov, *Soliton stability in plasmas and hydrodynamics*, Phys. Rep. **142** (1986), 103–165.

[LR]   D. Lebedev and A. Radul, *Generalized internal long wave equations: construction, Hamiltonian structure, and conservation laws*, Comm. Math. Phys. **91** (1983), 543–555.

[Li]   E. Lieb, *On the lowest eigenvalue of the Laplacian for the intersection of two domains*, Invent. Math. **74** (1983), 441–448.

[L1]   P. Lions, *The concentration compactness principle in the calculus of variations. The locally compact case, part 1*, Ann. Inst. H. Poincaré **1** (1984), 109–145.

[L2]   P. Lions, *The concentration compactness principle in the calculus of variations. The locally compact case, part 2*, Ann. Inst. H. Poincaré **4** (1984), 223–283.

[LKK]   A. Liu, T. Kubota, and D. Ko, *Resonant transfer of energy between nonlinear waves in neighboring pycnoclines*, Stud. Appl. Math. **63** (1980), 25–45.

[Lu]   D. Luenberger, *Optimization by vector space methods*, Wiley and Sons, New York, 1969.

[MS]   J. Maddocks and R. Sachs, *On the stability of KdV multi-solitons*, Comm. Pure Appl. Math. **46** (1993), 867–901.

[P]   W. Pritchard, *Solitary waves in rotating fluids*, J. Fluid Mech. **42** (1970), 61–83.

[PW]   R. Pego and M. Weinstein, *Asymptotic stability of solitary waves*, Comm. Math. Phys. **164** (1994), 305–349.

[R]   A. Radul, *The equation of a two-layer fluid is a completely integrable Hamiltonian system*, Soviet Math. Dokl. **32** (1985), 96–99.

[S]   R. Smith, *Non-linear Kelvin and continental-shelf waves*, J. Fluid Mech. **52** (1972), 379–391.

DEPARTMENT OF MATHEMATICS, UNIVERSITY OF OKLAHOMA, NORMAN, OKLAHOMA 73019
E-mail address: jalbert@ou.edu

Contemporary Mathematics
Volume **221**, 1999

Estimates for Green's Function in Terms of Asymmetry

Tilak Bhattacharya and Allen Weitsman

Abstract

Let Ω be a bounded plane region containing the origin, and having area π. If $G(z)$ denotes the Green's function for Ω, with pole at 0, and $G^*(r)$ its decreasing rearrangement, then it is shown that $G^*(r) < \log(1/r) - C\alpha^2$ for an interval of r, where α is the asymmetry of Ω.

1   Introduction

In this work we continue our study of obtaining bounds on various domain quantities, in terms of asymmetry. For a compact set Ω, in $I\!R^n$, let $V(\Omega)$ denote the volume of Ω and $B(x, \rho)$ be the ball of radius ρ, centered at x. Let ρ be such that $V(B(0, \rho)) = V(\Omega)$; then we define the asymmetry $\alpha = \alpha(\Omega)$ by

$$(1.1) \quad \alpha = \inf_x \frac{V(\Omega \backslash B(x, \rho))}{V(\Omega)}, \qquad \rho = (V(\Omega)/V(B(x, 1)))^{1/n}.$$

It is clear that $\alpha = 0$ when Ω is a ball. In $I\!R^2$, we shall use $A(\Omega)$ to denote the area of Ω.

The works, in [2], [6], [7], [8], and [9] relate asymmetry to various quantities such as the isoperimetric constant, capacity and the first eigenvalue of the Laplacian. In [8], a sharp lower bound for logarithmic capacity, in terms of asymmetry, was deduced by relating capacity to moment of inertia. Such a lower bound was also shown to hold for p-capacities of condensers in [2] by very different methods. The analysis in [2] was based on estimates for subcondensers, possessing special geometry, and was also shown to yield a lower bound for logarithmic capacity similar to the one in [8]. For a detailed survey of the field regarding capacity and asymmetry, and the work done in [7], [8] and [9], see [2]. Our effort in this work will be to deduce upper bounds,

---

[1]Mathematics Subject Classification. Primary 31A05, Secondary 30C20.

in terms of asymmetry, for Green's function of a bounded domain in $\mathbb{R}^2$. The connection between this and conformal mapping of simply connected regions is obvious, since the Green's function for such an Ω can be written as $-\log|f|$, where f is the conformal mapping of Ω onto the unit disk. On the other hand, the estimates on the Green's function are intimately connected to the eigenvalue problem which is still open. We shall give a brief discussion of this in §8

2   Notations and the main result

Given a Borel set $S \in \mathbb{R}^2$, let $A(S)$ denote its area and $L(\partial S)$ the perimeter of its boundary ∂S. Throughout this work, $\Omega \subset \mathbb{R}^2$ will be a bounded domain with boundary $\partial \Omega$ a finite union of rectifiable curves. We will also assume that the origin $0 \in \Omega$. Let $R > 0$ be such that $A(\Omega) = \pi R^2$; and $G(x)$, $x \in \Omega$, denote Green's function of Ω with pole at 0. That is

$$(2.1) \qquad G(x) = \log \frac{R}{|x|} + h(x), \quad x \in \Omega,$$

where $h(x)$ is harmonic in Ω and is such that $G(x)$ vanishes continuously on $\partial\Omega$.

Also, let Ω^* be the disk, centered at the origin 0, having the same area as Ω. Clearly, the radius of Ω^* is R. For $t > 0$, we set

$$(2.2) \qquad F(t) = \{x \in \Omega : G(x) > t\} \text{ and } A(t) = A(F(t)).$$

For $x \in \Omega^*$, the radial function $G^*(x) = G^*(|x|)$ will denote the symmetric decreasing rearrangement (Schwarz symmetrization) of $G(x)$. It is defined as

$$(2.3) \quad G^*(x) = G^*(|x|) = \inf\{t \geq 0 : A(t) < \pi|x|^2\} \quad 0 < |x| \leq R.$$

It is well known that [1; p 60],

$$(2.4) \qquad 0 \leq G^*(x) \leq \log \frac{R}{|x|}, \quad 0 < |x| \leq R, \quad x \in \Omega^*.$$

In this work we determine the effect of asymmetry on (2.4). More precisely, we prove

Theorem 1 : Let $\Omega \subset I\!\!R^2$ be a bounded domain with $0 \in \Omega$, and $\alpha = \alpha(\Omega)$ its asymmetry. Let $R > 0$ be such that $A(\Omega) = \pi R^2$. Then, for every $0 < \delta < 1$, exists a constant $C = C_\delta > 0$, depending only on δ, such that

$$(2.5) \qquad G^*(x) \leq \log \frac{R}{|x|} - C\,\alpha^2, \quad 0 < |x| < \sqrt{(1-\delta)}R.$$

We adapt the method developed in [2] to prove Theorem 1. In what follows, k will stand for a small positive constant to be determined in Sections 3 and 6. We will assume that

$$(2.6) \qquad 0 < k < \delta/500 < 10^{-3}.$$

Let $\partial F(t)$ denote the boundary of the set $F(t)$, and $L(\partial F(t))$ be its length . Since G is real analytic in $\Omega \backslash \{0\}$, the gradient DG vanishes only on a discrete set. Thus $\partial F(t)$ is an analytic curve except possibly for countably many t's, and $A(t)$ is decreasing and continuous.

As in [2], we will consider the following two possibilities.

Case 1 : For all t such that

$$(2.7) \qquad (1-\delta)\pi R^2 \leq A(t) \leq (1-\delta/2)\pi R^2.$$

we have
$$(2.8) \qquad L(\partial F(t))^2 \geq 4\pi(1 + k\alpha^2)A(t),$$

that is, asymmetry propagates inwards.

Case 2 : There exists a value T such that

$$(2.9) \qquad (1-\delta)\pi R^2 \leq A(T) \leq (1-\delta/2)\pi R^2.$$

and
$$(2.10) \qquad L(\partial F(T))^2 < 4\pi(1 + k\alpha^2)A(T).$$

When this situation occurs, we will say that asymmetry fails to propagate. We may take $\partial F(T)$ to be analytic.

We will also have occassion to use the Bonnesen formulas [11; pp 3-4]. Let D be a simply connected planar domain bounded by a rectifiable

Jordan curve. Let R_o and R_i denote the outradius and inradius of D respectively. Then

$$(2.11) \qquad L(\partial D)^2 - 4\pi A(D) \geq \pi^2 (R_o - R_i)^2,$$

$$(2.12) \qquad 2\pi R_o \leq \{L(\partial D) + \sqrt{L(\partial D)^2 - 4\pi A(D)}\},$$

and

$$(2.13) \qquad 2\pi R_i \geq \{L(\partial D) - \sqrt{L(\partial D)^2 - 4\pi A(D)}\}.$$

Our strategy will consist in proving the estimate (2.5) in Cases 1 and 2. We will make considerable use of the coarea formula and the isoperimetric inequality. Our proof in Case 2 will also employ a perturbation result for capacities of condensers. A version was first proven in [2].

Remark 2.1: We will prove Theorem 1 for $R = 1$ and $\delta = 0.2$. The proof undergoes only minor modifications for $\delta < 0.2$ and the full strength of the result in (2.5) for $R \neq 1$ may be recovered by scaling.

We have divided our work as follows. In Section 3, we prove Theorem 1 in Case 1. The proof in Case 2, when Ω is simply connected, is spread over Sections 4, 5 and 6. In Section 7, we fill in the details of the proof of Theorem 1 when Ω is multiply connected. Finally, in Section 8, we present a brief discussion regarding the connections of this work to the eigenvalue problem for the Laplacian.

3 Proof of Theorem 1 in Case 1

As noted above, for the rest of this work, we take $R = 1$ and $\delta = 0.2$ in (2.6), (2.7), (2.8) and (2.9).

We now prove Theorem 1, i.e., when asymmetry propagates inwards in the sense of (2.7) and (2.8). By applying the divergence theorem, the coarea formula, and Hölder's inequality we may conclude that outside a set of at most countably many $t's$,

$$(3.1) \quad \int_{\partial F(t)} |DG| = 2\pi, \quad \text{and} \quad \frac{dA}{dt} = -\int_{\partial F(t)} \frac{1}{|DG|}, \quad (t < \infty),$$

and

$$(3.2) \quad L(\partial F(t))^2 = \left(\int_{\partial F(t)} 1 \right)^2 \leq \left(\int_{\partial F(t)} |DG| \right) \left(\int_{\partial F(t)} \frac{1}{|DG|} \right)$$

$$= -2\pi \frac{dA}{dt},$$

where $A = A(t)$. From (2.7), (2.8), (3.1) and (3.2), it follows that for all t's such that $0.8\pi \le A(t) \le 0.9\pi$,

$$4\pi(1 + k\alpha^2)A(t) \le -2\pi \frac{dA}{dt}.$$

Thus

(3.3)
$$-\frac{1}{2(1 + k\alpha^2)A} \frac{dA}{dt} \ge 1.$$

Since $A(t)$ is decreasing and continuous, $-\log A(t)$ is increasing. Thus integrating (3.3), over the interval $[A, 0.9\pi]$ with $A \ge 0.8\pi$, we get

(3.4)
$$t(A) - t(0.9\pi) \le \frac{1}{2(1 + k\alpha^2)} \log \frac{0.9\pi}{A}.$$

Employing the usual isoperimetric inequality (instead of (2.8)) in (3.2), we obtain that

(i) whenever $0 < A(t) < 0.8\pi$,

(3.5)
$$t(A) - t(0.8\pi) \le \frac{1}{2} \log \frac{0.8\pi}{A(t)}$$

and

(ii) when $0.9\pi \le A(t) < \pi$,

(3.6)
$$t(A) - t(\pi) \le \frac{1}{2} \log \frac{\pi}{A(t)}.$$

Recall that $t(\pi) = 0$; taking $A = 0.9\pi$ in (3.6) and then adding (3.4) to the resulting inequality in (3.6) we find that for $0.8\pi \le A \le 0.9\pi$,

$$
\begin{aligned}
t(A) &\le \frac{1}{2} \log \frac{1}{0.9} + \frac{1}{2} \frac{1}{1 + k\alpha^2} \log \frac{0.9\pi}{A} \\
&= \frac{1}{2}\left(\log \frac{\pi}{A} - \frac{k\alpha^2}{1 + k\alpha^2} \log \frac{0.9\pi}{A} \right).
\end{aligned}
$$
(3.7)

Now take $A = 0.8\pi$ in (3.7). This gives

$$t(0.8\pi) \le \frac{1}{2}\left(\log \frac{1}{0.8} - \frac{k\alpha^2}{1 + k\alpha^2} \log \frac{0.9}{0.8} \right).$$

Employing this estimate for $t(0.8\pi)$ in (3.5) results in

$$t(A) \leq \frac{1}{2}\left(\log\frac{0.8\pi}{A} + \log\frac{1}{0.8} - \frac{k\alpha^2}{1+k\alpha^2}\log\frac{0.9}{0.8}\right)$$

$$(3.8) \qquad = \frac{1}{2}\left(\log\frac{\pi}{A} - \frac{k\alpha^2}{1+k\alpha^2}\log\frac{0.9}{0.8}\right),$$

for $0 < A < 0.8\pi$.

For $0 < A < 0.85\pi$ we may then deduce the following estimate from (3.7) and (3.8), namely,

$$(3.9) \qquad t(A) \leq \frac{1}{2}\left(\log\frac{\pi}{A} - \frac{k\alpha^2}{1+k\alpha^2}\log\frac{0.9}{0.85}\right).$$

Let $B(0,r)$ denote the disk with $\pi r^2 = A(t)$. Noting that $t(A) = G^*(r)$, we have

$$(3.10) \qquad G^*(r) \leq \log\frac{1}{r} - \frac{k\alpha^2}{1+k\alpha^2}\log\frac{0.9}{0.85},$$

for $0 < r < \sqrt{0.85}$. Theorem 1 will follow once a value for k is chosen. ∎

The remainder of the proof is devoted to Theorem 1 in Case 2, i.e., when (2.9) and (2.10) hold. We will make some preliminary reductions before presenting the proof. A part of the effort will be invested in describing the geometry of the level sets $F(t)$ involved in this case. The proof will follow from a perturbation result for capacities of condensers.

4 Preliminary Reductions

We will first assume that Ω is simply connected. The analysis continues to apply with minor modifications in the event that Ω is multiply connected. These details are presented in Section 7 .

The following lemma, though not difficult to prove, is essential for the construction of a condenser with special geometry. A lower bound for its 2-capacity will be key to the proof of Theorem 1 in Case 2.

Lemma 4.1 : Either there exists a $t = t_0 > 0$ with

$$(4.1) \qquad A(t_0) > \pi(1 - k\alpha^2)$$

and

(4.2) $$L(\partial F(t_0))^2 < 4\pi(1.01)A(t_0)$$

or Theorem 1 holds.

Proof : Suppose that (4.2) were false for all t's such that (4.1) holds, i.e.,

$$L(\partial F(t))^2 \geq 4\pi(1.01)A(t), \text{ whenever } \pi(1 - k\alpha^2) < A(t) < \pi.$$

Then using the above mentioned inequality, in (3.2), we get for $\pi(1 - k\alpha^2) \leq A(t) \leq \pi$,

(4.3) $$-\frac{1}{2(1.01)A}\frac{dA}{dt} \geq 1.$$

Integrating (4.3) over the interval $[\pi(1 - k\alpha^2), \pi]$ we get that

(4.4) $$t(\pi(1 - k\alpha^2)) \leq \frac{1}{2(1.01)} \log \frac{1}{1 - k\alpha^2}.$$

Now employing the usual isoperimetric inequality in (3.2) and integrating over the interval $[A, \pi(1 - k\alpha^2)]$ (for example, see (3.5)) we find that

(4.5) $$t(A) - t(\pi(1 - k\alpha^2)) \leq \frac{1}{2} \log \frac{\pi(1 - k\alpha^2)}{A}.$$

Adding (4.4) and (4.5) and recalling (2.6) we get that for $0 < A < 0.85\pi$,

$$t(A) \leq \frac{1}{2}\left(\log \frac{\pi}{A} - \frac{0.01}{1.01} \log \frac{1}{1 - k\alpha^2}\right),$$

which again implies Theorem 1.   ∎

We now continue the proof of Theorem 1 under the assumption that there exists a value of t such that both (4.1) and (4.2) hold. In what follows, we take t_0 to always denote such a value and this will stay fixed throughout. Let T be as in (2.9) and (2.10). It readily follows from (2.6) and (2.9) that

(4.6) $$T > t_0.$$

We now describe the geometry of the set $F(T)$. Since $G(x)$ has only one singularity, the level sets $F(t)$, for all t, are simply connected by the maximum principle. With $A(t) = A(F(t))$, set $R = \sqrt{A(t_0)/\pi}$ and $\rho = \sqrt{A(T)/\pi}$; it is easily seen from (2.6), (2.9) and (4.1) that

$$\sqrt{0.8} \le \rho \le \sqrt{0.9} \text{ and } \sqrt{0.999} < R < 1.$$

It follows from the Bonnesen formulas in (2.11) - (2.13) that $\partial F(T)$ is contained between two circles $C_1 = \{x : |x - x_o| = R_o\}$ and $C_2 = \{x : |x - x_i| = R_i\}$, where R_o and R_i are the outradius and the inradius of $F(T)$ respectively. Thus,

$$(4.7) \qquad 0 \le R_o - R_i \le 2\sqrt{\frac{kA(T)}{\pi}}\alpha = 2\sqrt{k}\alpha\rho \le 2\sqrt{k}\alpha,$$

$$\begin{aligned}
R_o \quad &\le \quad \frac{L(\partial F(T))}{2\pi} + \sqrt{\frac{kA(T)}{\pi}}\alpha \le (\sqrt{1 + k\alpha^2} + \sqrt{k}\alpha)\sqrt{\frac{A(T)}{\pi}} \\
&\le \quad (1 + 2\sqrt{k}\alpha)\sqrt{\frac{A(T)}{\pi}} = (1 + 2\sqrt{k}\alpha)\rho \\
R_i \quad &\ge \quad \frac{L(\partial F(T))}{2\pi} - \sqrt{\frac{kA(T)}{\pi}}\alpha \ge (1 - \sqrt{k}\alpha)\sqrt{\frac{A(T)}{\pi}} = (1 - \sqrt{k}\alpha)\rho,
\end{aligned}$$

where we have used (2.10) to calculate R_o and the usual isoperimetric inequality to estimate R_i. Also, from (4.7),

$$(4.8) \quad |x_o - x_i| \le R_o - R_i \le 2\sqrt{\frac{kA(T)}{\pi}}\alpha = 2\sqrt{k}\rho\alpha \le 2\sqrt{k}\alpha.$$

We set
$$(4.9) \qquad\qquad\qquad \varepsilon = 6\sqrt{k}\,\alpha.$$

Now, from (4.7), (4.8), (2.6) and (2.9), it follows that $|x_o - x_i| < R_i$. Since $x_i \in F(T)$, we have that $x_o \in F(T)$. Thus it is easily reasoned from (4.7) that $F(T) \subset B(x_i, \bar{R}_o)$ with

$$(4.10) \qquad\qquad \bar{R}_o \le 2R_o - R_i \le (1 + 5\sqrt{k}\alpha)\rho \le 1.11.$$

We will now use (2.12) to estimate the outradius $\hat{R}_o$ of $F(t_0)$. It follows from (4.1) and (4.2) that

$$\hat{R}_o \le (\sqrt{1.01} + \sqrt{0.01})\sqrt{\frac{A(t_0)}{\pi}} \le 1.11R.$$

If we use (2.6), (2.9), (4.1) and (4.7), we find that $F(t_0) \subset B(x_i, \tilde{R}_o)$ where

(4.11) $\qquad \tilde{R}_o \leq 2\hat{R}_o - R_i \leq 2.22R - (1 - \sqrt{k}\alpha)\rho \leq 1.40.$

The estimate in (4.11) follows quite easily from the observation that $B(x_i, R_i) \subset F(T) \subset F(t_0) \subset B(x_i, \tilde{R}_o)$. Again, with x_i as in (4.8), it follows from (4.1), (2.6) and (1.1) that if

(4.12) $\qquad \beta = \dfrac{A(F(t_0) \backslash B(x_i, R))}{A(t_0)}, \quad R = \sqrt{A(t_0)/\pi},$

then

$$\begin{aligned} \beta \quad &\geq \quad \frac{A(\Omega \backslash B(x_i, R)) - A(\Omega \backslash F(t_0))}{\pi} \\ &\geq \quad \frac{A(\Omega \backslash B(x_i, 1)) - A(\Omega \backslash F(t_0))}{\pi} \\ &\geq \quad \alpha - k\alpha^2 \end{aligned}$$

(4.13) $\qquad\qquad\qquad\quad \geq \quad \alpha/2$

Let Γ denote the condenser $\Gamma(\bar{F}(T), I\!\!R^2 \backslash F(t_0))$ whose inner set is $\bar{F}(T)$ and whose outer set is $I\!\!R^2 \backslash F(t_0)$. Here $\bar{S}$ stands for the closure of a set S. The following then summarizes the description of the geometry of the condenser Γ. From (2.6), (2.9) and (4.7) - (4.11), we see that

$\qquad$ (i) $\quad B(x_i, R_i) \subset F(T) \subset B(x_i, \bar{R}_o),$

$\qquad$ (ii) $\quad F(t_0) \subset B(x_i, \tilde{R}_o),$

$\qquad$ (iii) $\quad \bar{R}_o - R_i \leq 4\sqrt{k}\alpha\rho \leq 4\sqrt{k}\alpha \leq \varepsilon,$ where $\varepsilon = 6\sqrt{k}\alpha,$

(4.14) $\quad$ (iv) $\quad 4/5 \leq 1 - 6\sqrt{k}\alpha \leq R_i/\bar{R}_o \leq \rho/\bar{R}_o,$

$\qquad$ (v) $\quad 4/7 \leq (1 - \sqrt{k}\alpha)\rho/1.40 \leq R_i/\tilde{R}_o$

$$\leq \rho/\tilde{R}_o \leq \rho/R \leq \sqrt{\frac{100}{111}},$$

$\qquad$ (vi) $\quad \beta = A(F(t_0) \backslash B(x_i, R))/A(t_0) \geq \alpha/2.$

$\qquad$ (vii) $\quad R_i \leq \rho \leq \bar{R}_o$ and $R \leq \hat{R}_o \leq \tilde{R}_o.$

Let $\mathrm{Cap}(\Gamma)$ be the 2-capacity of Γ, i.e.,

(4.15) $\qquad \mathrm{Cap}(\Gamma) = \mathrm{Cap}_2(\Gamma) = \inf_w \int_{I\!\!R^2} |Dw|^2 dx dy,$

where w is absolutely continuous on $I\!\!R^2$ and takes the value 1 on $\bar{F}(T)$ and 0 on $I\!\!R^2 \backslash F(t_0)$. The minimizer v is harmonic, in $F(t_0)\backslash \bar{F}(T)$, and is given by $(G - t_0)/(T - t_0)$. With ε, x_i, β as above, and $v = (G - t_0)/(T - t_0)$ we now prove a perturbation result for the 2-capacity of the condenser Γ.

5   A general perturbation result.

Before proving the perturbation result for 2-capacities of condensers, we will first describe a new type of symmetrization first introduced in [2]. The version, we use in this work, differs from the one in [2] in that it is designed to handle inward propagation of asymmetry. In the current situation we redistribute the outer set instead of the inner set as was done in [2]. We first describe this symmetrization in a general setting and then apply this to obtain a perturbation type result for the 2-capacity of condensers with special geometry (see the last paragraph of §4).

Let O and Q be two bounded open sets in $I\!\!R^2$. We assume that (i) the origin 0 lies in O, (ii) $\bar{O} \subset Q$, and (iii) ∂O and ∂Q are unions of finitely many Lipschitz curves. The requirement in (i) is purely for the ease of presentation of details and bears no relation to the assumption in Theorem 1. Let $\rho = \sqrt{A(O)/\pi}$ and $R = \sqrt{A(Q)/\pi}$.

For each $\theta \in (-\pi, \pi]$, let $J(\theta) = \{re^{i\theta} : 0 \leq r\}$ be the ray from the origin making an angle θ with the positive x-axis. For a given value of θ, let

$$J(\theta) \cap Q = [r_0, r_1(\theta)) \cup_{j \geq 1} (r_{2j}(\theta), r_{2j+1}(\theta)), (r_0 = 0),$$

the intervals being pairwise disjoint. We now introduce the following quantities necessary to give a redistribution of the area of Q relative to $B(0, R)$.

Set

$$
\begin{aligned}
s(\theta) &= \sup\{r : re^{i\theta} \in J(\theta) \cap O\} \\
t(\theta) &= \inf\{r : re^{i\theta} \in J(\theta) \cap \partial Q\} \\
&= \sup\{r : [0, r) \subset J(\theta) \cap Q\}, \\
(5.1) \quad \hat{s}(\theta) &= \sup\{r : re^{i\theta} \in J(\theta) \cap O, r < t(\theta)\}, \\
\hat{t}(\theta) &= \inf\{r : re^{i\theta} \in J(\theta) \cap \partial Q, r > s(\theta)\}
\end{aligned}
$$

$$
\begin{aligned}
&= \sup\{r : [s(\theta), r) \subset Q\}, \\
N &= \{re^{i\theta} \in Q^c : s(\theta) > t(\theta), t(\theta) < r < s(\theta)\} \\
E &= \{\theta : J(\theta) \cap N \neq \emptyset\} = \{\theta : s(\theta) > t(\theta)\}.
\end{aligned}
$$

Here Q^c stands for the complement of Q. It is useful to observe that $s(\theta) \geq \hat{s}(\theta)$, $\hat{t}(\theta) \geq t(\theta)$ with equality iff $s(\theta) < t(\theta)$. We now distinguish two possibilities in the redistribution of Q.

Case A: Suppose first that $\hat{t}(\theta) \geq R$. Define $\xi(\theta) \geq \hat{t}(\theta)$ by

$$(5.2) \qquad \xi(\theta)^2 = \hat{t}(\theta)^2 + \sum_{i \in I}(r_{2i+1}(\theta)^2 - r_{2i}(\theta)^2),$$

where $I = \{i : r_{2i}(\theta) \geq \hat{t}(\theta)\}$. Then $\xi(\theta) \geq R$.

Case B: If $\hat{t}(\theta) < R$, we distinguish two subcases to define $\xi(\theta) \geq \hat{t}(\theta)$ and $\lambda(\theta) \geq R$.

(i) If $Re^{i\theta} \in J(\theta) \cap Q$, i.e., $r_{2j}(\theta) < R < r_{2j+1}(\theta)$ for some j, then

$$(5.3) \qquad \xi(\theta)^2 = \hat{t}(\theta)^2 + R^2 - r_{2j}(\theta)^2 + \sum_{i \in L}(r_{2i+1}(\theta)^2 - r_{2i}(\theta)^2),$$

where $L = \{i : \hat{t}(\theta) \leq r_{2i}(\theta) < r_{2i+1}(\theta) \leq r_{2j}(\theta)\}$, and

$$(5.4) \qquad \lambda(\theta)^2 = r_{2j+1}(\theta)^2 + \sum_{i \in M}(r_{2i+1}(\theta)^2 - r_{2i}(\theta)^2),$$

where $M = \{i : r_{2i}(\theta) \geq r_{2j+1}(\theta)\}$. Then $\xi(\theta) \leq R$ and $\lambda(\theta) > R$.

(ii) If $Re^{i\theta} \in J(\theta) \backslash Q$, we set

$$(5.5) \qquad \xi(\theta)^2 = \hat{t}(\theta)^2 + \sum_{i \in L'}(r_{2i+1}(\theta)^2 - r_{2i}(\theta)^2),$$

where $L' = \{i : \hat{t}(\theta) \leq r_{2i}(\theta) < r_{2i+1}(\theta) < R\}$, and

$$(5.6) \qquad \lambda(\theta)^2 = R^2 + \sum_{i \in M'}(r_{2i+1}(\theta)^2 - r_{2i}(\theta)^2),$$

where $M' = \{i : r_{2i}(\theta) \geq R\}$. Then $\xi(\theta) \leq R$ and $\lambda(\theta) \geq R$.

Now suppose that $0 < R_i \leq \rho \leq \bar{R}_o$, and $0 < R \leq \tilde{R}_o$ are such that $B(0, R_i) \subset O \subset \bar{O} \subset B(0, \bar{R}_o)$ and $\bar{Q} \subset B(0, \tilde{R}_o)$. Then (5.1)-(5.6) imply

$$
\begin{aligned}
(i) \quad & R_i \leq \hat{s}(\theta) \leq s(\theta) \leq \bar{R}_o, \\
(ii) \quad & R_i \leq t(\theta) \leq \hat{t}(\theta) \leq \xi(\theta) \leq \tilde{R}_o, \\
(iii) \quad & s(\theta) < \hat{t}(\theta) \leq \xi(\theta), \\
(iv) \quad & R_i \leq \hat{s}(\theta) < t(\theta) < s(\theta) \leq \bar{R}_o, \theta \in E, \\
(v) \quad & \text{if } \hat{t}(\theta) < R \text{ then } \xi(\theta) < R \text{ and } \lambda(\theta) \geq R, \\
(vi) \quad & \text{if } \hat{t}(\theta) \geq R \text{ then } \xi(\theta) \geq R, \\
(vii) \quad & R_i \leq \rho \leq \min(R, \bar{R}_o) \text{ and } R \leq \tilde{R}_o. \\
(viii) \quad & \hat{s}(\theta) \leq s(\theta) \text{ and } t(\theta) \leq \hat{t}(\theta) \text{ with equality iff } s(\theta) < t(\theta).
\end{aligned}
$$

(5.7)

It is helpful to note that $R_i \leq \inf_\theta \{\sup\{r : [0,r) \subset J(\theta) \cap O\}\}$ and $\bar{R}_o \geq \sup_\theta s(\theta)$.

Based on (5.1)-(5.7), we now make some easy but useful observations. Suppose that $\beta = A(Q \backslash B(0,R))/A(Q)$. Then by consideration of $B(0,R) \backslash Q$, (5.7) (v) and (vi),

$$
\begin{aligned}
0 < 2\pi R^2 \left(\beta - \frac{A(N \cap B(0,R))}{\pi R^2}\right) & \leq \int_{\{\xi(\theta) \leq R\}} R^2 - \xi(\theta)^2 d\theta \\
& \leq 2\pi R^2 \beta.
\end{aligned}
$$

(5.8)

By consideration of $Q \backslash B(0,R)$, (5.7) (v) and (vi),

$$
\begin{aligned}
2\pi R^2 \beta & \leq \int_{\{\xi(\theta) \geq R\}} \xi(\theta)^2 - R^2 d\theta + \int_{\{\hat{t}(\theta) < R\}} \lambda(\theta)^2 - R^2 d\theta \\
& \leq 2\pi R^2 \left(\beta + \frac{A(N \backslash B(0,R))}{\pi R^2}\right).
\end{aligned}
$$

(5.9)

Subtracting (5.8) from (5.9), we get

$$
(5.10) \quad 0 \leq \int_{-\pi}^{\pi} \xi(\theta)^2 - R^2 d\theta + \int_{\{\hat{t}(\theta) < R\}} \lambda(\theta)^2 - R^2 d\theta \leq 2A(N).
$$

By adding (5.8) and (5.9), and employing (5.7) (v), we get

$$
(5.11) \quad \int_{-\pi}^{\pi} |\xi(\theta)^2 - R^2| d\theta \leq 4\pi R^2 \left(\beta + \frac{A(N)}{\pi R^2}\right).
$$

Also define μ and $\bar{\mu}$ by

$$
(5.12) \quad 2\pi R^2 \mu = \int_{-\pi}^{\pi} \xi(\theta)^2 - R^2 d\theta, \quad 2\pi R^2 \bar{\mu} = \int_{-\pi}^{\pi} s(\theta)^2 - \rho^2 d\theta.
$$

By (5.1), (5.7) (v) and (5.10),

$$(5.13) \qquad \mu \leq \frac{A(N)}{\pi R^2} \text{ and } \bar{\mu} \geq 0.$$

We now prove a perturbation result for 2-capacity.

Let O and Q be as before, satisfying $B(0, R_i) \subset O \subset \bar{O} \subset B(0, \bar{R}_o)$, and
$O \subset \bar{O} \subset Q \subset \bar{Q} \subset B(0, \tilde{R}_o)$. Let $\rho = \sqrt{A(O)/\pi}$ and $R = \sqrt{A(Q)/\pi}$.
Then
$0 < R_i \leq \rho \leq \bar{R}_o$ and $R \leq \tilde{R}_o$. Assume further that

$$(5.14) \qquad \begin{array}{ll} (i) & \text{for a fixed } \varepsilon, \ 0 < \varepsilon \leq 1/2, \ (1 - \varepsilon)\bar{R}_o < R_i \leq \rho \leq \bar{R}_o, \\ (ii) & 1/2 \leq \min\{R_i/\tilde{R}_o, R_i/\bar{R}_o, \rho/R\} \\ & \qquad \leq \max\{R_i/\tilde{R}_o, R_i/\bar{R}_o, \rho/R\} \leq 1, \\ (iii) & \text{for a fixed } \delta, 0 < \delta \leq 1/2, \ 1/4 \leq (\rho/R)^2 \leq 1/(1 + \delta) < 1. \end{array}$$

Let Γ denote the condenser $\Gamma(\bar{O}, \mathbb{R}^2 \backslash Q)$; set

$$I = \mathrm{Cap}_2(\Gamma) = \inf_u \int_{Q \backslash \bar{O}} |Du|^2 dx dy,$$

where u is absolutely continuous and takes the value 1 on $\mathbb{R}^2 \backslash Q$ and 0 on $\bar{O}$. Let v denote the minimizer. Then v is harmonic in $Q \backslash \bar{O}$ and assumes the appropriate boundary values. Let $\beta = A(Q \backslash B(0, R))/A(Q)$, where $R = \sqrt{A(Q)/\pi}$ (see line preceding (5.8)). We prove

Lemma 5.1: Let $O, Q, \rho, R, R_i, \bar{R}_o, \tilde{R}_o, \beta, \varepsilon$ and v be as described above. Assume that (5.14) holds. Then for all sufficiently small $\varepsilon > 0$, we have

$$I = \mathrm{Cap}_2(\Gamma) = \int_{Q \backslash \bar{O}} |Dv|^2 dx dy \geq \frac{2\pi}{\log R/\rho} + K_0 \beta^2 - K_1 \varepsilon^2 - K_2 \varepsilon \beta,$$

where K_0, K_1 and K_2 are positive constants depending only on δ.

Proof: Throughout this proof C_j's will denote positive constants which are either absolute or depend only on δ. We shall employ the symmetrization described above with the same notations as in (5.1)-(5.6).

From (5.7) and (5.14) we see that

$$(i) \quad 0 < t(\theta) - \hat{s}(\theta) \leq \bar{R}_o - R_i \leq \varepsilon \bar{R}_o, \theta \in E,$$
$$(5.15)(ii) \quad (1/e)^2 < 1/4 \leq \min(s(\theta)^2/\xi(\theta)^2, \rho^2/R^2, R_i^2/\bar{R}_o^2, R_i^2/\tilde{R}_o^2),$$
$$(iii) \quad |\rho^2 - s(\theta)^2| \leq 2\varepsilon \bar{R}_o^2,$$
$$(iv) \quad 1 - \varepsilon \leq R_i/\bar{R}_o \leq s(\theta)/\bar{R}_o \leq 1.$$

Now

$$I = \int_{Q\setminus\bar{O}} v_r^2 + \frac{1}{r^2} v_\theta^2 \, r dr d\theta$$

$$\geq \int_{Q\setminus\bar{O}} v_r^2 \, r dr d\theta$$

$$(5.16) \qquad \geq \int_{-\pi}^{\pi} \left(\inf_z \int_{J(\theta)\cap\{Q\setminus\bar{O}\}} z_r^2 \, r dr \right) d\theta,$$

where the infimum is taken over $z = z(r, \theta)$ such that $z = 1$ on $J(\theta) \cap (\mathbb{R}^2\setminus Q)$ and $z = 0$ on $J(\theta) \cap \bar{O}$. The minimizer $\bar{z}$ satisfies the one variable Euler equation $(r\bar{z}')' = 0$ in $J(\theta) \cap \{Q\setminus\bar{O}\}$. We will estimate I by employing the symmetrization described above and obtaining a lower bound for the inner integral on the right hand side of (5.16). We do this by first solving for $\bar{z}$ from the o.d.e. over the disjoint intervals $(s(\theta), \hat{t}(\theta))$ and $(\hat{s}(\theta), t(\theta))$, the latter occuring whenever $s(\theta) > t(\theta)$, i. e., when $\theta \in E$ (see (5.1)). Note that $\bar{z}$ vanishes on the left end points of these intervals and takes the value 1 on the right end points. Also see (5.7). Thus a lower bound for I is obtained by calculating the inner integral for this function $\bar{z}$ over the above mentioned intervals. Recalling the definition of E from (5.1), it follows from (5.16), (5.7) and (5.1) that

$$I \geq \int_{-\pi}^{\pi} \frac{1}{\log(\hat{t}(\theta)/s(\theta))} d\theta + \int_E \frac{1}{\log(t(\theta)/\hat{s}(\theta))} d\theta$$

$$(5.17) \qquad \geq \int_{-\pi}^{\pi} \frac{1}{\log(\xi(\theta)/s(\theta))} d\theta + \int_E \frac{1}{\log(t(\theta)/\hat{s}(\theta))} d\theta.$$

If the second integral, on the right hand side of (5.17), is larger than $4\pi/\log(R/\rho)$, then Lemma 5.1 follows trivially from (5.15) (ii). Otherwise,

$$\int_E \frac{1}{\log(t(\theta)/\hat{s}(\theta))} d\theta \leq \frac{4\pi}{\log(R/\rho)}.$$

But $\log(t(\theta)/\hat{s}(\theta)) \leq (t(\theta)/\hat{s}(\theta) - 1)$; it then follows from (5.15) (i), (5.7) (iv), (5.15) (ii), (5.14) (iii) that

$$meas_\theta E \leq \frac{16\pi\varepsilon}{\log(1+\delta)} = C_1\varepsilon.$$

Recalling the definition of N from (5.1), (5.7) (iv) and (5.14) (i) yield

$$(5.18) \qquad A(N) \leq \frac{32\pi\varepsilon^2}{\log(1+\delta)}\bar{R}_o^2 = C_2\varepsilon^2\bar{R}_o^2.$$

Now from (5.17),

$$I \geq \int_{-\pi}^{\pi} \frac{1}{\log(\xi(\theta)/s(\theta))}d\theta$$

$$(5.19) \qquad = 2\int_{-\pi}^{\pi} \frac{-1}{\log(s(\theta)^2/\xi(\theta)^2)}d\theta.$$

To estimate (5.19) we note that the function $f(x) = -1/\log x$ satisfies

$$(5.20) \qquad \begin{array}{ll} (i) & f(x) > 0, (0 < x < 1), \\ (ii) & f'(x) > 0, (0 < x < 1), \\ (iii) & f''(x) > 0, (1/e^2 < x < 1). \end{array}$$

We shall use (5.20) in the form

$$(5.21) \qquad f(x) = f(\bar{x}) + f'(\bar{x})(x - \bar{x}) + \frac{f''(\zeta)}{2}(x - \bar{x})^2,$$

for some $\zeta \in (x, \bar{x})$ or $(\bar{x}, x)$. From (5.7) (iii) and (5.14) (iii), it follows that $s(\theta)/\xi(\theta) < 1$ and $1/2 \leq \rho/R \leq 1/\sqrt{1+\delta} < 1$. Then with $\bar{x} = \rho^2/R^2$, it follows from (5.19), (5.20) and (5.21) that

$$I - \frac{2\pi}{\log R/\rho} \geq 2\int_{-\pi}^{\pi} \frac{-1}{\log(s(\theta)^2/\xi(\theta)^2)} + \frac{1}{\log(\rho^2/R^2)}d\theta$$

$$(5.22) \qquad \geq 2f'(\rho^2/R^2)\int_{-\pi}^{\pi}\left(\frac{s(\theta)^2}{\xi(\theta)^2} - \frac{\rho^2}{R^2}\right)d\theta$$

$$+ C_3\int_{-\pi}^{\pi}\left(\frac{s(\theta)^2}{\xi(\theta)^2} - \frac{\rho^2}{R^2}\right)^2 d\theta.$$

The positive constant C_3, in (5.22), results from the fact that (5.15)(ii) implies that $\zeta \geq \min(s(\theta)^2/\xi(\theta)^2, \rho^2/R^2) \geq 1/4 > 1/e^2$.

Next we estimate the quantities

$$B = \int_{-\pi}^{\pi} \frac{s(\theta)^2}{\xi(\theta)^2} - \frac{\rho^2}{R^2} d\theta, \text{ and } D = \int_{-\pi}^{\pi} \left(\frac{s(\theta)^2}{\xi(\theta)^2} - \frac{\rho^2}{R^2}\right)^2 d\theta.$$

We may rewrite B as

$$
\begin{aligned}
B &= \int_{-\pi}^{\pi} s(\theta)^2 \left(\frac{1}{\xi(\theta)^2} - \frac{1}{R^2}\right) + \frac{s(\theta)^2 - \rho^2}{R^2} d\theta \\
&= \int_{-\pi}^{\pi} (s(\theta)^2 - \rho^2) \left(\frac{1}{\xi(\theta)^2} - \frac{1}{R^2}\right) \\
&\quad + \rho^2 \left(\frac{1}{\xi(\theta)^2} - \frac{1}{R^2}\right) + \frac{s(\theta)^2 - \rho^2}{R^2} d\theta.
\end{aligned}
$$

(5.23)

By (5.12) and (5.13),

(5.24)
$$\int_{-\pi}^{\pi} \frac{s(\theta)^2 - \rho^2}{R^2} d\theta \geq 0.$$

Also by (5.15) (iii), (5.7) (ii), (5.14) (ii) and (5.11),

$$
\begin{aligned}
\left| \int_{-\pi}^{\pi} (s(\theta)^2 - \rho^2) \left(\frac{1}{\xi(\theta)^2} - \frac{1}{R^2}\right) d\theta \right| &\leq \int_{-\pi}^{\pi} |s(\theta)^2 - \rho^2| \left|\frac{\xi(\theta)^2 - R^2}{\xi(\theta)^2 R^2}\right| d\theta \\
&\leq \frac{C_4 \varepsilon}{R^2} \int_{-\pi}^{\pi} |\xi(\theta)^2 - R^2| d\theta \\
&\leq C_5 \varepsilon \left(\beta + \frac{A(N)}{\pi R^2}\right).
\end{aligned}
$$

(5.25)

By (5.13),

$$
\begin{aligned}
\rho^2 \int_{-\pi}^{\pi} \left(\frac{1}{\xi(\theta)^2} - \frac{1}{R^2}\right) d\theta &= \rho^2 \int_{-\pi}^{\pi} \left\{\frac{R^2 - \xi(\theta)^2}{R^2 \xi(\theta)^2} + \frac{\xi(\theta)^2 - R^2}{R^4} - \frac{\mu}{R^2}\right\} d\theta \\
&= \rho^2 \int_{-\pi}^{\pi} \frac{(R^2 - \xi(\theta)^2)^2}{R^4 \xi(\theta)^2} d\theta - \frac{2\pi \mu \rho^2}{R^2} \\
&\geq \frac{-2A(N)\rho^2}{R^4}.
\end{aligned}
$$

(5.26)

Putting together (5.23), (5.24), (5.25) and (5.26) we have

(5.27)
$$B \geq \frac{-2A(N)\rho^2}{R^4} - C_5 \varepsilon \left(\beta + \frac{A(N)}{\pi R^2}\right).$$

We now estimate D. It is easy to see that

$$(5.28) \quad \frac{1}{2}\left(\frac{s(\theta)^2}{\xi(\theta)^2} - \frac{s(\theta)^2}{R^2}\right)^2 \leq \left(\frac{\rho^2}{R^2} - \frac{s(\theta)^2}{R^2}\right)^2 + \left(\frac{s(\theta)^2}{\xi(\theta)^2} - \frac{\rho^2}{R^2}\right)^2.$$

Integrating with respect to θ and using (5.15) (iii) and (5.14), we have

$$(5.29) \quad \int_{-\pi}^{\pi}\left(\frac{\rho^2 - s(\theta)^2}{R^2}\right)^2 d\theta \leq \frac{8\pi \bar{R}_o^4 \varepsilon^2}{R^4} = 8\pi\varepsilon^2\left(\frac{\bar{R}_o}{R_i}\frac{R_i}{\rho}\frac{\rho}{R}\right)^4 \leq C_6\varepsilon^2.$$

Using Hölder's inequality,

$$(5.30) \quad \left(\int_{\{\xi(\theta)\leq R\}} R^2 - \xi(\theta)^2 d\theta\right)^2 \leq \left(\int_{-\pi}^{\pi}|\xi(\theta)^2 - R^2|d\theta\right)^2$$
$$\leq 2\pi \int_{-\pi}^{\pi}(\xi(\theta)^2 - R^2)^2 d\theta.$$

Next

$$\int_{-\pi}^{\pi}\left(\frac{s(\theta)^2}{\xi(\theta)^2} - \frac{s(\theta)^2}{R^2}\right)^2 d\theta = \int_{-\pi}^{\pi}\frac{s(\theta)^4}{\xi(\theta)^4 R^4}\left(R^2 - \xi(\theta)^2\right)^2 d\theta$$
$$(5.31) \qquad\qquad\qquad \geq \inf\frac{s(\theta)^4}{\xi(\theta)^4 R^4}\int_{-\pi}^{\pi}\left(R^2 - \xi(\theta)^2\right)^2 d\theta.$$

First employing (5.8) in (5.30) and (5.15) (ii) in (5.31), and then combining the two estimates we see that

$$(5.32) \quad \int_{-\pi}^{\pi}\left(\frac{s(\theta)^2}{\xi(\theta)^2} - \frac{s(\theta)^2}{R^2}\right)^2 d\theta \geq C_7\left(\beta - \frac{A(N)}{\pi R^2}\right)^2.$$

Thus (5.28), together with the estimates in (5.29) and (5.32), yields

$$(5.33) \quad D = \int_{-\pi}^{\pi}\left(\frac{s(\theta)^2}{\xi(\theta)^2} - \frac{\rho^2}{R^2}\right)^2 d\theta \geq C_8\beta^2 - C_9\varepsilon^2 - C_{10}\frac{A(N)}{\pi R^2}.$$

By the assumptions in (5.14) and the conclusions of (5.15), the positive constants $C_1 - C_{10}$ are either absolute or depend only on δ. The estimates in (5.18), (5.27) and (5.33) when used in (5.22) yield the estimate of the lemma, namely,

$$I \geq \frac{2\pi}{\log R/\rho} + K_0\beta^2 - K_1\varepsilon\beta - K_2\varepsilon^2,$$

where K_0, K_1, K_2, are positive constants depending only on δ and become absolute once a value for δ is chosen. ∎

Remark 5.1: We intend to use Lemma 5.1 with $F(T) = O, F(t_0) = Q$ and $x_i = 0$ (see Section 4 and (4.15)). It is easily verified, with $\varepsilon = 6\sqrt{k}\alpha$ (see (4.9)) and $\delta = 0.11$, that (4.14) together with (2.6) implies (5.14). Thus, with $\beta = A(F(t_0)\backslash B(x_i, R))/A(t_0) \geq \alpha/2$ (see (4.13)), it is easily seen that there are absolute constants $K > 0$ and $k_1 > 0$ such that for $k \leq k_1$ we have the following lower bound in (4.15), namely,

$$(5.34) \qquad \mathrm{Cap}(\Gamma) \geq \frac{2\pi}{\log R/\rho} + K\alpha^2.$$

In Section 6, we work out the proof of Theorem 1 in Case 2.

6   Proof of Theorem 1 in Case 2

By (4.1), it follows that

$$(6.1) \qquad \frac{1}{2}\log\frac{A(\Omega)}{A(T)} = \frac{1}{2}\log\frac{\pi}{A(T)} \geq \frac{1}{2}\log\frac{A(t_0)}{A(T)} = \log\frac{R}{\rho}.$$

Thus it is seen from (6.1), (5.34) and (4.15) with $v = (G - t_0)/(T - t_0)$ (see the last paragraph of §4) and $k \leq k_1$, that

$$(6.2) \quad \int_{F(t_0)\backslash\bar{F}(T)} |DG|^2 dxdy \;=\; (T - t_0)^2 \mathrm{Cap}_2(\Gamma)$$

$$\geq \; (T - t_0)^2 \left\{ \frac{4\pi}{\log(\pi/A(T))} + K\alpha^2 \right\}.$$

We now estimate t_0 in (4.1); recall that $G(x) = t_0$ on $\partial F(t_0)$ (also see (4.2)). Let $\bar{\Gamma}$ be the condenser $\bar{\Gamma}(\bar{F}(t_0), \mathbb{R}^2\backslash\Omega)$. By symmetrization [4; p. 3] we find

$$\mathrm{Cap}_2(\bar{\Gamma}) = \frac{1}{t_0^2} \int_{\Omega\backslash\bar{F}(t_0)} |DG|^2 dxdy \geq \frac{4\pi}{\log(\pi/A(t_0))}.$$

It follows from (4.1) that

$$(6.3) \qquad t_0^2 \leq \frac{1}{4\pi}\log\frac{1}{1 - k\alpha^2}\int_{\Omega\backslash\bar{F}(t_0)} |DG|^2 dxdy.$$

An application of the divergence theorem yields

$$(6.4) \int_{\Omega\backslash\bar{F}(t_0)} |DG|^2 dxdy \;=\; t_0\int_{\partial F(t_0)} \frac{\partial G}{\partial n} = -t_0\int_{|x|=r} \frac{\partial G}{\partial r}$$

$$= \; 2\pi r t_0(\frac{1}{r} + O(1)) \to 2\pi t_0 \text{ as } r \to 0^+.$$

Employing (6.4) in (6.3) and noting (2.6), we find that

$$(6.5) \qquad\qquad t_0 \le k\alpha^2 \equiv M.$$

We see from (6.2), (4.13) and (2.9) that there is an absolute constant C such that

$$\int_{\Omega \setminus \bar{F}(T)} |DG|^2 dxdy \ge \int_{F(t_0) \setminus \bar{F}(T)} |DG|^2 dxdy$$

$$(6.6) \qquad\qquad \ge \frac{4\pi(T - t_0)^2}{\log(\pi/A(T))}(1 + C\alpha^2).$$

It is easy to check that (see (6.4)),

$$(6.7) \qquad\qquad \int_{\Omega \setminus \bar{F}(T)} |DG|^2 dxdy = 2\pi T.$$

We study two cases, namely, (i) $T > M$ and (ii) $T \le M$, where M is as in (6.5).

Case(i): Let $T > M$. Then from (6.6) and (6.7) we find that

$$\frac{4\pi(T - k\alpha^2)^2}{\log(\pi/A(T))}(1 + C\alpha^2) \le 2\pi T.$$

Simplifying,

$$(T - k\alpha^2)^2 \le \frac{T}{1 + C\alpha^2} \frac{1}{2} \log \frac{\pi}{A(T)}.$$

On expanding the left hand side and simplifying (we may take $C < 1$), we have

$$(6.8) \qquad\qquad T \le \frac{1}{2}\left(1 - \frac{C}{2}\alpha^2\right) \log \frac{\pi}{A(T)} + 2k\alpha^2.$$

Recalling (2.9) we see that by taking $k \le k_2$, k_2 small enough, (6.8) yields

$$(6.9) \qquad\qquad T \le \frac{1}{2}(1 - \bar{C}\alpha^2) \log \frac{\pi}{A(T)},$$

where $\bar{C}$ is an absolute constant. Again with $\rho = \sqrt{A(T)/\pi} \le \sqrt{0.9}$, we get from (6.9) that

$$T \le (1 - \bar{C}\alpha^2) \log \frac{1}{\rho},$$

i.e.,

(6.10) $$G^*(\rho) \leq (1 - \bar{C}\alpha^2) \log \frac{1}{\rho}.$$

We proceed now as in Section 3. We apply the usual isoperimetric inequality to obtain (see (3.2) and (3.3))

$$\frac{dt}{dA} \geq -\frac{1}{2A}.$$

For $t > T$,

$$t - T \leq \frac{1}{2} \log \frac{A(T)}{A(t)}.$$

With $r = \sqrt{A(t)/\pi}$, we find

(6.11) $$G^*(r) - G^*(\rho) \leq \log \frac{\rho}{r}.$$

Combining (6.10) and (6.11), for $0 < r < \sqrt{0.8} \leq \sqrt{A(T)/\pi}$, we deduce that

$$\begin{aligned} G^*(r) &\leq \log \frac{\rho}{r} + (1 - \bar{C}\alpha^2) \log \frac{1}{\rho} \\ &= \log \frac{1}{r} - \bar{C}\alpha^2 \log \frac{1}{\rho}. \end{aligned}$$

Recalling (2.9) we obtain

(6.12) $$G^*(r) \leq \log \frac{1}{r} - \tilde{C}\alpha^2, \quad 0 < r < \sqrt{0.8}.$$

Thus (6.12) implies Theorem 1 in Case (i).

Case (ii): We now consider the case $T \leq M$, where M is as in (6.5). Again, (6.11) holds for $t > T$, i.e., for $0 < r < \rho$. Thus

$$\begin{aligned} t \leq \log \frac{\rho}{r} + T &\leq \log \frac{\rho}{r} + k\alpha^2 \\ &= \log \frac{1}{r} - \log \frac{1}{\rho} + k\alpha^2 \\ &\leq \log \frac{1}{r} - \frac{1}{2} \log \frac{1}{0.9} + k\alpha^2. \end{aligned}$$

Once again the estimate (2.5) in Theorem 1 holds for k small.

    Thus Theorem 1 is completely proven for domains Ω that are simply connected. ∎

7   The multiply connected case.

We now present details of the proof of Theorem 1 when Ω is multiply connected. The proof in Case 1 undergoes no change. However, as the study of the geometry of sets in Case 2 involves the use of Bonnesen inequalities, we need to modify the calculations somewhat. The basic aim is to ensure that the estimates of Section 4 continue to hold, with minor modifications, and that the geometry of the condenser Γ is not significantly altered (see (4.15)). Thus the result of Lemma 5.1 can be applied even when Ω is multiply connected. This will in turn ensure that the analysis in Section 6 continues to hold thus yielding the estimate (2.5) in Case 2.

Let us then begin by recalling the definitions of T and t_0 (see (2.9), (2.10), (4.1) and (4.2)). Firstly, T is such that

(7.1) $0.8\pi \leq A(T) \leq 0.9\pi,$

and
(7.2) $L(\partial F(T))^2 < 4\pi(1 + k\alpha^2)A(T).$

Secondly, t_0 is such that

(7.3) $(1 - k\alpha^2)\pi \leq A(t_0) \leq \pi,$

and
(7.4) $L(\partial F(t_0))^2 < 4\pi(1.01)A(t_0).$

Noting that Green's function $G(x)$ has only one pole in Ω, it follows from the maximum principle that $F(t)$ has only one component. However, in general, $F(t)$ would be multiply connected except, perhaps when t is large. With some abuse of notation, let $H(t)$ denote the holes of $F(t)$, i.e., the set $F(t) \cup H(t)$ is simply connected. Since $T > t_0$, $F(T) \subset F(t_0)$. We now prove a lemma that will provide us with an estimate for the perimeter and the area of the holes $H(T)$ of $F(T)$.

Lemma 7.1: Let $D \subset \mathbb{R}^2$ be a bounded domain whose boundary ∂D consists of a finite union of rectifiable curves. Let H denote the holes of D, i.e., the set $D \cup H$ is simply connected. Suppose that for some small $\delta > 0$, we have

$$L(\partial D)^2 \leq 4\pi(1 + \delta)A(D).$$

Then

(7.5) $$L(\partial H) \leq \frac{\delta}{2} L(\partial D) \text{ and } A(H) \leq \frac{\delta^2}{4} A(D).$$

Proof: Let $S = D \cup H$; then $L(\partial S) + L(\partial H) = L(\partial D)$, and $A(S) = A(D) + A(H)$. Thus,

$$L(\partial D)^2 = \{L(\partial S) + L(\partial H)\}^2 \leq 4\pi(1 + \delta)A(D).$$

Expanding, we obtain

(7.6) $$L(\partial S)^2 + 2L(\partial S)L(\partial H) + L(\partial H)^2 \leq 4\pi(1 + \delta)A(D).$$

Applying the usual isoperimetric inequality for S and H, employing $A(D) \leq A(S)$ and simplifying (7.6), we see that

$$2\sqrt{(A(D) + A(H))A(H)} + A(H) \leq \delta A(D).$$

Thus,

(7.7) $$A(H) \leq \frac{\delta^2}{4} A(D).$$

Again using $A(D) \leq A(S)$, (7.6) yields

$$2L(\partial S)L(\partial H) + L(\partial H)^2 \leq 4\pi \delta A(S) \leq \delta L(\partial S)^2.$$

Finally,

$$L(\partial H) \leq \frac{\delta}{2} L(\partial S) \leq \frac{\delta}{2} L(\partial D).$$

Thus (7.5) holds.   ∎

Our strategy for multiply connected domains is as follows. Set $B = F(T) \cup H(T)$ and $H = B \cap H(t_0)$, i.e., H denotes the holes of $F(t_0)$ that lie in B. Note that $H \subset H(T)$. It is clear that B is simply connected. Also set $D = H \cup F(t_0)$. The motivation for this choice of H follows from the observation that (7.2) and Lemma 7.1 imply that $A(H) \leq A(H(T)) \leq k^2\alpha^4$ (see (7.8)). However, it is not clear that $A(H(t_0))$ can be bounded by such a term (see (7.4)). With these modifications, we employ the methods of Section 4 to describe the geometry of the condenser $\Gamma' = \Gamma'(\bar{B}, I\!\!R^2 \backslash D)$ and conclude (4.14). Now, if Γ is the condenser $\Gamma(\bar{F}(T), I\!\!R^2 \backslash F(t_0))$ then $\text{Cap}(\Gamma) \geq \text{Cap}(\Gamma')$. We use Lemma 5.1 to get a lower bound for $\text{Cap}(\Gamma')$. The desired estimate for $\text{Cap}(\Gamma)$ will then be shown to follow from this estimate for $\text{Cap}(\Gamma')$.

We now proceed with the calculations to estimate the inradius and outradius of the set $B = F(T) \cup H(T)$. Recall (7.1) and (7.2). By an application of Lemma 7.1,

(7.8) $A(H(T)) \leq \dfrac{k^2\alpha^4}{4} A(T)$ and $L(\partial H(T)) \leq \dfrac{k\alpha^2}{2} L(\partial F(T)).$

Call $S = \partial F(T) \backslash \partial H(T) = \partial B$. Note that $L(\partial B) = L(S) \leq L(\partial F(T))$ and $A(T) \leq A(B)$. If R_i and R_o denote the inradius and the outradius of B respectively then (2.11), (2.12), (2.13), (7.1), (7.2) and (7.8) yield

$$\pi^2 (R_o - R_i)^2 \leq L(\partial B)^2 - 4\pi A(B)$$

$$\leq L(\partial F(T))^2 - 4\pi A(T) \leq 4\pi k \alpha^2 A(T).$$

Thus,

(7.9) $$0 \leq R_o - R_i \leq 2\sqrt{\dfrac{kA(T)}{\pi}}\, \alpha \leq 2\sqrt{k}\alpha.$$

Also,

$$\begin{aligned}
R_o &\leq \dfrac{1}{2\pi}\left\{ L(\partial B) + \sqrt{L(\partial B)^2 - 4\pi A(B)} \right\} \\
&\leq \dfrac{1}{2\pi}\left\{ L(\partial F(T)) + \sqrt{L(\partial F(T))^2 - 4\pi A(T)} \right\} \\
&\leq \dfrac{1}{2\pi} L(\partial F(T)) + \alpha\sqrt{\dfrac{kA(T)}{\pi}}
\end{aligned}$$

(7.10) $\leq (1 + 2\sqrt{k}\alpha)\sqrt{\dfrac{A(T)}{\pi}},$

and using the usual isoperimetric inequality for $F(T)$, we see

$$\begin{aligned}
R_i &\geq \dfrac{1}{2\pi}\left\{ L(\partial B) - \sqrt{L(\partial B)^2 - 4\pi A(B)} \right\} \\
&\geq \dfrac{1}{2\pi}\left\{ L(\partial F(T)) - L(\partial H(T)) - \sqrt{L(\partial F(T))^2 - 4\pi A(T)} \right\} \\
&\geq \left\{ (1 - k\alpha^2/2) - \sqrt{k}\alpha \right\}\sqrt{\dfrac{A(T)}{\pi}}
\end{aligned}$$

(7.11) $\geq (1 - 1.05\sqrt{k}\alpha)\sqrt{\dfrac{A(T)}{\pi}}.$

Clearly, there exist x_o and x_i in $\mathbb{R}^2$ such that S is contained in the region between the two circles $C_o = \{x : |x - x_o| = R_o\}$ and $C_i = \{x : |x - x_i| = R_i\}$. It is easy to see from (7.9) that

(7.12) $$|x_o - x_i| \leq R_o - R_i \leq 2\sqrt{k}\alpha.$$

Set
(7.13) $\varepsilon = 6.5\sqrt{k}\alpha.$

It is necessary to estimate the outradius $\hat{R}_o$ of $F(t_0)$, which is same
as the outradius of $F(t_0) \cup H(t_0)$. From (7.4) and Lemma 7.1, we see
that $A(H(t_0)) \leq 0.0001A(t_0)$ and $L(\partial F(t_0)) \leq 0.005L(\partial F(t_0))$. Set
$E = F(t_0) \cup H(t_0)$. Using $L(\partial E) \leq L(\partial F(t_0))$, $A(t_0) \leq A(E)$ and (7.4)
in (2.12), we find

$$\begin{aligned}
\hat{R}_o &\leq \frac{1}{2\pi}\left\{L(\partial E) + \sqrt{L(\partial E)^2 - 4\pi A(E)}\right\} \\
&\leq \frac{1}{2\pi}\left\{L(\partial F(t_0)) + \sqrt{L(\partial F(t_0)^2 - 4\pi A(t_0)}\right\} \\
&\leq \{\sqrt{1.01} + \sqrt{0.01}\}\sqrt{\frac{A(t_0)}{\pi}} \\
&\leq 1.11\sqrt{\frac{A(t_0)}{\pi}}.
\end{aligned}$$

Let $H = B \cap H(t_0)$, i.e., H denotes the holes of $F(t_0)$ which lie in
B. Clearly, $H \subset H(T)$, and (7.8) implies

(7.14) $A(H) \leq A(H(T)) \leq \dfrac{k^2\alpha^4}{4}A(T) \leq \dfrac{k^2\alpha^4}{4}A(t_0) \leq \dfrac{\pi}{4}k^2\alpha^4.$

The set H need not contain all the holes of $F(t_0)$, and so $(\mathbb{R}^2\backslash B)\cap H(t_0)$
may be nonempty. Now set $R = \sqrt{A(t_0)/\pi}$, $\rho = \sqrt{A(T)/\pi}$ and $D = F(t_0) \cup H$. Define

(7.15) $R' = \sqrt{\dfrac{A(D)}{\pi}}$, $\rho' = \sqrt{\dfrac{A(B)}{\pi}}$, and $\beta' = \dfrac{A(D\backslash B(x_i, R'))}{A(D)}$,

where x_i is as in (7.12). In the event that Ω is multiply connected, x_o
may not lie in $F(t_0)$. Observe that (7.1), (7.3) and (7.14) imply the
following, namely,

$$\sqrt{0.8} \leq \rho \leq \rho' = \sqrt{\frac{A(B)}{\pi}} = \sqrt{\frac{A(T) + A(H(T))}{\pi}}$$

(7.16) $\leq \rho\sqrt{1 + (k^2\alpha^4)/4},$

and

$$\sqrt{0.999} \leq \sqrt{1 - k\alpha^2} \leq R \leq R' = \sqrt{\frac{A(t_0) + A(H)}{\pi}}$$

(7.17) $\leq R\sqrt{1 + (k^2\alpha^4)/4}.$

We may now carry out the calculations in (4.10), (4.11) and conclude (4.14) (i) - (v) for the sets B and D. We list these for easy reference. First note that (i) $B(x_i, R_i) \subset B \subset B(x_i, \bar{R}_o)$, where $\bar{R}_o \leq 2R_o - R_i \leq (1 + 5.05\sqrt{k}\alpha)\rho \leq 1.12$, and (ii) $B \subset \bar{B} \subset D \subset B(x_i, \tilde{R}_o)$, where $\tilde{R}_o \leq 2\hat{R}_o - R_i \leq 2.22R - (1 - 1.05\sqrt{k}\alpha)\rho \leq 1.40$.

Furthermore, (a) $R_i \leq \rho' \leq \bar{R}_o$ and $R' \leq \hat{R}_o \leq \tilde{R}_o$, (b) $\bar{R}_o - R_i \leq 2(R_o - R_i) \leq 4\sqrt{k}\alpha \leq \varepsilon$, where $\varepsilon = 6.5\sqrt{k}\alpha$, (c) $(1 - \varepsilon) \leq (1 - 6.1\sqrt{k}\alpha) \leq R_i/\bar{R}_o \leq \rho'/\bar{R}_o \leq 1$, (d) $4/7 \leq (1 - 1.05\sqrt{k}\alpha)\rho/1.40 \leq R_i/\tilde{R}_o \leq \rho'/\tilde{R}_o \leq \rho'/R' \leq \sqrt{0.9(1 + k^2\alpha^4/4)/0.999} \leq \sqrt{10001/11100}$.

Let $\Gamma = \Gamma(\bar{F}(T), I\!\!R^2 \backslash F(t_0))$ denote the condenser consisting of the pair $\bar{F}(T)$ and $I\!\!R^2 \backslash F(t_0)$, and let Γ' be the condenser $\Gamma(\bar{B}, I\!\!R^2 \backslash D)$. Then
$$(7.18) \qquad \mathrm{Cap}(\Gamma) \geq \mathrm{Cap}(\Gamma').$$

Now let β' be as in (7.15). We will now derive a lower bound for $\mathrm{Cap}(\Gamma)$ of the type given by Lemma 5.1 by first calculating a lower bound for $\mathrm{Cap}(\Gamma')$. The final estimate for $\mathrm{Cap}(\Gamma)$ will follow from (7.18). We apply now the symmetrization of Section 5 to the set D, relative to $B(x_i, R')$. Taking $\delta = 0.1$ we may verify (5.14) and (5.15) for $\beta', \varepsilon, \rho', R', R_i, \bar{R}_o$ and $\tilde{R}_o$. Thus we may conclude the estimate in Lemma 5.1 for $\mathrm{Cap}(\Gamma')$, namely,

$$(7.19) \qquad \mathrm{Cap}(\Gamma') \geq \frac{2\pi}{\log(R'/\rho')} + K_1\beta'^2 - K_2\varepsilon^2 - K_3\varepsilon\beta',$$

where K_1, K_2, and K_3 are absolute constants. Our intention is to express the right hand side of (7.19) in terms of ρ, R, and α. By (7.16) and (7.17), we see that

$$(7.20) \qquad \log\frac{R'}{\rho'} \leq \log\frac{R\sqrt{1 + (k^2\alpha^4)/4}}{\rho} \leq \log\frac{R}{\rho} + \frac{k^2\alpha^4}{8}.$$

Recalling that $\sqrt{0.8} \leq \rho \leq \sqrt{0.9}$, $\sqrt{0.999} \leq R \leq 1$ (see (7.1) and (7.3)) and $\log(1 + x) \geq 2x/3$ $(0 < x < 0.5)$, we see from (7.17) that for $k < k_0$, small, there is an absolute constant K_4 such that

$$(7.21) \qquad \frac{2\pi}{\log(R'/\rho')} \geq \frac{2\pi}{\log(R/\rho)} - K_4 k^2\alpha^4.$$

Recall that the quantity β, in (4.12), is bounded below by $\alpha/2$ (see (4.13)). We now estimate β'; set $\bar{R} = \sqrt{A(\Omega \cup H)/\pi} > 1$. Note that

$D \subset \Omega \cup H$ and so $\bar{R} > R'$; thus (1.1), (7.1) and (7.14) yield

$$
\begin{aligned}
\beta' &= \frac{A(D \backslash B(x_i, R'))}{A(D)} \\
&\geq \frac{A((\Omega \cup H) \backslash B(x_i, R')) - A((\Omega \cup H) \backslash (D))}{A(\Omega \cup H)} \\
&\geq \frac{A(\Omega \backslash B(x_i, \bar{R})) - A(\Omega \backslash F(t_0))}{B(x_i, \bar{R})} \\
&\geq \frac{A(\Omega \backslash B(x_i, 1)) - A(B(x_i, \bar{R}) \backslash B(x_i, 1)) - A(\Omega \backslash F(t_0))}{\pi(1 + k^2 \alpha^4)}
\end{aligned}
$$

$$
(7.22) \quad \geq \frac{\alpha - k^2 \alpha^4 - k \alpha^2}{1 + k^2 \alpha^4} \geq \alpha/2,
$$

where k is small. This gives us (4.14) (vi) for β'.

It is clear from (7.18), (7.19), (7.20), (7.21) and (7.22) that there is a $k_1 > 0$, and an absolute constant $K > 0$, such that

$$
(7.23) \qquad \mathrm{Cap}(\Gamma) \geq \frac{2\pi}{\log(R/\rho)} + K\alpha^2.
$$

The rest of the proof now follows from the analysis in Section 6. ∎

8   The eigenvalue problem.

The problem of ascertaining the dependence on asymmetry of the first eigenvalue λ_1 for the Laplacian

$$
(8.1) \qquad \Delta u + \lambda u = 0 \quad \text{in } \Omega, \qquad u = 0 \quad \text{on } \partial\Omega,
$$

seems more difficult than the corresponding problem for capacity, but can be reformulated in terms of the Green's function.

Conjecture: Let $\lambda = \lambda_1$ be the first eigenvalue for (8.1), where Ω is a bounded plane region. Then there exists a constant C such that

$$
(8.2) \qquad \lambda_1 \geq (1 + C\alpha(\Omega)^2)\lambda_1^*,
$$

where $\lambda_1^* = \lambda(\Omega^*)$ is the first eigenvalue for the disk having the same area as Ω.

The ellipse shows that this conjecture, if true, is best possible. This can be deduced from the expansion in [10; p.326]. At present, the best result is due to Hansen and Nadirashvili [9], who showed that (8.2) is true with exponent 3 in place of 2.

Using the fact that the solution to (8.1) can be written

$$u(z) = \frac{\lambda}{2\pi} \int_\Omega G_\Omega(z, \zeta) u(\zeta) d\zeta,$$

where G_Ω is the Green's function for Ω, it can easily be shown that (8.2) would follow from the hypothetical inequality

(8.3) $\qquad \int_\Omega \int_\Omega \cdots \int_\Omega G_\Omega(0, \zeta_1) G_\Omega(\zeta_1, \zeta_2) \ldots G_\Omega(\zeta_{n-1}, \zeta_n) d\zeta_1 \ldots d\zeta_n$

$\leq (1 - c\alpha^2)^n \int_{\Omega^*} \int_{\Omega^*} \cdots \int_{\Omega^*} G_{\Omega^*}(0, \zeta_1) G_{\Omega^*}(\zeta_1, \zeta_2) \ldots G_{\Omega^*}(\zeta_{n-1}, \zeta_n) d\zeta_1 \ldots d\zeta_n.$

where Ω^* is the disk of the same area as Ω. The case $n = 1$ in (8.3) follows from Theorem 1.

References

1. C. Bandle, Isoperimetric Inequalities and Applications, Pitman Monographs and Studies in Math., 7, Boston, 1980.

2. T. Bhattacharya and A. Weitsman, Bounds for capacities in terms of asymmetry, To appear in Revista Matematica Iberoamericana.

3. H. Federer, Geometric Measure Theory, Springer-Verlag, Berlin, 1969.

4. H. Federer and W. Fleming, Normal and integral currents, Ann. of Math. 72 (1960).

5. F. Gehring, Inequalities for condensers, hyperbolic capacity, and extremal length, Mich. Math. J. 18 (1971), 1-21.

6. R. Hall, A quantitative isoperimetric inequality in n dimensional space, J. Reine angew. Math 428 (1992), 61-76.

7. R. Hall, W. Hayman and A. Weitsman, On capacity and asymmetry, J. d'Analyse Math. 56(1991), 87-123.

8. W. Hansen and N. Nadirashvili, Isoperimetric inequalities for capacities, Harmonic Analysis and Discrete Potential Theory, ed. M. Picardello, Plenum Press, 1992, 193-206.

9. W. Hansen and N. Nadirashvili, Isoperimetric inequalities in potential theory, Potential Analysis 3(1994), 1-14.

10. D. Joseph, Parameter and domain dependence of eigenvalues of elliptic partial differential equations, Arch. Rat. Mech. Anal. 24 (1967), 326-351.

11. R. Osserman, Bonnesen-style isoperimetric inequalities, Am. Math. Monthly 86(1972), 1-29.

Indian Statistical Institute   Department of Mathematics
7 SJSS Marg   Purdue University
New Delhi 110 016   W. Lafayette, IN 47907
India   USA

Contemporary Mathematics
Volume **221**, 1999

A Generalized Korteweg-de Vries
Equation in a Quarter Plane

Jerry L. Bona and Laihan Luo

ABSTRACT. An initial- and boundary-value problem for the nonlinear wave
equation

$$u_t + P(u)_x + u_{xxx} = 0 \qquad (*)$$

is considered in the quarter plane $\{(x,t) : x \geq 0,\, t \geq 0\}$ with initial data and
boundary data specified at $t = 0$ and on $x = 0$, respectively. Such problems
arise in the modelling of open-channel flows where the waves are generated by a
wavemaker mounted at one end of a flume, and in other situations where waves
propagate into an undisturbed patch of the dispersive medium. Equation $(*)$,
which is a generalized version of the classical Korteweg-de Vries equation, fea-
tures a general form of nonlinearity in gradient form. With suitable restrictions
on P and with conditions imposed on the initial data and boundary data which
are quite reasonable with regard to potential applications, the aforementioned
initial-boundary-value problem for $(*)$ is shown to be well posed.

Key words: Generalized Korteweg-de Vries equation; quarter-plane problem;
initial-boundary-value problem; nonlinear, dispersive, wave equations.

1. Introduction

This paper is concerned with the initial- and boundary-value problem

$$u_t + P(u)_x + u_{xxx} = 0, \qquad \text{for } x,\, t \geq 0, \qquad (1.1a)$$

$$u(x,0) = f(x), \qquad \text{for } x \geq 0, \qquad (1.1b)$$

$$u(0,t) = g(t), \qquad \text{for } t \geq 0, \qquad (1.1c)$$

where $u = u(x,t)$ is real-valued function of the two real variables x and t, P is a real-
valued function of a real variable and subscripts adorning a function connote partial
differentiation. Equations like (1.1a) are mathematical models for the unidirectional

---

1991 *Mathematics Subject Classification.* Primary 35Q53; Secondary 35Q51, 45G10, 76B15,
76B25.

This work was partially supported by both the National Science Foundation and the W. M.
Keck Foundation.

propagation of small-amplitude long waves in nonlinear dispersive systems. In such applications, u is typically an amplitude or a velocity, x is often proportional to distance in the direction of propagation and t is proportional to elapsed time.

The well-developed theory for the pure initial-value problem for (1.1a) wherein $u(x, 0)$ is specified on the entire real line with zero boundary conditions at $x = \pm\infty$ insures that there exists a unique smooth solution u corresponding to given, smooth initial data f, at least over some time interval $[0, T^*)$, where $T^* = T^*(f) > 0$ [15, 16, 19, 24, 25, 28-31, 33, 39]. If $P(u) = u^p$, for $p < 5$, for example, then T^* may be taken to be $+\infty$ because of certain *a priori* bounds that are available in this case [29]. However, the question of whether or not T^* can be taken to be $+\infty$ in case $p \geq 5$ is open. Numerical results [6, 7, 8] seem to indicate that for $p \geq 5$, solutions of (1.1a) corresponding to significant classes of smooth initial data form singularities in finite time.

The pure initial-value problem is often not practically convenient if one attempts to assess the performance of equations like (1.1a) as models for waves, or to use them predictively. There will usually be difficulty associated with determining the entire wave profile accurately at a given instant of time. Indeed, a much more common situation arises when some sort of wavemaker is used to generate waves at the edge of an undisturbed stretch of the medium in question, which then propagate into the medium. This corresponds to the special case of (1.1) in which $f \equiv 0$. Guided by experimental studies on water waves in channels see [11, 26, 27, 44], Bona and Winther [17, 18] considered the Korteweg-de Vries-equation (KdV-equation henceforth)

$$u_t + u_x + uu_x + u_{xxx} = 0 \tag{1.2}$$

with initial- and boundary-conditions implemented as in (1.1) and proved that such a quarter-plane problem is well posed (see also [20, 22] for theory involving nonlinearities P having the general form in (1.1), but still restricted to grow at most quadratically and with more restricted initial data). Earlier, Bona and Bryant [4] had studied the same quarter-plane problem for the regularized long-wave equation

$$u_t + u_x + uu_x - u_{xxt} = 0 \tag{1.3}$$

and proved it to be well posed. (This latter quarter-plane formulation of the wavemaker problem with appropriate dissipative terms appended was later used to test (1.3) against experimentally determined water-wave data in [11].)

For physical situations other than wave motion on the surface of a perfect fluid, simple models sometimes yield nonlinearities that are somewhat more complex that the quadratic one appearing in (1.2) or (1.3). Examples include internal wave motion and waves in crystalline lattices [37, 38, 41]. This fact gives impetus to the present generalization of the earlier theory.

In this paper, we study the quarter-plane problem (1.1) and show it possesses a unique, global classical solution which depends continuously on variations of the data f and g within their respective function classes. Of course the nonlinearity must be restricted for these results to obtain. Other than being smooth, the non-linearity P will be required to satisfy a one-sided growth condition of the form $\Lambda(u) \leq |u|^\rho$ for all large values of $|u|$ and suitable values of ρ, where $\frac{d\Lambda(u)}{du} = P(u)$. (It is worth note that in our companion paper [9] on the quarter-plane problem for a generalized version

$$u_t + u_x + P(u)_x - \nu u_{xx} - \alpha^2 u_{xxt} = 0 \tag{1.4}$$

of (1.3), a similar, but less restrictive condition appears on P. In this latter reference, initial- and two-point boundary-value problems for (1.4) are also studied. Such two-point boundary-value problems seem rather complicated for (1.1a) since they require the imposition of an extra boundary condition [5, 10, 17].)

The paper is organized as follows. Section 2 outlines briefly the notation and terminology to be used subsequently and presents a statement of the principal result. In Section 3 the regularized problem

$$u_t + P(u)_x + u_{xxx} - \epsilon u_{xxt} = 0, \qquad \text{for } x, \, t \geq 0, \tag{1.5a}$$

$$u(x,0) = f(x), \qquad \text{for } x \geq 0, \tag{1.5b}$$

$$u(0,t) = g(t), \qquad \text{for } t \geq 0, \tag{1.5c}$$

is considered, and shown to admit a satisfactory theory when ϵ is fixed and positive. *A priori* adduced, ϵ-independent bounds for solutions of the regularized problem (1.5) are derived in Section 4 and Section 5. Passage to the limit as $\epsilon \downarrow 0$ in the weak-star topology is effected in Section 6, where smooth solutions of the initial- and boundary-value problem (1.1) are shown to exist. In Section 7, these solutions are shown to lie in more restricted spaces and to depend continuously on the initial- and boundary-conditions.

2. Notation and Statement of the Main Results

We begin with a review of terminology and notation. For an arbitrary Banach space X, the associated norm will be denoted $||\cdot||_X$. If $\Omega = (a, b)$ is a bounded open interval in $\mathbb{R}^+ = (0, +\infty)$ and k a non-negative integer, we denote by $C^k(\bar{\Omega}) = C^k(a, b)$ the functions that, along with their first k derivatives, are continuous on $[a, b]$ with the norm

$$||f||_{C^k(\bar{\Omega})} = \sup_{\substack{x \in \bar{\Omega} \\ 0 \leq j \leq k}} |f^{(j)}(x)|. \tag{2.1}$$

If Ω is an unbounded interval, $C_b^k(\bar{\Omega})$ is defined just as when Ω is bounded except that $f, f', \cdots, f^{(k)}$ are required to be bounded as well as continuous on $\bar{\Omega}$. The norm is defined as in (2.1). Similar definitions apply if Ω is an open set in $\mathbb{R}^N$. The space $C^\infty(\bar\Omega) = \cap_j C^j(\bar\Omega)$ will appear tangentially, but its Frechet-space topology will not be needed. $\mathcal{D}(\Omega)$ is the usual subspace of $C^\infty(\bar\Omega)$ consisting of functions with compact support in Ω. Its dual space $\mathcal{D}'(\Omega)$ is the space of Schwartz distributions on Ω. For $1 \le p < \infty$, $L_p(\Omega)$ connotes those functions f which are pth-power absolutely integrable on Ω with the usual modification in case $p = \infty$. If $s \ge 0$ is an integer and $1 \le p \le \infty$, let $W^{s,p}(\Omega)$ be the Sobolev space consisting of those $L_p(\Omega)$-functions whose first s generalized derivatives lie in $L_p(\Omega)$, with the usual norm,

$$\|f\|_{W^{s,p}(\Omega)}^p = \sum_{k=0}^{s} \|f^{(k)}\|_{L_p(\Omega)}^p.$$

If $p = 2$ we write $H^s(\Omega)$ for $W^{s,2}(\Omega)$. In the analysis of the quarter-plane problem, the spaces $H^s(\Omega)$ will occur often with s a positive integer and $\Omega = \mathbb{R}^+$ or $\Omega = (0, T)$. Because of their frequent occurrence, it is convenient to abbreviate their norms, thusly;

$$\| \cdot \|_s = \| \cdot \|_{H^s(\mathbb{R}^+)} \qquad \text{and} \qquad | \cdot |_{s,T} = \| \cdot \|_{H^s(0,T)}. \qquad (2.2a)$$

If $s = 0$, the subscript s will be omitted altogether, so that

$$\| \cdot \| = \| \cdot \|_{L_2(\mathbb{R}^+)} \qquad \text{and} \qquad | \cdot |_T = \| \cdot \|_{0,T}. \qquad (2.2b)$$

Similarly, $C_b^k(\mathbb{R}^+)$ appears frequently and will be denoted simply C_b^k. In case $k = 0$, we will systematically drop the superscript and so the class of bounded continuous functions on $\mathbb{R}^+$ is written C_b. The notation $H^\infty(\Omega) = \cap_j H^j(\Omega)$ will be used for the C^∞-functions on Ω, all of whose derivatives lie in $L_2(\Omega)$. For $s \ge 1$, $H_0^s(\mathbb{R}^+)$ is the closed linear subspace of $H^s(\mathbb{R}^+)$ of functions f such that $f(0) = f'(0) = \cdots = f^{(s-1)}(0) = 0$. $H_{loc}^s(\Omega)$ is the set of real-valued functions f defined on Ω such that, for each $\varphi \in \mathcal{D}(\Omega)$, $\varphi f \in H^s(\Omega)$. This space is equipped with the weakest topology such that all of the mappings $f \to \varphi f$, for $\varphi \in \mathcal{D}(\Omega)$, are continuous from H_{loc}^s into $H^s(\Omega)$. With this topology, $H_{loc}^s(\Omega)$ is a Fréchet space. If X is a Banach space, T a positive real number and $1 \le p \le +\infty$, denote by $L_p(0, T; X)$ the Banach space of all measurable functions $u : (0, T) \to X$, such that $t \to \|u(t)\|_X$ is in $L_p(0, T)$, with norm

$$\|u\|_{L_p(0,T;X)} = \left(\int_0^T \|u(t)\|_X^p dt \right)^{\frac{1}{p}}, \qquad \text{if } 1 \le p < +\infty,$$

and if $p = \infty$, then

$$\|u\|_{L_\infty(0,T;X)} = \operatorname*{essential\ supremum}_{0<t<T}\{\|u(t)\|_X\}.$$

Similarly, if k is a positive integer, then $C^k(0, T; X)$ denotes the space of all continuous functions $u : [0, T] \to X$, such that their derivatives up to the kth order exist and are continuous. The space $L_{loc}^\infty(\bar{\mathbb{R}}^+; X)$ is the class of measurable maps

$u \colon \bar{\mathbb{R}}^+ \to X$ which are essentially bounded on any compact subset of $\bar{\mathbb{R}}^+$. The abbreviation $\mathcal{B}_T^{k,l}$ will be employed for the functions $u \colon \mathbb{R}^+ \times [0,T] \to \mathbb{R}$ such that $\partial_t^i \partial_x^j u \in C(0,T;C_b)$ for $0 \le j \le k$, and $0 \le i \le l$. This space of functions will carry the norm

$$\|u\|_{\mathcal{B}_T^{k,l}} = \sum_{\substack{0 \le j \le k \\ 0 \le i \le l}} \|\partial_t^i \partial_x^j u\|_{C(0,T;C_b)}.$$

The space $\mathcal{B}_T^{0,0}$ will be abbreviated simply $\mathcal{B}_T$, so that

$$\|u\|_{\mathcal{B}_T} = \sup_{0 \le x,\, 0 \le t \le T} |u(x,t)|.$$

The next few sections are somewhat technical, and it seems useful to state at the outset a sample of one of our principal results so the reader may keep in mind the overall goal of the paper. Throughout the development of our theory, it will be assumed that the nonlinearity P appearing in the differential equation is at least locally Lipschitz. If B is a bounded subset of the real line, then $\gamma(B)$ will denote the Lipschitz constant for P on B, so that $\gamma(B)$ is the smallest number for which

$$|P(z_1) - P(z_2)| \le \gamma(B)|z_1 - z_2|, \qquad \text{for all} \qquad z_1, z_2 \in B. \tag{H1}$$

It will also be presumed that $P(0) = 0$, an assumption that entails no loss of generality since P appears differentiated in the equation.

Main Result. *Let there be given $T > 0$, initial data $f \in H^3(\mathbb{R}^+)$, boundary data $g \in H^2(0,T)$, and assume the compatibility condition $f(0) = g(0)$ to be satisfied. Suppose that in addition to being locally Lipschitz, P satisfies the one-sided growth condition*

$$\limsup_{|s| \to \infty} \frac{\Lambda(s)}{|s|^{10/3}} \le 0, \tag{$**$}$$

where $\frac{d\Lambda(s)}{ds} = P(s)$ and $\Lambda(0) = 0$. Then the initial-boundary-value problem (1.1) has a unique solution $u \in C(0,T;H^3(\mathbb{R}^+))$ which depends continuously on the auxiliary data f and g. If the auxiliary data (f,g) is further restricted by the requirement $f \in H^4(\mathbb{R}^+)$ and the compatibility conditions

$$f(0) = g(0) \qquad and \qquad g'(0) + P'(f(0))f'(0) + f'''(0) = 0$$

are satisfied, the solution u lies in $C(0,T;H^4(\mathbb{R}^+))$.

Remark. The appellation *solution* of (1.1) means a distributional solution of (1.1a) for which the auxiliary conditions (1.1b) and (1.1c) can be given a well-defined sense.

3. The Regularized Problem

In this section attention will be given to the following regularized initial- and boundary-value problem:

$$u_t + P(u)_x + u_{xxx} - \epsilon u_{xxt} = 0, \qquad \text{for } x, t \geq 0, \tag{3.1a}$$

$$u(x,0) = f(x), \qquad \text{for } x \geq 0, \tag{3.1b}$$

$$u(0,t) = g(t), \qquad \text{for } t \geq 0, \tag{3.1c}$$

with the compatibility condition $u(0,0) = f(0) = g(0)$. The positive parameter ϵ will be treated as fixed in this section. Following the development in [17], let

$$v(x,t) = \epsilon^{\frac{1}{p}} u(\epsilon^{\frac{1}{2}}(x-t), \epsilon^{\frac{3}{2}}t),$$

where p is a positive number to be specified later. If $P(u) = cu^{r+1}$, then $p = r$. The function u is a smooth solution of (3.1) if and only if v is a smooth solution of the problem

$$v_t + \epsilon v_x + \epsilon^{\frac{p+1}{p}} P(\epsilon^{-\frac{1}{p}} v)_x - v_{xxt} = 0, \qquad \text{in } \Omega, \tag{3.2a}$$

$$v(x,0) = F(x), \qquad \text{for } x \geq 0, \tag{3.2b}$$

$$v(t,t) = G(t), \qquad \text{for } t \geq 0. \tag{3.2c}$$

Here $\Omega = \{(x,t) : t > 0 \text{ and } x > t\}$, $F(x) = \epsilon^{\frac{1}{p}} f(\epsilon^{\frac{1}{2}} x)$, and $G(t) = \epsilon^{\frac{1}{p}} g(\epsilon^{\frac{3}{2}} t)$. The dependence of F and G on ϵ is suppressed, since ϵ is viewed as fixed for the nonce. The compatibility of f and g at the origin implies and is implied by the relation $F(0) = G(0)$.

By converting the differential equation (3.2a) with initial condition (3.2b) and boundary condition (3.2c) into an integral equation and applying the contraction-mapping theorem to this new equation, a small-time existence theory can be established. The argument closely parallels that worked out in detail in [9, 17], and we therefore content ourselves with a sketch. First regard equation (3.2a) as an ordinary differential equation for the independent variable v_t by considering $\epsilon v_x + \epsilon^{\frac{p+1}{p}} P(\epsilon^{\frac{1}{p}} v)_x$ as a given external force. Solving this second-order equation, performing a formal integration by parts, and following that by an integration from 0 to t leads to the equation

$$v(x,t) = F(x) + (G(t) - F(t))e^{-(x-t)} + \mathbb{B}(v)(x,t), \tag{3.3}$$

where

$$\mathbb{B}(v)(x,t) = \int_t^{+\infty} K(x-t, \xi-t) \int_0^t \left[\epsilon^{\frac{p+1}{p}} P(\epsilon^{-\frac{1}{p}} v(\xi,\tau)) + \epsilon v(\xi,\tau) \right] d\tau d\xi \tag{3.4}$$

and

$$K(x,\xi) = \frac{1}{2} \left[\exp(-(x+\xi)) + \text{sgn}(x-\xi) \exp(-|x-\xi|) \right]. \tag{3.5}$$

For $v \in \mathcal{B}_T$, define the function $\mathbb{A}v$ by

$$(\mathbb{A}v)(x,t) = F(x) + (G(t) - F(t))e^{-(x-t)} + \mathbb{B}(v)(x,t). \tag{3.6}$$

Assuming that F and G are bounded and continuous, it follows that $\mathbb{A}$ is an operator mapping $v \in \mathcal{B}_T$ into itself since K is integrable. Define the quantity $R(T)$ by

$$\frac{1}{2}R(T) = ||\mathbb{A}\vartheta||_{\mathcal{B}_T} \le 2||F||_{C_b(\bar{\mathbb{R}}+)} + ||G||_{C(0,T)}, \tag{3.7}$$

where $\vartheta(x,t) \equiv 0$, and let

$$\mathbf{B}_T = \{w \in \mathcal{B}_T : ||w||_{\mathcal{B}_T} \le R(T)\},$$

and

$$\lambda(T) = T(\epsilon + \gamma(R)), \tag{3.8}$$

where $\gamma(R) = \gamma([-R, R])$ is the Lipschitz constant for P on the set $[-R, R]$. Then for $u, v \in \mathbf{B}_T$, it transpires that

$$||\mathbb{A}u - \mathbb{A}v||_{\mathcal{B}_T} = \sup_{0 \le x} \sup_{0 \le t \le T} |\mathbb{A}u - \mathbb{A}v|$$

$$\le T(\epsilon + \gamma(R))||u - v||_{\mathcal{B}_T}$$

$$\le \lambda(T)||u - v||_{\mathcal{B}_T},$$

and

$$||\mathbb{A}v||_{\mathcal{B}_T} = ||\mathbb{A}v - \mathbb{A}\vartheta||_{\mathcal{B}_T} + ||\mathbb{A}\vartheta||_{\mathcal{B}_T}$$

$$= \lambda(T)||v||_{\mathcal{B}_T} + \frac{1}{2}R(T) \le (\lambda(T) + \frac{1}{2})R(T).$$

If T is chosen small enough so $\lambda(T) \le \frac{1}{2}$, then it follows from the last two inequalities that $\mathbb{A}$ is a contractive mapping of $\mathbf{B}_T$ into itself. These remarks together with the contraction-mapping theorem suffice to establish the following result.

Proposition 3.1. *Let $T > 0$, $F \in C_b(\mathbb{R}^+)$ and $G \in C(0,T)$ be given. Suppose that P is locally Lipschitz continuous. Then there is a positive constant T' depending only on $||F||_{C_b(\mathbb{R}+)}$, $||G||_{C(0,T)}$ and the Lipschitz constant γ such that if $T_0 = \min(T', T)$, then there is a unique solution of (3.6) in $\mathcal{B}_{T_0}$.*

Remark: Uniqueness follows readily from the sort of inequalities displayed above. A detailed view of the uniqueness may be found in [**17**]. Notice that the size of the time interval T' depends only upon the maximum value of F and G.

It will be important in subsequent sections to have smooth solutions, up to the boundaries, of the regularized problem (3.1) at our disposal. This amounts to the program of relating solutions of the integral equation (3.3) to solutions of the regularized initial-boundary-value problem (3.2). The following result will be sufficient for later developments.

Proposition 3.2. *Suppose that $F \in C_b^k(\bar{\mathbb{R}}^+)$ and $G \in C^m(0,T)$, where $k \geq 2$, $m \geq 1$, and $k \geq m$, and that $F(0) = G(0)$. Further suppose $P \in C^{k+m-1}(\mathbb{R})$. Let v be a solution in $\mathcal{B}_{T_0}$ of the integral equation (3.3), where T_0 lies in the interval $(0,T]$. Then it follows that*

$$\partial_x^i \partial_t^j v \subset \mathcal{B}_{T_0}, \qquad \text{for } 0 \leq j \leq m \text{ and } 0 \leq i \leq k+j. \tag{3.9}$$

Conversely, if v is a classical solution of the transformed problem (3.2) in $\bar{\Omega}_T$, then v is a solution of the integral equation (3.3) over $\bar{\Omega}_T$, and so v satisfies (3.9).

The proof follows from the integral equation (3.3) as in [**4, 16, 17**] and so is omitted here. The partial derivatives in (3.9) may be defined at the boundary of Ω_T via the obvious one-sided difference quotients. In case $j > 0$ in (3.9), the condition $\partial_x^i \partial_t^j v \in \mathcal{B}_T$ connotes that this partial derivative exists classically in $\bar{\Omega}_T \backslash \{(0,0)\}$, is bounded and continuous there, and that it may be extended continuously to $\bar{\Omega}_T$.

Suppose a classical solution v of (3.2), defined on $\bar{\Omega}_T$ for some $T > 0$, is in hand, and suppose the boundary data G is defined at least on $[0, T_1]$, where $T_1 > T$. As soon as an *a priori* bound on the L_∞-norm of a solution defined on $\bar{\Omega}_{T_1}$ is provided, it follows from the Remark below Proposition 3.1 that the solution can be extended to $\bar{\Omega}_{T_1}$ by a finite number of iterations of the local existence result propounded in Proposition 3.1. Moreover, if F and G possess the regularity assumed in Proposition 3.2, it follows that the extended solution does as well.

Provision of the relevant *a priori* bound is now considered. Additional conditions on F and G seem to be needed at this stage, namely that the initial data be suitably evanescent at infinity. This condition is quite reasonable from the point of view of the physical situations for which (1.1) serves as a model.

Lemma 3.3. *Let $F \in C_b^k(\bar{\mathbb{R}}^+)$ and $G \in C^m(0,T)$ with $F(0) = G(0)$, where $k \geq 2$, $m \geq 1$ and $k \geq m$. Let v be a solution of (3.2) in $\mathcal{B}_{T_0}$. Let r lie in the range $0 \leq r \leq k$ and suppose that*

$$\partial_x^j F(x) \to 0 \qquad \text{as } x \to +\infty,$$

for $0 \leq j \leq r$. Then it follows that

$$\partial_x^j \partial_t^i v(x,t) \to 0 \qquad \text{as } x \to +\infty,$$

uniformly for $0 \leq t \leq T_0$, for i, j such that $0 \leq i \leq m$ and $0 \leq j \leq r+i$.

The proof of this technical result follows from the representation (3.3) just as in [**3, 4**] and we may safely skip the details.

Attention is now given to the derivation of the *a priori* bounds needed to guarantee the local solutions provided in Proposition 3.1 admit global extensions to solutions of the initial-boundary-value problem (3.2). According to the remarks above, it suffices to show the following. Suppose to be given suitably restricted initial data F defined on $\mathbb{R}^+$ and boundary data G defined on $[0, T]$ for some positive T. Suppose also that u is a correspondingly smooth solution of (3.2) defined on Ω_{T_0} for some $0 < T_0 \leq T$. If it is demonstrated in these circumstances that there is a constant C dependent only on F, G and T such that

$$\|u\|_{C_b(\bar{\Omega}_T)} \leq C, \tag{3.10}$$

then it follows that u can be extended to a solution of (1.1) defined on Ω_T. In particular, if G is given for all $t \geq 0$ and lies in the function class that allows the derivation of (3.10) on any bounded time interval, it will follow that u can be extended as a solution of (3.2) defined on the entire quarter plane $\mathbb{R}^+ \times \mathbb{R}^+$.

A bound on solutions of (3.2) that implies (3.10) is the subject of the next proposition. The proof can be found in [9].

Proposition 3.4. *Let Λ be defined by $\Lambda(0) = 0$ and $\Lambda' = P$. Suppose $F \in C_b^3(\bar{\mathbb{R}}^+) \cap H^1(\mathbb{R}^+)$, $G \in C^1(0, T)$ with $F(0) = G(0)$, and that Λ is at least a C^4- function satisfying the one-sided growth condition*

$$\limsup_{|s| \to \infty} |s|^{-4} \Lambda(s) \leq 0. \tag{H2}$$

Then for any $T > 0$, the system (3.2) with initial and boundary data F and G has a unique solution $v \in \mathcal{B}_T^{3,1} \cap C(0, T; H^1)$. Moreover, if $F \in C_b^3(\bar{\mathbb{R}}^+) \cap H^2(\mathbb{R}^+)$ and $G \in C^1(0, T)$ with $F(0) = G(0)$, and

$$\limsup_{|s| \to \infty} |s|^{-2} |P''(s)| \leq c^*, \tag{H3}$$

for some finite constant c^*, then the initial-boundary-value problem (3.2) has a unique solution $v \in \mathcal{B}_T^{3,1} \cap C(0, T; H^2)$ for arbitrary $T > 0$.

In any case, the solution depends continuously on the initial- and boundary data in the sense that the mapping $(F, G) \longmapsto v$ is continuous from $C_b^3(\bar{\mathbb{R}}^+) \cap H^1(\mathbb{R}^+) \times C^1(0, T)$ to $\mathcal{B}_T^{3,1} \cap C(0, T; H^1)$, or from $C_b^3(\bar{\mathbb{R}}^+) \cap H^2(\mathbb{R}^+) \times C^1(0, T)$ to $\mathcal{B}_T^{3,1} \cap C(0, T; H^2)$.

Corollary 3.5. *Let $f \in C_b^3(\bar{\mathbb{R}}^+) \cap H^2(\mathbb{R}^+)$ and $g \in C^1(0, T)$, where $f(0) = g(0)$. Let u be a classical solution of (3.1), up to the boundary, on $\bar{\mathbb{R}}^+ \times [0, T_0]$. Suppose P satisfies the conditions in Proposition 3.4. Then there exists a constant C dependent on $\|f\|_2$ and on $|g|_{1,T}$ such that any classical solution u of (3.1) defined on $\bar{\Omega}_{T_0}$, for $T_0 \leq T$, satisfies*

$$\|u\|_{C_b(\bar{\Omega}_{T_0})} \leq C.$$

Proof: Let u be a classical solution of (3.1) on $\bar{\Omega}_{T_0}$ for some $T_0 \leq T$. Then since

$$v(x, t) = \epsilon^{\frac{1}{p}} u\left(\epsilon^{\frac{1}{2}}(x - t), \ \epsilon^{\frac{3}{2}} t\right), \tag{3.11}$$

v is a classical solution of (3.2a) on $\bar{\mathbb{R}}^+ \times [0, T_0']$, where $T_0' = \epsilon^{\frac{3}{2}} T_0$, which satisfies the auxiliary conditions (3.8) where

$$F(x) = \epsilon^{\frac{1}{p}} f(\epsilon^{\frac{1}{2}} x), \quad \text{and} \quad G(t) = \epsilon^{\frac{1}{p}} g(\epsilon^{\frac{3}{2}} t).$$

Here ϵ is fixed, and so F and G satisfy the hypotheses of Proposition 3.4. Hence the $H^2(\bar{\mathbb{R}}^+)$-norm of v is bounded on $[0, T_0']$ by a constant that depends on $||F||_2$, and on $|G|_{1,T'}$, say, where $T' = \epsilon^{\frac{3}{2}} T$. By an elementary inequality, one has

$$||v(\cdot, t)||_{C_b(\mathbb{R}^+)} \le \sqrt{2} (||v(\cdot, t)|| \, ||v_x(\cdot, t)||)^{\frac{1}{2}}.$$

It follows that v is bounded on $\bar{\mathbb{R}}^+ \times [0, T_0']$ by a constant C dependent only on $||F||_2$ and $|G|_{1,T'}$. In particular, C does not depend on T_0' for T_0' in the range $[0, T']$. Since u is defined from v by

$$u(x, t) = \epsilon^{-\frac{1}{p}} v\left(\epsilon^{-\frac{1}{2}} x + \epsilon^{-\frac{3}{2}} t, \ \epsilon^{-\frac{3}{2}} t\right), \tag{3.12}$$

the desired result follows. $\square$

Here is one more result about the transformed problem (3.2). The proof follows easily from the integral equation (3.3).

Proposition 3.6. *Let* $F \in C_b^k(\bar{\mathbb{R}}^+) \cap H^k(\mathbb{R}^+), G \in C^m(0, T)$, *where* $F(0) = G(0)$ *and* $k \ge 3, m \ge 1$ *and* $k \ge m$. *Let* v *be the solution of (3.2) in* $\mathcal{B}_{T_0}(\Omega)$, *up to the boundary, where* $\Omega = \{(x, t) : t > 0 \text{ and } x > t\}$. *Suppose* P *satisfies (H3) in the Proposition 3.4 and is in* $C^{k+m-1}(\mathbb{R})$. *Then there exists a constant* C, *dependent on* $||F||_2$ *and* $|G|_{1,T}$ *such that, for each* $t \in [0, T_0]$,

$$||\partial_x^i \partial_t^j v(\cdot, t)||_{L_2((t,\infty))} \le C$$

provided that $0 \le j \le m$ *and* $0 \le i \le k + j$.

It is worth summarizing the accomplishments of the present section. As the transformed problem (3.2) is only of transient interest, the theory is recapitulated in terms of the regularized problem (3.1). Thus the results stated now are consequences of the propositions established above together with the transformation (3.12) taking (3.2) to (3.1).

Theorem 3.7. *Let* $f \in C_b^k(\bar{\mathbb{R}}^+)$ *and* $g \in C^m(0, T)$, *with* $f(0) = g(0)$, *where* $k \ge 3, m \ge 1$ *and* $k \ge m$. *Let* $\epsilon > 0$ *and suppose* P *to lie in* $C^{k+m-1}(\mathbb{R})$. *Then there exists* $T_0 > 0$ *and a unique function* u *in* $C_b(\bar{\mathbb{R}}^+ \times [0, T_0])$ *which is a classical solution of the regularized problem (3.1). Additionally,*

$$\partial_x^i \partial_t^j u \in C_b(\bar{\mathbb{R}}^+ \times [0, T_0]),$$

for i *and* j *such that* $0 \le j \le m$, $0 \le i \le k$ *and* $i + j \le k$. *Suppose* $f \in H^r(\mathbb{R}^+)$, *where* P *satisfies hypothesis (H2) if* $r \ge 1$ *or hypothesis (H3) if* $r \ge 2$ *in Proposition*

3.4. Then u may be extended to a solution of (3.1) on $\bar{\mathbb{R}}^+\times[0,T]$. In this case, there is a constant C which depends on ϵ such that, for $0 \leq t \leq T$,

$$\|\partial_x^i \partial_t^j u(\cdot, t)\| \leq C,$$

for i and j such that $0 \leq j \leq \min\{r, m\}$, $0 \leq i \leq r$ and $i + j \leq r$.

Corollary 3.8. *Let $f \in H^\infty(\mathbb{R}^+)$ and $g \in C^\infty(\mathbb{R}^+)$, with $f(0) = g(0)$. Suppose P satisfies the growth condition (H3) in the Proposition 3.4 and is a C^∞-function. Then there exists a unique solution u of (3.1) on the entire quarter-plane $\bar{\mathbb{R}}^+ \times \bar{\mathbb{R}}^+$ corresponding to the data f and g. Moreover the solution u is bounded on any finite time interval, lies in $C^\infty(\bar{\mathbb{R}}^+ \times \bar{\mathbb{R}}^+)$ and for each $k \geq 0$,*

$$\partial_x^i \partial_t^j u \in C(\bar{\mathbb{R}}^+; H^k(\mathbb{R}^+)),$$

for all $i, j \geq 0$

Proof: The existence of global solutions follows immediately from Theorem 3.7 together with the uniqueness result. Also, for any $i, j \geq 0$, $k > 0$, and $T > 0$, $\partial_x^i \partial_t^j u$ is uniformly bounded in $H^k(\mathbb{R}^+)$, for $0 \leq t \leq T$. Since $u \in L_\infty(0, T; H^k(\mathbb{R}^+))$ and $u_t \in L_\infty(0, T; H^k(\mathbb{R}^+))$, it follows immediately (cf. [35]) that $u \in C(0, T; H^k(\mathbb{R}^+))$. □

4. *A priori* **Bounds in H^3 for the Regularized Problem**

In Section 3, the bounds obtained in H^1 or H^2 for solutions of (3.2) do not appear to yield ϵ-independent bounds on solutions of (3.1) because the transformation (3.12) that takes solutions of (3.2) to solutions of (3.1) is singular at $\epsilon = 0$. In this and the next section, ϵ-independent *a priori* bounds are derived for solutions of the regularized initial- and boundary-value problem (3.1) which, for any fixed $T > 0$, are independent of $t \in [0, T]$.

Throughout this section it will be assumed that $f \in H^\infty(\mathbb{R}^+)$, $g \in C^\infty(0, T)$, and $f(0) = g(0)$. From Corollary 3.8 it is inferred that there is a classical solution $u = u_\epsilon$ of (3.1) corresponding to the auxiliary data f and g which is such that

$$u \in C^\infty(\bar{\mathbb{R}}^+ \times [0, T]),$$

and, for integers $j, k \geq 0$,

$$\partial_t^j u \in C(0, T; H^k(\mathbb{R}^+)).$$

Some preliminary relations, established via energy-type arguments, will be derived in a sequence of technical lemmas. These prefatory results will be combined to obtain ϵ-independent bounds for u within the function class $C(0, T; H^3(\mathbb{R}^+))$ under the assumption

$$\limsup_{|s| \to \infty} |s|^{-\frac{10}{3}} \Lambda(s) \leq 0 \qquad (**)$$

on P, where as before, $\Lambda' = P$ and $\Lambda(0) = 0$. Besides the condition (**), it is also assumed that $P \in C^\infty(\mathbb{R}^+)$, though it will be clear that weaker differentiability suffices for most of the results below. Because of (**) and the fact that $P(0) = 0$, it follows that for any $\delta > 0$, there is a constant $C = C_\delta$ such that

$$\Lambda(s) \leq Cs^2 + \delta s^{\frac{10}{3}}$$

for all $s \geq 0$. Note that, because (**) is a one-sided condition, high growth rates at infinity are not excluded. For example, if $P(s) = -(2k+2)s^{2k+1}$, then $\Lambda(s) = -s^{2k+2}$ satisfies (**) no matter how large the positive integer k.

At various times, constants will arise in our considerations that depend only on the data f and g. Many of these will be denoted simply by C, and this symbol's occurence in different formulae is not taken to connote the same constant.

Lemma 4.1. *Let $f \in H^\infty(\mathbb{R}^+)$, $g \in C^\infty(\mathbb{R}^+)$, with $f(0) = g(0)$. Suppose P satisfies the growth condition (**) and suppose $0 < \epsilon \leq 1$. There exists a positive constant*

$$a_1 = a_1(||f||_1, |g|_{1,T}), \tag{4.1}$$

such that the solution u of (3.1) corresponding to the data f and g satisfies

$$||u(\cdot,t)||_1^2 + \int_0^t \left[u_x^2(0,s) + (u_{xx}(0,s) - \epsilon u_{xt}(0,s))^2\right]ds \leq a_1, \tag{4.2}$$

for $0 \leq t \leq T$, and uniformly for $\epsilon \in (0,1]$.

Proof: Multiply (3.1a) by $2u$ and integrate the resulting relation over $\mathbb{R}^+ \times (0,t)$. After integrations by parts, in which the fact that u and various of its derivatives vanish at $+\infty$ is used repeatedly, it is verified that

$$||u(\cdot,t)||^2 + \epsilon||u_x(\cdot,t)||^2 + \int_0^t u_x^2(0,s)ds$$

$$= \int_0^t \left[2g(s)(u_{xx}(0,s) - \epsilon u_{xt}(0,s)) + 2Q(g(s))\right]ds + ||f||^2 + \epsilon||f'||^2$$

$$\leq C(||f||_1, |g|_{1,T}) + C(|g|_T)\left(\int_0^t \left(u_{xx}(0,s) - \epsilon u_{xt}(0,s)\right)^2 ds\right)^{\frac{1}{2}}, \tag{4.3}$$

where $Q'(\lambda) = \lambda P'(\lambda)$ and $Q(0) = 0$. In particular, we have

$$||u(\cdot,t)||^4 \leq C(||f||_1, |g|_{1,T}) + C(|g|_T)\int_0^t \left(u_{xx}(0,s) - \epsilon u_{xt}(0,s)\right)^2 ds.$$

Next multiply the regularized equation (3.1a) by the combination $2\epsilon u_{xt} - 2u_{xx} - 2P(u)$ and integrate the result over $\mathbb{R}^+ \times (0,t)$. After several integrations by parts, it is seen that

$$\|u_x(\cdot,t)\|^2 + \int_0^t \left[u_{xx}(0,s) - \epsilon u_{xt}(0,s) + P(g(s))\right]^2 ds$$

$$= \|f'\|^2 + 2\int_0^{+\infty} \Lambda(u(x,t))dx - 2\int_0^{+\infty} \Lambda(f(x))dx$$

$$+ \int_0^t \left(\epsilon g_t^2(s) - 2g_t(s)u_x(0,s)\right)ds$$

$$\leq C(\|f\|_1, |g|_{1,T}) \left(1 + \int_0^t u_x^2(0,s)ds\right)^{\frac{1}{2}} + \|u\|^2 \tilde{E}(\|u\|^{\frac{1}{2}} \|u_x\|^{\frac{1}{2}}),$$

where

$$\Lambda'(\lambda) = P(\lambda), \quad \Lambda(\lambda) = \lambda^2 E(\lambda), \quad \text{and} \quad \tilde{E}(r) = \sup_{|\lambda| \leq r} E(\lambda).$$

By using (∗∗), the elementary inequality

$$\|f\|_{C_b(\mathbb{R}^+)}^2 \leq 2\|f\| \|f'\| \tag{4.4}$$

and Young's inequality, the last inequality may be put in the form

$$\|u_x(\cdot,t)\|^2 + \int_0^t \left[u_{xx}(0,s) - \epsilon u_{xt}(0,s) + P(g(s))\right]^2 ds$$

$$\leq C(\|f\|_1, |g|_{1,T}, \frac{1}{\delta}) + \delta\left[\|u(\cdot,t)\|^{\frac{8}{3}} \|u_x(\cdot,t)\|^{\frac{2}{3}} + \int_0^t u_x^2(0,s)ds\right], \tag{4.5}$$

for any choice of $\delta > 0$. By a further use of Young's inequality, it is adduced that

$$\frac{1}{2}\|u_x(\cdot,t)\|^2 + \int_0^t \left[u_{xx}(0,s) - \epsilon u_{xt}(0,s)\right]^2 ds$$

$$\leq C(\|f\|_1, |g|_{1,T}, \frac{1}{\delta}) + \delta\|u(\cdot,t)\|^4. \tag{4.6}$$

Substitute (4.3) into (4.6) to obtain

$$\|u(\cdot,t)\|^4 + \|u_x(\cdot,t)\|^2 + \left(\int_0^t u_x^2(0,s)ds\right)^2$$

$$+ \int_0^t \left[u_{xx}(0,s) - \epsilon u_{xt}(0,s)\right]^2 ds \leq C(\|f\|_1, |g|_{1,T}, \frac{1}{\delta}), \tag{4.7}$$

for a suitable $\delta > 0$. Inequality (4.2) now follows, and the proposition is proved.
□

Remarks: Note that if the boundary data is small in the sense of the norm $|g|_T$, then the condition (∗∗) can be improved. In particular, if the boundary data is zero, then the L_2-bound depends only on the initial data. Hence, following the steps above, if

$$\limsup_{|s|\to\infty} |s|^{-6}\Lambda(s) \leq 0, \qquad\qquad (**')$$

then one derives ϵ-independent H^1-bounds. The restriction $(**')$ is the same restriction imposed on P when pure initial-value problems for equation (1.1a) are considered [**28, 29**]. Note also that if the initial data f and the boundary data g are small enough in $H^1(\mathbb{R}^+)$ and $H^1(0,T)$, respectively, Λ is unrestricted in sign. In fact, from (4.3) and the estimate of $||u_x(\cdot,t)||^2$, one shows that

$$||u(\cdot,t)||_1^2 \big(1 - \delta\tilde{E}(||u||_1)\big) \leq C(||f||_1, |g|_{1,T}),$$

where $\delta = ||f||_1^2 + |g|_{1,T}^2$. If δ is small enough relative to $||f||_1$ and $|g|_{1,T}$, one obtains a global H^1-bound from the above estimate. Furthermore, the bound obtained in this way depends only on the auxiliary data and not explicitly on T. In this situation, $||u(\cdot,t)||_1$ grows at most linearly with the energy supplied by the wavemaker.

From (4.4) it follows that

$$||u||_{C_b(\bar{\mathbb{R}}^+\times[0,T])}^2 \leq 2 \sup_{0\leq t\leq T} \{||u_x(\cdot,t)||||u(\cdot,t)||\} \leq a_1, \qquad (4.8)$$

for all t in $[0,T]$. Using the differential equation (3.1a) and the fact $\int_0^t u_x^2(0,s)ds \leq a_1$, it follows that

$$\int_0^t \big[u_{xxx}(0,s) - \epsilon u_{xxt}(0,s)\big]^2 ds = \int_0^t \big[g_t(s) + a(g(s))u_x(0,s)\big]^2 ds$$
$$\leq C(||f||_1, |g|_{1,T}),$$

where $a(\lambda) = P'(\lambda)$. These conclusions are formalized in the following corollary.

Corollary 4.2. *Let f, g and P satisfy the conditions in Lemma 4.1. Then there is a constant a_1 depending only on $||f||_1$ and $|g|_{1,T}$ such that*

$$||u||_{C_b(\bar{\mathbb{R}}^+\times[0,T])}^2 \leq a_1(||f||_1, |g|_{1,T})$$

and

$$\int_0^t \big(u_{xxx}(0,s) - \epsilon u_{xxt}(0,s)\big)^2 ds \leq a_1,$$

for $0 \leq t \leq T$, and uniformly for $\epsilon \in (0,1]$.

Next we obtain an $H^3(\mathbb{R}^+)$-bound on solutions of (3.1). It will be shown that $||u(\cdot,t)||_3$ is bounded on $[0,T]$, independently of ϵ small enough. First define $A(t)$

and $B(t)$ by

$$A^4(t) = \sup_{0 \le s \le t} \{||u(\cdot, s)||_2^4\} + \left(\int_0^t u_{xxx}^2(0, s) ds \right)^2$$

$$+ \left(\int_0^t u_{xx}^2(0, s) ds \right)^2 + \left(\int_0^t \epsilon u_{xt}^2(0, s) ds \right)^2$$

(4.9a)

and

$$B^2(t) = \sup_{0 \le s \le t} \{||u(\cdot, s)||_3^2\} + \epsilon ||u_{xxxx}(\cdot, t)||^2 + \int_0^t \epsilon ||u_{xt}(\cdot, s)||^2 ds$$

$$+ \int_0^t u_{xxxx}^2(0, s) ds + \int_0^t u_{xt}^2(0, s) ds + \epsilon \int_0^t u_{xxt}^2(0, s) ds.$$

(4.9b)

In fact, it will be demonstrated below that $A(t)$ and $B(t)$ are bounded on $[0, T]$, independently of ϵ small enough. The next lemma gives an $H^2(\mathbb{R}^+)$-estimate not directly effective in bounding $||u(\cdot, t)||_2$, independently of ϵ, but which will prove useful later.

Remark: It seems that for the boundary-value problems (1.1), the H^2-bound is difficult to obtain alone, in the same way that the $L_2(\mathbb{R}^+)$-bound was not derived on its own. A similar problem occurs when a two-point, nonhomogeneous boundary-value problem for the KdV equation is considered (see [**10**]).

Lemma 4.3. *Let $T > 0$, $f \in H^\infty(\mathbb{R}^+)$, $g \in H^\infty(0, T)$, with $f(0) = g(0)$. There exist constants a_2, ϵ_1, C_1 and C_2, where*

$$a_2 = a_2(||f||_2 + \epsilon^{\frac{1}{2}}||f||_3, |g|_{1,T}), \quad \epsilon_1 = \epsilon_1(||f||_1, |g|_{1,T}),$$
$$C_1 = C_1(||f||_1, |g|_{1,T}), \quad and \quad C_2 = C_2(||f||_1, |g|_{1,T}, T),$$

such that the solution of (3.1) corresponding to the data f and g satisfies

$$A^4(t) \le a_2(||f||_2 + \epsilon^{\frac{1}{2}}||f||_3, |g|_{1,T}) + C_1 \int_0^t u_{xt}^2(0, s) ds$$

$$+ C_2 \int_0^t [||u_{xx}(\cdot, s)||^4 + \epsilon^2 ||u_{xxx}(\cdot, s)||^4 + \epsilon ||u_{xx}(\cdot, s)||^8] ds,$$

(4.10)

provided that $t \in [0, T]$ and $\epsilon \in (0, \epsilon_1]$.

Proof: Multiply (3.1a) by $10(u_{xt} - a(u)u_{xx})$ where $a(u) = P'(u)$ as before, differentiate (3.1a) once with respect to x and multiply the result by $2u_{xxx}$, add the equations thus obtained, and integrate their sum over $\mathbb{R}^+ \times (0,t)$. Several integrations by parts and using the relation

$$\int_0^t \int_0^{+\infty} u_{xt} u_x a(u)\,dx\,ds = \int_0^{+\infty} \left[u_x^2(x,t)a(u(x,t)) - (f'(x))^2 a(f(x))\right]dx$$
$$+ \int_0^t g'(s)a(g(s))u_x(0,s)\,ds + \int_0^t \int_0^{+\infty} u_{xx}u_t a(u)\,dx\,ds,$$

yield

$$6\|u_{xx}(\cdot,t)\|^2 + \epsilon\|u_{xxx}(\cdot,t)\|^2 + \int_0^t u_{xxx}^2(0,s)\,ds$$

$$= 6\|f''\|^2 + \epsilon\|f'''\|^2 + \int_0^{+\infty} \left[10u_x^2(x,t)a(u(x,t)) + 5\epsilon u_{xx}^2(x,t)a(u(x,t))\right.$$
$$\left. - 10(f'(x))^2 a(f(x)) - 5\epsilon(f''(x))^2 a(f(x))\right]dx$$

$$+ \int_0^t \left[5\epsilon u_{xt}^2(0,s) - 12u_{xt}(0,s)u_{xx}(0,s) + \frac{1}{2}u_x^4(0,s)a''(g(s))\right.$$

$$\qquad - 2a'(g(s))u_x^2(0,s)u_{xx}(0,s) + 4a(g(s))u_{xx}^2(0,s)$$

$$\qquad + 10g'(s)a(g(s))u_x(0,s) + 5a^2(g(s))u_x^2(0,s) - 5(g'(s))^2\Big]ds$$

$$+ \int_0^t \int_0^{+\infty} \left[\frac{1}{2}a'''(u)u_x^5 + 10a(u)a'(u)u_x^3 - 5\epsilon a'(u)u_t u_{xx}^2\right]dx\,ds.$$

(4.11)

Take $\epsilon_1 \leq 1$ small enough that

$$5\epsilon\|a(u(x,t))\|_{C_b(\bar{\mathbb{R}}^+ \times [0,T])} \leq 3,$$

say, for any $\epsilon \leq \epsilon_1$. Then the first six terms on the right-hand side of (4.11) can be controlled by the terms on the left-hand side, the constant a_1 in Lemma 4.1 and a constant depending on $\|f\|_2 + \epsilon^{\frac{1}{2}}\|f\|_3$ and $|g|_{1,T}$. By using (4.2) and choosing ϵ_1 small, say, there obtains

$$\int_0^t \left[u_{xx}^2(0,s) + \epsilon^2 u_{xt}^2(0,s)\right]ds \leq a_1 + 2\epsilon \int_0^t u_{xx}(0,s)u_{xt}(0,s)\,ds$$

$$\leq C(\|f\|_1, |g|_{1,T}) + \epsilon^{\frac{3}{2}}\int_0^t u_{xt}^2(0,s)\,ds. \quad (4.12)$$

The last three boundary terms in the second integral on the right-hand side of (4.11) may be bounded by a suitable multiple C* of the left-hand side of (4.12). Use of Lemma 4.1 allows the estimation of the first two boundary terms in the second integral on the right-hand side of (4.11) as follows:

$$\int_0^t \left[5\epsilon u_{xt}^2(0,s) - 12\epsilon u_{xt}(0,s)u_{xx}(0,s) \right] ds$$

$$= \int_0^t \left[12 u_{xt}(0,s)(\epsilon u_{xt}(0,s) - u_{xx}(0,s)) \right] ds - \int_0^t 7\epsilon u_{xt}^2(0,s) ds$$

$$\leq C(\|f\|_1, |g|_{1,T}) \left(\int_0^t u_{xt}^2(0,s) ds \right)^{\frac{1}{2}} - \int_0^t 7\epsilon u_{xt}^2(0,s) ds. \tag{4.13}$$

Thus, if it is supposed that ϵ_1 is small enough that $6\epsilon_1 \geq C^*(|g|_{1,T})\epsilon_1^{\frac{3}{2}}$, then (4.12) can be controlled in the form shown on the right-hand side of (4.13), so one has

$$\int_0^t \left[u_{xx}^2(0,s) + \epsilon u_{xt}^2(0,s) \right] ds$$

$$\leq C(\|f\|_1, |g|_{1,T}) + C(\|f\|_1, |g|_{1,T}) \left(\int_0^t u_{xt}^2(0,s) ds \right)^{\frac{1}{2}}. \tag{4.14}$$

For any $\delta > 0$, apply (4.4) to $u_x(x,t)$ to adduce

$$\|u_x\|_{C_b(\bar{\mathbb{R}}^+ \times [0,t])}^2 \leq 2 \sup_{0 \leq s \leq t} \{ \|u_x(\cdot,s)\| \|u_{xx}(\cdot,s)\| \}$$

$$\leq C\delta^{-1} + \delta \sup_{0 \leq s \leq t} \{ \|u_{xx}(\cdot,s)\|^2 \}. \tag{4.15}$$

One shows straightforwardly that for any $\delta > 0$,

$$\left| \int_0^t \frac{1}{2} u_x^4(0,s) a''(g(s)) ds \right| \leq C(\|f\|_1, |g|_{1,T})\delta^{-1} + \delta \sup_{0 \leq s \leq t} \{ \|u_{xx}(\cdot,s)\|^2 \}.$$

Similarly, one shows that

$$\left| \int_0^t 2a'(g(s)) u_x^2(0,s) u_{xx}(0,s) ds \right|$$

$$\leq C(|g_{1,T}|) \left(\int_0^t u_x^4(0,s) ds \right)^{\frac{1}{2}} \left(\int_0^t u_{xx}^2(0,s) ds \right)^{\frac{1}{2}}$$

$$\leq C(\|f\|_1, |g|_{1,T})\delta^{-3} + \delta \left(\sup_{0 \leq s \leq t} \{ \|u_{xx}(\cdot,s)\|^2 \} + \int_0^t u_{xx}^2(0,s) ds \right).$$

Thus in summary, the boundary terms in the second integral on the right-hand side of (4.11) are bounded above by

$$C(\|f\|_1, |g|_{1,T})\delta^{-3} + \delta \sup_{0 \leq s \leq t} \{ \|u_{xx}(\cdot,s)\|^2 \} + C(\|f\|_1, |g|_{1,T}) \left(\int_0^t u_{xt}^2(0,s) ds \right)^{\frac{1}{2}}$$

for any $\delta > 0$. By using (4.15) and Lemma 4.1, the first two terms of the third integral in (4.11) can be estimated as follows:

$$\left| \int_0^t \int_0^{+\infty} \left[\frac{1}{2}a'''(u)u_x^5 + 10a(u)a'(u)u_x^3\right] dx ds \right|$$

$$\leq C(\|f\|_1, |g|_{1,T}) \left(\int_0^t \|u_{xx}(\cdot, s)\|^4 ds \right)^{1/2}.$$

Multiplying (3.1a) by u_t and integrating the resulting expression over $\mathbb{R}^+$, there appears

$$\|u_t(\cdot, t)\|^2 + \epsilon \|u_{xt}(\cdot, t)\|^2$$

$$= -\epsilon g'(t)u_{xt}(0, t) - \int_0^{+\infty} \left[u_t u_{xxx} + u_t P(u)_x\right] dx.$$

Hence it is deduced that

$$\|u_t(\cdot, t)\|^2 + 2\epsilon \|u_{xt}(\cdot, t)\|^2$$

$$\leq C(\|f\|_1, |g|_{1,T}) + 2|\epsilon g'(t)u_{xt}(0, t)| + \|u_{xxx}(\cdot, t)\|^2. \tag{4.16}$$

Applying (4.4) to u_t and using (4.16) shows that the last term in the last integral on the right-hand side of (4.11) is bounded above in the following way:

$$\left| \int_0^t \int_0^{+\infty} 5\epsilon a'(u)u_t u_{xx}^2 dx ds \right|$$

$$\leq C(\|f\|_1, |g|_{1,T})\epsilon \int_0^t \left[\|u_{xx}(\cdot, s)\|^2 \|u_t(\cdot, s)\|^{\frac{1}{2}} \|u_{xt}(\cdot, s)\|^{\frac{1}{2}}\right] ds$$

$$\leq \epsilon^{\frac{1}{2}} C(\|f\|_1, |g|_{1,T}) \int_0^t \|u_{xx}(\cdot, s)\|^4 ds + \epsilon \left(\int_0^t \|u_t(\cdot, s)\|^2 ds + \epsilon \int_0^t \|u_{xt}(\cdot, s)\|^2 ds \right)$$

$$\leq \epsilon C(\|f\|_1, |g|_{1,T}, T) + \epsilon^{\frac{1}{2}} C(\|f\|_1, |g|_{1,T}) \int_0^t \|u_{xx}(\cdot, s)\|^4 ds$$

$$+ \epsilon^2 C(\|f\|_1, |g|_{1,T}) \left(\int_0^t u_{xt}^2(0, s) ds \right)^{\frac{1}{2}} + \epsilon \int_0^t \|u_{xxx}(\cdot, s)\|^2 ds.$$

If δ and ϵ_1 are chosen small enough, the above estimates show that (4.11) is reduced to

$$A^2(t) \leq C(\|f\|_2 + \epsilon \|f\|_3, |g|_{1,T}) + C(\|f\|_1, |g|_{1,T}) \left(\int_0^t u_{xt}^2(0, s) ds \right)^{\frac{1}{2}}$$

$$+ C(\|f\|_1, |g|_{1,T}) \int_0^t \left[\|u_{xx}(\cdot, s)\|^2 + \epsilon^{\frac{1}{2}} \|u_{xx}(\cdot, s)\|^4 + \epsilon \|u_{xxx}(\cdot, s)\|^2\right] ds.$$

The desired inequality now follows.   □

The estimate of the $H^2(\mathbb{R}^+)$-norm for the solution u of (3.1) given in Lemma 4.3 will be used in determining the following bound for $A^4(t) + B^2(t)$. When in hand, this bound implies one on $H^3(\mathbb{R}^+)$.

Lemma 4.4. *Let* $T > 0$, $f \in H^\infty(\mathbb{R}^+)$, $g \in H^\infty(0,T)$, *with* $f(0) = g(0)$. *There exist* a_3 *and* ϵ_2 *where*

$$a_3 = a_3(||f||_3 + \epsilon^{\frac{1}{2}}||f^{(4)}||, |g|_{2,T}, T)$$

and

$$\epsilon_2 = \epsilon_2(||f||_3 + \epsilon^{\frac{1}{2}}||f^{(4)}||, |g|_{2,T})$$

such that the solution of (3.1), corresponding to the data f *and* g, *satisfies*

$$A^4(t) + B^2(t) \le a_3,$$

provided that $t \in [0,T]$ *and* $\epsilon \in (0, \epsilon_2]$.

Proof: First multiply (3.1a) by $2u_{xxxt} + 2u_{xxx}u_x a'(u) + a'(u)u_{xx}^2 + u_x^2 u_{xx} a''(u)$, differentiate (3.1a) twice with respect to x and multiply this by $u_x^2 a'(u)$, then integrate their sum over $\mathbb{R}^+ \times (0,t)$. Many integrations by parts leads to

$$||u_{xxx}||^2 + \int_0^t u_{xt}^2(0,s)ds + \int_0^t \epsilon u_{xxt}^2(0,s)ds$$

$$= ||f'''||^2 + \int_0^{+\infty} \Big[2f'''(x)f'(x)a(f(x)) - 2u_{xxx}(x,t)u_x(x,t)a(u(x,t))$$

$$+ (f''(x))^2 a(f(x)) - u_{xx}^2(x,t)a(u(x,t)) + \epsilon u_{xxx}^3(x,t)a'(u(x,t))$$

$$- \epsilon(f''(x))^3 a'(f(x)) - u_{xx}(x,t)u_x^2(x,t)a'(u(x,t)) + f''(x)(f'(x))^2 a'(f(x))$$

$$- 3\epsilon(f'(x))^2(f''(x))^2 a''(f(x)) + 3\epsilon u_x^2(x,t)u_{xx}^2(x,t)a''(u(x,t))$$

$$+ \epsilon u_x^4(x,t)u_{xx}(x,t)a'''(u(x,t)) - \epsilon(f'(x))^4 f''(x)a'''(f(x)) \Big] dx$$

$$+ \int_0^t \Big[2u_{xxt}(0,s)g'(s) - 2u_x(0,s)u_{xx}(0,s)a'(g(s))\big[u_{xxx}(0,s) - \epsilon u_{xxt}(0,s)\big]$$

$$+ u_{xx}^3(0,s)a'(g(s)) + u_x^2(0,s)a'(g(s))\big[u_{xxxx}(0,s) - \epsilon u_{xxxt}(0,s)\big] \qquad (4.17)$$

$$- 2u_{xx}(0,s)u_{xt}(0,s)a(g(s)) - u_x^3(0,s)a''(g(s))\big[u_{xxx}(0,s) - \epsilon u_{xxt}(0,s)\big] \Big] ds$$

$$+ \int_0^t \int_0^{+\infty} \Big[4\epsilon u_{xxx}u_{xxt}u_x a'(u) - 4u_{xxx}^2 u_x a'(u) - 3u_{xxx}u_x^2 a(u)a'(u)$$

$$- u_x^3 u_{xx}a(u)a''(u) + u_x u_{xx}^3 a''(u) - u_x u_{xx}^2 a(u)a'(u)$$

$$- u_x^4 u_{xxx}a'''(u) - 3u_x^3 u_{xx}\big(a'(u)\big)^2 - u_x^5 a'(u)a''(u)$$

$$- 6u_{xxx}u_{xx}u_x^2 a''(u) - 6\epsilon u_x u_{xx}^2 u_{xt}a''(u) - 3\epsilon u_x^2 u_{xx}^2 u_t a'''(u)$$

$$- \epsilon u_{xx}^3 u_t a''(u) - 4\epsilon u_x^3 u_{xx}u_{xt}a'''(u) - \epsilon u_x^4 u_{xx}u_t a^{(4)}(u) \Big] dx ds.$$

Note that the following integration by parts has been used in deriving (4.17):

$$-\int_0^t \int_0^{+\infty} u_{xxx}u_{xt}a(u)\,dx\,ds = \int_0^t u_{xx}(0,s)u_{xt}(0,s)a(g(s))$$

$$+ \int_0^t \int_0^{+\infty} \Big[u_{xx}u_{xt}u_x a'(u) + u_{xx}u_{xxt}a(u)\Big]\,dx\,ds.$$

Then multiply (3.1a) by $4u_{xxxt}+4u_{xxx}u_x a'(u)$, differentiate (3.1a) once with respect to x and multiply this by $4u_{xxx}a(u)$, then integrate their sum over $\mathbb{R}^+ \times (0,t)$. After integrations by parts, one obtains

$$2||u_{xxx}||^2 + \int_0^t 2u_{xt}^2(0,s)\,ds + \int_0^t 2\epsilon u_{xxt}^2(0,s)\,ds$$

$$= 2||f'''||^2 + \int_0^{+\infty} \Big[2\epsilon u_{xxx}^2(x,t)a(u(x,t)) - 2\epsilon(f'''(x))^2 a(f(x))$$

$$+ 4f'''(x)f'(x)a(f(x)) - 4u_{xxx}(x,t)u_x(x,t)a(u(x,t))\Big]\,dx$$

$$+ \int_0^t \Big[4u_{xxt}(0,s)g'(s) + 2u_{xxx}^2(0,s)a(g(s))\Big]\,ds \qquad (4.18)$$

$$+ \int_0^t \int_0^{+\infty} \Big[4\epsilon u_{xxx}u_x a'(u)u_{xxt} - 8u_{xxx}u_x^2 a'(u)a(u)$$

$$- 2u_{xxx}^2 u_x a'(u) - 4u_{xxx}u_{xx}a^2(u) - 2\epsilon u_{xxx}^2 a'(u)u_t\Big]\,dx\,ds.$$

Note also that the relation

$$-u_t u_{xxx}u_x a'(u) = u_{xxx}u_x a'(u)(a(u)u_x + u_{xxx} - \epsilon u_{xxt}),$$

which is obtained by using (1.2a), has been used in deriving (4.18). Finally differentiate (3.1a) twice with respect to x and multiply this by $2u_{xxxx}$, then integrate the result over $\mathbb{R}^+ \times (0,t)$. After integrations by parts, there appears

$$||u_{xxx}||^2 + \epsilon||u_{xxxx}||^2 + \int_0^t u_{xxxx}^2(0,s)\,ds$$

$$= ||f'''||^2 + \epsilon||f^{(4)}||^2 - \int_0^t \Big[6u_x(0,s)u_{xx}(0,s)u_{xxx}(0,s)a'(g(s))$$

$$+ 2u_{xxt}(0,s)u_{xxx}(0,s) + u_{xxx}^2(0,s)a(g(s))$$

$$+ 2u_x^3(0,s)u_{xxx}(0,s)a''(g(s)) - 2u_{xx}^3(0,s)a'(g(s))\Big]\,ds \qquad (4.19)$$

$$- \int_0^t \int_0^{+\infty} \Big[7u_{xxx}^2 u_x a'(u) + 12u_{xxx}u_{xx}u_x^2 a''(u)$$

$$+ 2u_x^4 u_{xxxx}a'''(u) - 2u_{xx}^3 u_x a''(u)\Big]\,dx\,ds.$$

Subtract (4.17) from (4.18), multiply the result by $\frac{7}{2}$ and then add the result to (4.19). One ends up with

$$\frac{9}{2}\|u_{xxx}\|^2 + \epsilon\|u_{xxxx}\|^2 + \int_0^t \left[\frac{7}{2}u_{xt}^2(0,s) + \frac{7}{2}\epsilon u_{xxt}^2(0,s) + u_{xxxx}^2(0,s)\right]ds$$

$$= \frac{9}{2}\|f'''\|^2 + \epsilon\|f^{(4)}\|^2 + \int_0^{+\infty}\Big[7f'''(x)f'(x)a(f(x)) - 7u_{xxx}(x,t)u_x(x,t)a(u(x,t))$$

$$+ 7\epsilon u_{xxx}^2(x,t)a(u(x,t)) - 7\epsilon(f'''(x))^2 a(f(x)) - \frac{7}{2}(f''(x))^2 a(f(x))$$

$$+ \frac{7}{2}u_{xx}^2(x,t)a(u(x,t)) - \frac{7}{2}\epsilon u_{xx}^3(x,t)a'(u(x,t)) + \frac{7}{2}\epsilon(f''(x))^3 a'(f(x))$$

$$+ \frac{7}{2}u_{xx}(x,t)u_x^2(x,t)a'(u(x,t)) - \frac{7}{2}f''(x)(f'(x))^2 a'(f(x))$$

$$+ \frac{21}{2}\epsilon(f'(x))^2(f''(x))^2 a''(f(x)) - \frac{21}{2}\epsilon u_x^2(x,t)u_{xx}^2(x,t)a''(u(x,t))$$

$$- \frac{7}{2}\epsilon u_x^4(x,t)u_{xx}(x,t)a'''(u(x,t)) + \frac{7}{2}\epsilon(f'(x))^4 f''(x)a'''(f(x))\Big]dx$$

$$+ \int_0^t \Big[2u_{xxt}(0,s)\Big[\frac{7}{2}g'(s) - u_{xxx}(0,s)\Big] + 6u_{xxx}^2(0,s)a(g(s))$$

$$+ 7u_{xx}(0,s)u_{xt}(0,s)a(g(s)) + \frac{3}{2}u_x^3(0,s)u_{xxx}(0,s)a''(g(s)) \tag{4.20}$$

$$- \frac{7}{2}\epsilon u_x^3(0,s)u_{xxt}(0,s)a''(g(s)) + u_x(0,s)u_{xx}(0,s)u_{xxx}(0,s)a'(g(s))$$

$$- 7\epsilon u_x(0,s)u_{xx}(0,s)u_{xxt}(0,s)a'(g(s)) - \frac{3}{2}u_{xx}^3(0,s)a'(g(s))$$

$$- \frac{7}{2}u_x^2(0,s)a'(g(s))\big[u_{xxxx}(0,s) - \epsilon u_{xxxt}(0,s)\big]\Big]ds$$

$$+ \int_0^t\int_0^{+\infty}\Big[-14u_{xxx}u_{xx}a^2(u) - \frac{35}{2}u_{xxx}u_x^2 a(u)a'(u) + \frac{7}{2}u_x^3 u_{xx}a(u)a''(u)$$

$$- \frac{3}{2}u_x u_{xx}^3 a''(u) + \frac{7}{2}u_x u_{xx}^2 a(u)a'(u) + 9u_{xxx}u_{xx}u_x^2 a''(u)$$

$$+ \frac{3}{2}u_x^4 u_{xxx}a'''(u) + \frac{21}{2}u_x^3 u_{xx}\big(a'(u)\big)^2 + \frac{7}{2}u_x^5 a'(u)a''(u)$$

$$+ \frac{7}{2}\epsilon u_{xx}^3 u_t a''(u) + 21\epsilon u_x u_{xx}^2 u_{xt}a''(u) + \frac{21}{2}\epsilon u_x^2 u_{xx}^2 u_t a'''(u)$$

$$+ 14\epsilon u_x^3 u_{xx}u_{xt}a'''(u) + \frac{7}{2}\epsilon u_x^4 u_{xx}u_t a^{(4)}(u) - 7\epsilon u_{xxx}^2 u_t a'(u)\Big]dx\,ds.$$

We recall again the convention that constants dependent only on the data f and g will generally be denoted simply by C, and that this symbol's occurrence in different formulae is not taken to connote the same constant.

First, an argument analogous to that leading to (4.15) shows that

$$\|u_{xx}\|^2_{C_b(\bar{\mathbb{R}}^+ \times [0,t])} \le C\delta^{-3} + \delta \sup_{0 \le s \le t} \{\|u_{xxx}(\cdot, s)\|^2\}. \tag{4.21}$$

By (4.12), (4.15) and (4.21), the terms in (4.20) that feature integration with respect to x only are bounded in the terms of a suitable small multiple of $\|u_{xxx}(\cdot, t)\|^2 + \|u_{xx}(\cdot, t)\|^4$ provided ϵ is small. Note that equation (3.1a) implies

$$-\int_0^t u_{xxx}(0, s)u_{xxt}(0, s)ds = \int_0^t \left[u_{xxt}(0, s)[g'(s) + a(g(s))u_x(0, s) - \epsilon u_{xxt}(0, s)]\right]ds.$$

Integration by parts with respect to t yields

$$\int_0^t \left[u_{xxt}(0, s)a(g(s))u_x(0, s)\right]ds = u_{xx}(0, s)a(g(s))u_x(0, s)\Big|_{s=0}^{s=t}$$

$$-\int_0^t \left[u_{xx}(0, s)[a'(g(s))g'(s)u_x(0, s) + a(g(s))u_{xt}(0, s)]\right]ds.$$

Similarly, one shows that

$$\int_0^t u_{xxt}(0, s)g'(s)ds = u_{xx}(0, s)g'(s)\Big|_{s=0}^{s=t} - \int_0^t u_{xx}(0, s)g''(s)ds.$$

Then due to (4.15) and (4.21), the first boundary term can be estimated as

$$\int_0^t 2u_{xxt}(0, s)\left[\frac{7}{2}g'(s) - u_{xxx}(0, s)\right]ds$$

$$= \int_0^t 2u_{xxt}(0, s)\left[\frac{7}{2}g'(s) + g'(s) + a(g(t))u_x(0, s) - \epsilon u_{xxt}(0, s)\right]ds$$

$$\le C\delta^{-1} - 2\epsilon \int_0^t u^2_{xxt}(0, s)ds + \delta \int_0^t u^2_{xt}(0, s)ds$$

$$+ \delta\left(\int_0^t u^2_{xx}(0, s)ds\right)^2 + \delta\|u_{xxx}(\cdot, t)\|^2 + \delta\|u_{xx}(\cdot, t)\|^4$$

for any $\delta > 0$, where C depends on $\|f\|_1$ and $|g|_{2,T}$. Elementary inequalities show that there is a positive constant C depending on $\|f\|_1$ and $|g|_{2,T}$ such that for any $\delta > 0$,

$$\int_0^t \left[6u^2_{xxx}(0, s)a(g(s)) + 7u_{xx}(0, s)u_{xt}(0, s)a(g(s))\right]ds$$

$$\le C\delta^{-3} + \delta\left(\int_0^t u^2_{xxx}(0, s)ds\right)^2 + \delta\int_0^t u^2_{xt}(0, s)ds + \delta\left(\int_0^t u^2_{xx}(0, s)ds\right)^2.$$

By using (4.14), there is another constant C depending on $||f||_1$ and $|g|_{1,T}$ such that for any $\delta > 0$,

$$\int_0^t \left[\frac{3}{2} u_x^3(0,s) u_{xxx}(0,s) a''(g(s)) - \frac{7}{2} \epsilon u_x^3(0,s) u_{xxt}(0,s) a''(g(s)) \right.$$

$$\left. + u_x(0,s) u_{xx}(0,s) u_{xxx}(0,s) a'(g(s)) - 7\epsilon u_x(0,s) u_{xx}(0,s) u_{xxt}(0,s) a'(g(s)) \right] ds$$

$$\leq C||u_x||^2_{C_b(\mathbb{R}^+ \times [0,T])} \left(\int_0^t u_x^2(0,s) ds \right)^{\frac{1}{2}} \left[\left(\int_0^t u_{xxx}^2(0,s) ds \right)^{\frac{1}{2}} + \epsilon \left(\int_0^t u_{xxt}^2(0,s) ds \right)^{\frac{1}{2}} \right]$$

$$+ C||u_x||_{C_b(\mathbb{R}^+ \times [0,T])} \left(\int_0^t u_{xx}^2(0,s) ds \right)^{\frac{1}{2}} \left[\left(\int_0^t u_{xxx}^2(0,s) ds \right)^{\frac{1}{2}} + \epsilon \left(\int_0^t u_{xxt}^2(0,s) ds \right)^{\frac{1}{2}} \right]$$

$$\leq C(\delta^{-3}) + \delta \left(\epsilon \int_0^t u_{xxt}^2(0,s) ds + \left(\int_0^t u_{xx}^2(0,s) ds \right)^2 + \sup_{0 \leq s \leq t} \{||u_{xx}(\cdot,s)||^4\} \right)$$

$$+ \delta \left(\int_0^t u_{xxx}^2(0,s) ds \right)^2.$$

Similarly, one shows that

$$- \int_0^t \frac{3}{2} u_{xx}^3(0,s) a'(g(s)) ds \leq C||u_{xx}||_{C_b(\mathbb{R}^+ \times [0,T])} \int_0^t u_{xx}^2(0,s) ds$$

$$\leq C \left(\sup_{0 \leq s \leq t} \{||u_{xx}(\cdot,s)|| ||u_{xxx}(\cdot,s)||\} \right)^{\frac{1}{2}} \int_0^t u_{xx}^2(0,s) ds$$

$$\leq C\delta^{-5} + \delta \left(\int_0^t u_{xx}^2(0,s) ds \right)^2 + \delta \sup_{0 \leq s \leq t} \{||u_{xx}(\cdot,s)||^4 + ||u_{xxx}(\cdot,s)||^2\}.$$

By using (3.1a), the last boundary term in (4.20) is seen to satisfy the inequality

$$- \frac{7}{2} \int_0^t u_x^2(0,s) a'(g(s)) \left[u_{xxxx}(0,s) - \epsilon u_{xxxt}(0,s) \right] ds$$

$$= \frac{7}{2} \int_0^t u_x^2(0,s) a'(g(s)) \left[u_{xt}(0,s) + u_{xx}(0,s) a(g(s)) + u_x^2(0,s) a'(g(s)) \right] ds$$

$$\leq C\delta^{-3} + \delta \left(\int_0^t u_{xx}^2(0,s) ds \right)^2 + \delta \int_0^t u_{xt}^2(0,s) ds + \delta \sup_{0 \leq s \leq t} \{||u_{xx}(\cdot,s)||^4\}.$$

The use of (4.15), (4.16) and (4.21) shows that the entire set of double integrals in (4.20) except the last one can be controlled by the quantity

$$\int_0^t \left(A^4(s) + B^2(s) + \epsilon A^8(s) + \epsilon B^4(s) \right) ds.$$

Using (4.4) to bound $||u_{xxx}(\cdot, t)||_{C_b(\mathbb{R}^+)}$ and applying (4.16) shows that the last double integral in (4.20) is bounded above by

$$\left| - 7 \int_0^t \int_0^{+\infty} \epsilon u_{xxx}^2 u_t a'(u) dx ds \right|$$

$$\leq \int_0^t C\epsilon ||u_{xxx}(\cdot, s)||_{C_b(\mathbb{R}^+)} ||u_{xxx}(\cdot, s)|| ||u_t(\cdot, s)|| ds$$

$$\leq C \int_0^t \epsilon ||u_{xxx}(\cdot, s)||^{\frac{3}{2}} ||u_{xxxx}(\cdot, s)||^{\frac{1}{2}} ||u_t(\cdot, s)|| ds$$

$$\leq C \int_0^t \epsilon ||u_{xxx}(\cdot, s)||^2 ||u_t(\cdot, s)||^{\frac{4}{3}} ds + C \int_0^t \epsilon ||u_{xxxx}(\cdot, s)||^2 ds$$

$$\leq C\delta^{-\frac{7}{3}} + \delta \epsilon^{\frac{1}{3}} \left(\epsilon \int_0^t u_{xt}^2(0, s) ds \right)^2 + C \int_0^t \epsilon ||u_{xxx}(\cdot, s)||^4 ds + C \int_0^t \epsilon ||u_{xxxx}(\cdot, s)||^2 ds.$$

From the preceding estimates, it is deduced that there exist positive constants a_3 and C_3, where

$$a_3 = a_3(||f||_3 + \epsilon_1^{\frac{1}{2}} ||f^{(4)}||, |g|_{2,T}) \quad \text{and} \quad C_3 = C_3(||f||_1, |g|_{1,T}),$$

such that the solution of (3.1) corresponding to f and g satisfies

$$||u_{xxx}(\cdot, t)||^2 + \epsilon ||u_{xxxx}(\cdot, t)||^2 + \int_0^t [u_{xt}^2(0, s) + \epsilon u_{xxt}^2(0, s) + u_{xxxx}^2(0, s)] ds$$

$$\leq a_3 + \delta A^4(t) + C_3 \int_0^t \left(A(s)^4 + ||u_{xxx}(\cdot, s)||^2 + \epsilon ||u_{xxxx}(\cdot, s)||^2 \right. \tag{4.22}$$

$$\left. + \epsilon[||u_{xxx}(\cdot, s)||^2 + \epsilon ||u_{xxxx}(\cdot, s)||^2]^2 \right) ds$$

for any $\delta > 0$. By adding an appropriate multiple of (4.22) to (4.10), it is adduced that the functionals $A(t)$ and $B(t)$ associated with the solution u are restricted by the inequality

$$A^4(t) + B^2(t) \leq \alpha + \beta \int_0^t [A^4(s) + B^2(s) + \epsilon(A^8(s) + B^4(s))] ds,$$

where α and β are positive constants that depend only on initial data f and boundary data g. Define $\bar{A}$ to be the maximal solution of the system

$$\bar{A}(t) = \alpha + \beta \int_0^t [\bar{A}(s) + \epsilon \bar{A}^2(s)] ds.$$

Then, $\bar{A}(t) \geq A^4(t) + B^2(t)$ for all t for which $\bar{A}(t)$ is finite. Moreover, $\bar{A}(t)$ may be determined explicitly as

$$\bar{A}(t) = \frac{\alpha e^{\beta t}}{1 + \epsilon \alpha - \epsilon \alpha e^{\beta t}}$$

so long as $\epsilon \alpha e^{\beta t} < 1$, say. The desired result thus follows by choosing ϵ small enough. In fact, if ϵ_2 is chosen so that

$$1 + \epsilon_2 \alpha - \epsilon_2 \alpha e^{\beta T} \geq \frac{1}{2},$$

then the desired result is established.   □

Corollary 4.5. *Let $T > 0$, $f \in H^{\infty}(\mathbb{R}^+)$, $g \in H^{\infty}(0,T)$, with $f(0) = g(0)$. There exists a constant a_4 with*

$$a_4 = a_4(\|f\|_3 + \epsilon^{\frac{1}{2}}\|f^{(4)}\|, |g|_{2,T})$$

such that the solution of (3.1) corresponding to the data f and g satisfies

$$\|u_t(\cdot,t)\| + \epsilon^{\frac{1}{2}}\|u_{xt}(\cdot,t)\| \leq a_4$$

provided that $t \in [0,T]$ and $\epsilon \in (0,1]$.

Proof: Using equation (3.1a), one shows that

$$\|u_t(\cdot,t) - \epsilon u_{xxt}(\cdot,t)\|^2 \leq C(\|u(\cdot,t)\|_3)$$

$$\leq C(\|f\|_3 + \epsilon^{\frac{1}{2}}\|f^{(4)}\|, |g|_{2,T}).$$

Integrating by parts, one derives from the above inequality that

$$\|u_t(\cdot,t)\|^2 + 2\epsilon\|u_{xt}(\cdot,t)\|^2 + \epsilon^2\|u_{xxt}(\cdot,t)\|^2$$

$$\leq C(\|f\|_3 + \epsilon^{\frac{1}{2}}\|f^{(4)}\|, |g|_{2,T}) + 2\epsilon|g'(t)u_{xt}(0,t)|$$

$$\leq C(\|f\|_3 + \epsilon^{\frac{1}{2}}\|f^{(4)}\|, |g|_{2,T}) + 4\epsilon|g|_{2,T}\|u_{xt}(\cdot,t)\|^{\frac{1}{2}}\|u_{xxt}(\cdot,t)\|^{\frac{1}{2}}$$

$$\leq C(\|f\|_3 + \epsilon^{\frac{1}{2}}\|f^{(4)}\|, |g|_{2,T}) + \epsilon\|u_{xt}(\cdot,t)\|^2 + \frac{1}{2}\epsilon^2\|u_{xxt}(\cdot,t)\|^2,$$

or, by elementary means, that

$$\|u_t(\cdot,t)\|^2 + \epsilon\|u_{xt}(\cdot,t)\|^2 + \frac{1}{2}\epsilon^2\|u_{xxt}(\cdot,t)\|^2$$

$$\leq C(\|f\|_3 + \epsilon^{\frac{1}{2}}\|f^{(4)}\|, |g|_{2,T}).$$

The corollary is established.   □

The bounds established in this section would be sufficient to conclude an existence theory set in the space $L_{\infty}(\mathbb{R}^+; H^3(\mathbb{R}^+))$ for the quarter-plane problem (3.1). If the further compatibility condition

$$g'(0) + a(f(0))f'(0) + f'''(0) = 0$$

is posited, where $f \in H^4(\mathbb{R}^+)$ and $g \in H^2(\mathbb{R}^+)$, then it will follow from the next two lemmas that the quarter-plane problem has a solution $u \in L_{\infty}(0,T; H^4(\mathbb{R}^+))$ with $u_t \in L_{\infty}(0,T; H^1(\mathbb{R}^+))$. These preliminary results will be improved in Section 7 when the issue of continuous dependence is considered.

Lemma 4.6. *Let* $T > 0$, $f \in H^\infty(\mathbb{R}^+)$, $g \in H^\infty(0,T)$, *with* $f(0) = g(0)$. *There exists a constant* a_5 *with*

$$a_5 = a_5(\|u_t(\cdot,0)\|_1, \|f\|_4, |g|_{2,T})$$

such that the solution of (3.1) corresponding to the data f *and* g *satisfies*

$$\|u_t(\cdot,t)\|_1^2 + \int_0^t \left[u_{xxt}(0,s) - \epsilon u_{xtt}\right]^2 ds \le a_5,$$

provided that $t \in [0,T]$ *and* $\epsilon \in (0,1]$.

Proof: Let $v(x,t) = u_t(x,t)$ so that v satisfies the partial-differential equation

$$v_t + \left(a(u)v\right)_x + v_{xxx} - \epsilon v_{xxt} = 0, \quad \text{for } (x,t) \in \bar{\mathbb{R}}^+ \times [0,T]. \tag{4.23}$$

An L_2-bound for $u_t(\cdot,t)$ has been established in Corollary 4.5. Now we derive an H^1-bound for $u_t(\cdot,t)$. Multiplying (4.23) by $2(\epsilon v_{xt} - a(u)v - v_{xx})$, integrating the results over $\mathbb{R}^+ \times (0,t)$ and then using some elementary inequalities including

$$\left| \int_0^{+\infty} v^3(x,t)dx \right| \le \|v(\cdot,t)\|^2 \|v(\cdot,t)\|_{C_b(\bar{\mathbb{R}}^+)}$$

$$\le \sqrt{2}\|v(\cdot,t)\|^{\frac{5}{2}} \|v_x(\cdot,t)\|^{\frac{1}{2}}$$

$$\le C\|v(\cdot,t)\|^{\frac{10}{3}} + \|v_x(\cdot,t)\|^3,$$

one comes to

$$\|v_x(\cdot,t)\|^2 + \epsilon \int_0^t (g''(s))^2 ds + \int_0^t [v_{xx}(0,s) - \epsilon v_{xt}(0,s)]^2 ds$$

$$= \|v_x(\cdot,0)\|^2 + \int_0^{+\infty} [a(u(x,t))v^2(x,t) - a(f(x))v^2(x,0)]dx$$

$$- \int_0^t a^2(g(s))(g''(s))^2 ds - \int_0^t 2a(g(s))g'(s)[v_{xx}(0,s) - \epsilon v_{xt}(0,s)]ds$$

$$- \int_0^t 2g'(s)v_x(0,s)ds - \int_0^t \int_0^{+\infty} a'(u)v^3 dx ds \tag{4.24}$$

$$\le C(\|u_t(\cdot,0)\|_1, |g|_{2,T}) + C\|v(\cdot,t)\|^2 + C\int_0^t v_x^2(0,s)ds$$

$$+ \frac{1}{2}\int_0^t [v_{xx}(0,s) - \epsilon v_{xt}(0,s)]^2 ds + C\int_0^t [\|v_x(\cdot,s)\|^2 + \|v(\cdot,s)\|^{\frac{10}{3}}]ds.$$

The use of Lemma 4.4, Corollary 4.5 and Gronwall's lemma in (4.24) shows that there is a constant a_5 depending on $\|u_t(\cdot,0)\|_1$ and $|g|_{2,T}$ such that

$$\|v(\cdot,t)\|_1^2 + \int_0^t [v_{xx}(0,s) - \epsilon v_{xt}(0,s)]^2 ds \le a_5$$

for all $t \in [0, T]$.   □

The constant a_5 in Lemma 4.6 depends on $||u_t(\cdot, 0)||_1$, a quantity about which there is currently no information. In order to estimate usefully solutions of (3.1), some control of $||u_t(\cdot, 0)||_1$ must be obtained in terms of the data f and g. An appropriate bound is forthcoming if the data satisfies an additional compatibility condition.

Lemma 4.7. *Let $T > 0$, $f \in H^\infty(\mathbb{R}^+)$, $g \in H^\infty(0, T)$, with $f(0) = g(0)$ and*

$$g'(0) = -[a(f(0))f'(0) + f'''(0)]. \tag{4.25}$$

There exists a constant a_6 where

$$a_6 = a_6(||f||_4, |g|_{2,T}),$$

such that the solution of (3.1) corresponding to the data f and g satisfies

$$||u_t(\cdot, 0)||_1^2 \le a_6,$$

for $t \in [0, T]$ and $\epsilon \in (0, \epsilon_2]$.

Proof: First note that by Corollary 4.5, $||u_t(\cdot, 0)||^2$ is controlled by a constant of the form $a_4(||f||_3 + \epsilon^{\frac{1}{2}}||f||_4, |g|_{2,T})$. Let

$$\phi(x) = -[a(f(x))f'(x) + f'''(x)].$$

Then $u_t(\cdot, 0)$ is a solution of the boundary-value problem

$$u_t(\cdot, 0) - \epsilon u_{xxt}(\cdot, 0) = \phi, \tag{4.26}$$

$$u_t(0, 0) = g'(0), \quad \lim_{x \to +\infty} u_t(x, 0) = 0.$$

Differentiate (4.26) with respect to x, multiply the result by $u_{xt}(\cdot, 0)$ and integrate the result over $\mathbb{R}^+$. After integrations by parts, there appears the equation

$$||u_{xt}(\cdot, 0)||^2 + \epsilon||u_{xxt}(\cdot, 0)||^2 = \int_0^{+\infty} u_{xt}(x, 0)\phi_x dx,$$

from which one obtains

$$||u_{xt}(\cdot, 0)||^2 + \epsilon||u_{xxt}(\cdot, 0)||^2 \le C||\phi_x||^2 \le C||f||_4^2,$$

where C is a constant independent of ϵ. Note that we have used $u_{xxt}(0, 0) = 0$ which is obtained by using the compatibility condition (4.25) and equation (4.26). The proof of the lemma is then finished.   □

The following lemmas will be helpful in Section 5 when higher-order estimates are considered.

Lemma 4.8. *Let $T > 0$, $f \in H^\infty(\mathbb{R}^+)$, $g \in H^\infty(0,T)$, with $f(0) = g(0)$. There exists a constant $a_7 = a_7(\|u_t(\cdot,0)\|_3, |g|_{3,T})$, such that the solution of (3.1) corresponding to the data f and g satisfies*

$$\|u_t(\cdot,t)\|_3^2 + \int_0^t \left\{ [u_{xxxt}(0,s)]^2 + [u_{xtt}(0,s)]^2 + \epsilon[u_{xxtt}(0,s)]^2 \right\} ds \leq a_7,$$

provided that $t \in [0,T]$ and $\epsilon \in (0, \epsilon_2]$.

Proof: Let $v = u_t$, where u is the solution of the regularized initial- and boundary-value problem (3.1) corresponding to the given smooth and compatible data f and g. For t in $[0,T]$, define

$$A^2(t) = \|v(\cdot,t)\|_3^2 + \int_0^t \left\{ v_{xxx}^2(0,s) + v_{xt}^2(0,s) + \epsilon v_{xxt}^2(0,s) \right\} ds.$$

The Lemma 4.4 and Lemma 4.6 imply that

$$\|u\|_{L_\infty(0,T;H^3(\mathbb{R}^+))}, \qquad \|v\|_{L_\infty(0,T;H^1(\mathbb{R}^+))} \leq C,$$

$$\|u\|_{L_\infty(0,T;W^{2,\infty}(\mathbb{R}^+))}, \qquad \|v\|_{L_\infty(\mathbb{R}^+\times[0,T])} \leq C, \qquad (4.27)$$

$$\int_0^t v_x^2(0,s)ds \quad \text{and} \quad \int_0^t (v_{xx}(0,s) - \epsilon v_{xt}(0,s))^2 ds \leq C,$$

where here, and in the remainder of this proof, C will denote various constants which all depend on the same variables as the constant a_7 in the statement of the lemma, but which will always be independent of ϵ. Note that v satisfies equation (4.23). Differentiate (4.23) once with respect to x, multiply by $-2v_{xxx}$ and integrate the resulting expression over $\mathbb{R}^+ \times (0,t)$. There appears the equation

$$\|v_{xx}(\cdot,t)\|^2 + \epsilon\|v_{xxx}(\cdot,t)\|^2 + \int_0^t v_{xxx}^2(0,s)ds$$

$$= \|v_{xx}(\cdot,0)\|^2 + \epsilon\|v_{xxx}(\cdot,0)\|^2 - 2\int_0^t v_{xt}(0,s)v_{xx}(0,s)ds \qquad (4.28)$$

$$+ 2\int_0^t \int_0^{+\infty} \left(a(u)v \right)_{xx} v_{xxx} dx ds.$$

Inequalities (4.27) imply that

$$\int_0^t \int_0^{+\infty} \left(a(u)v \right)_{xx} v_{xxx} dx ds \leq C\left(1 + \int_0^t \|v(\cdot,s)\|_3^2 ds \right).$$

Note that

$$-2\int_0^t v_{xt}(0,s)v_{xx}(0,s)ds$$

$$= -2\int_0^t v_{xt}(0,s)[v_{xx}(0,s) - \epsilon v_{xt}(0,s)]ds - 2\epsilon\int_0^t v_{xt}^2(0,s)ds.$$

Then by inequalities (4.27), one shows that for any $\delta > 0$, (4.28) can be estimated in the form

$$||v_{xx}(\cdot, t)||^2 + \epsilon||v_{xxx}(\cdot, t)||^2 + \int_0^t v_{xxx}^2(0, s)ds + 2\epsilon \int_0^t v_{xt}^2(0, s)ds$$

$$\leq C_\delta + \delta \int_0^t v_{xt}^2(0, s)ds + C \int_0^t A^2(s)ds. \tag{4.29}$$

Next, multiply (4.23) by $2v_{xxxt}$ and integrate the results over $\mathbb{R}^+ \times (0, t)$. After integrations by parts, we obtain the equation

$$||v_{xxx}(\cdot, t)||^2 + \int_0^t [v_{xt}^2(0, s) + \epsilon v_{xxt}^2(0, s)]ds$$

$$= ||v_{xxx}(\cdot, 0)||^2 + \int_0^t 2v_{xxt}(0, s)v_t(0, s)ds$$

$$- \int_0^{+\infty} \left[2v_{xxx}(x, t)\big(a(u(x, t)v(x, t)\big)_x - 2v_{xxx}(x, 0)\big(a(f(x))v(x, 0)\big)_x\right]dx \tag{4.30}$$

$$+ \int_0^t \int_0^{+\infty} 2v_{xxx}\left[\big(a(u)v_{xt} + a'(u)u_x v_t\big) + \big(a'(u)v^2\big)_x\right]dxds.$$

Because

$$||v||_{L_\infty(0, t; W^{2, \infty}(\mathbb{R}^+))} \leq C_\delta + \delta||v(\cdot, t)||_3^2, \tag{4.31}$$

and because of the relations $v_{tt}(0, s) = g'''(s)$ and $v_t(0, s) = g''(s)$, it is adduced that corresponding to any $\delta > 0$ there is a constant C_δ for which

$$\int_0^t 2v_{xxt}(0, s)v_t(0, s)ds = 2v_{xx}(0, s)v_t(0, s)\big|_{s=0}^{s=t} - \int_0^t 2v_{xx}(0, s)v_{tt}(0, s)ds$$

$$\leq C_\delta + \delta||v(\cdot, t)||_3^2 + C \int_0^t ||v(\cdot, s)||_3^2 ds.$$

One can easily obtain the inequality

$$\left|\int_0^{+\infty} v_{xxx}(x, t)\big(a(u(x, t))v(x, t)\big)_x dx\right| \leq C_\delta + \delta||v(\cdot, t)||_3^2,$$

valid for any $\delta > 0$. Differentiating equation (4.23) with respect to x yields

$$-v_{xt} = \big(a(u)v\big)_{xx} + v_{xxxx} - \epsilon v_{xxxt}.$$

Using the above equation leads to the inequality

$$\int_0^t \int_0^{+\infty} a(u)v_{xt}v_{xxx}\,dxds$$

$$= -\int_0^t \int_0^{+\infty} a(u)v_{xxx}\Big[\big(a(u)v\big)_{xx} + v_{xxxx} - \epsilon v_{xxxt}\Big]dxds$$

$$\leq C + \epsilon\|a(u)\|_{C_b(\mathbb{R}^+ \times [0,T])}\|v_{xxx}(\cdot,t)\|^2$$

$$+ C\int_0^t v_{xxx}^2(0,s)ds + C\int_0^t (\|v_{xxx}(\cdot,s)\|^2 + \|v_{xx}(\cdot,s)\|^2)ds$$

$$\leq C + C\int_0^t v_{xxx}^2(0,s)ds + C\epsilon\|v_{xxx}(\cdot,t)\|^2 + C\int_0^t A^2(s)ds.$$

By using equation (4.23) and (4.27) again, one shows that

$$\|v_t(\cdot,t) - \epsilon v_{xxt}(\cdot,t)\|^2 \leq \|v_{xxx}(\cdot,t)\|^2 + C\|v(\cdot,t)\|_2^2.$$

Expanding the norm on the left-hand side of this inequality and integrating by parts the mixed term gives

$$\|v_t(\cdot,t)\|^2 + 2\epsilon\|v_{xt}(\cdot,t)\|^2 + \epsilon^2\|v_{xxt}(\cdot,t)\|^2$$
$$\leq C\|v(\cdot,t)\|_3^2 + 2\epsilon|v_t(0,t)v_{xt}(0,t)|$$
$$\leq C\|v(\cdot,t)\|_3^2 + 4\epsilon|g^{(k+1)}(t)|\|v_{xt}(\cdot,t)\|^{\frac{1}{2}}\|v_{xxt}(\cdot,t)\|^{\frac{1}{2}}$$
$$\leq C\|v(\cdot,t)\|_3^2 + C\epsilon^{\frac{1}{2}}|g^{(k+1)}(t)|^2 + \epsilon\|v_{xt}(\cdot,t)\|^2 + \frac{1}{2}\epsilon^2\|v_{xxt}(\cdot,t)\|^2,$$

from which there obtains

$$\|v_t(\cdot,t)\|^2 + \epsilon\|v_{xt}(\cdot,t)\|^2 + \frac{1}{2}\epsilon^2\|v_{xxt}(\cdot,t)\|^2 \leq C\|v(\cdot,t)\|_3^2 + C\epsilon^{\frac{1}{2}}|g^{(k+1)}(t)|^2.$$

Using this last information yields

$$\left|\int_0^t \int_0^{+\infty} a'(u)u_x v_t v_{xxx}\,dxds\right| \leq C\int_0^t \|v_t(\cdot,s)\|\,\|v_{xxx}(\cdot,s)\|ds$$

$$\leq C + C\int_0^t A(s)^2 ds.$$

Taking recourse to (4.27) again, we see that

$$\int_0^t \int_0^{+\infty} v_{xxx}\big(a'(u)v^2\big)_x\,dxds \leq \int_0^t C\|v_{xxx}(\cdot,s)\|\|v_x(\cdot,s)\|ds$$

$$\leq C\int_0^t A^2(s)ds.$$

If ϵ is chosen small enough, (4.30) can be estimated as

$$||v_{xxx}(\cdot,t)||^2 + \int_0^t [v_{xt}^2(0,s) + \epsilon v_{xxt}^2(0,s)]ds$$

$$\leq C + C \int_0^t v_{xxx}^2(0,s)ds + C \int_0^t A^2(s)ds. \qquad (4.32)$$

Multiply (4.29) by a suitable constant and add the result to (4.32). Then applying Gronwall's lemma shows that $A(t)$ is bounded by a constant a_7, as advertised.   □

Lemma 4.9. *Let $T > 0$, $f \in H^\infty(\mathbb{R}^+)$, $g \in H^\infty(0,T)$, with $f(0) = g(0)$ and assume (4.25) also holds. Then there exist constants a_8 and a_9 where*

$$a_8 = a_8(||f||_6, |g|_{2,T}) \qquad and \qquad a_9 = a_9(||f||_6, |g|_{3,T}),$$

such that the solution of (3.1) corresponding to the data f and g satisfies

$$||u_t(\cdot,0)||_3^2 + \epsilon^2 ||u_{xxxxt}(\cdot,0)||^2 \leq a_8(||f||_6, |g|_{2,T}),$$

and in consequence,

$$||u_{tt}(\cdot,0)||^2 + \epsilon^2 ||u_{xtt}(\cdot,0)||^2 \leq a_9(||f||_6, |g|_{3,T}),$$

for $t \in [0,T]$ and $\epsilon \in (0,\epsilon_2]$.

Proof: Differentiate (4.26) with respect to x, multiply by $u_{xxxt}(\cdot,0)$ and integrate the result over $\mathbb{R}^+$. Since $u_{xxt}(0,0) = 0$, there appears

$$||u_{xxt}(\cdot,0)||^2 + \epsilon ||u_{xxxt}(\cdot,0)||^2 = \int_0^{+\infty} u_{xxt}(x,0)\phi_{xx}(x)dx. \qquad (4.33)$$

Differentiate (4.26) with respect to x twice, multiply the result by $u_{xxxxt}(\cdot,0)$ and integrate the result over $\mathbb{R}^+$ to obtain the equation

$$||u_{xxxt}(\cdot,0)||^2 + \epsilon ||u_{xxxxt}(\cdot,0)||^2 = \int_0^{+\infty} u_{xxxt}(x,0)\phi_{xxx}(x)dx, \qquad (4.34)$$

by again using the fact $u_{xxt}(0,0) = 0$. Applying some elementary inequalities to (4.33) and (4.34), one immediately obtains the first result in the statement of the lemma.

From equation (3.1a), one shows that u_{tt} is the solution of the boundary-value problem

$$u_{tt}(\cdot,0) - \epsilon u_{xxtt}(\cdot,0) = \psi\big(u_t(\cdot,0), f(x), u_{xt}(\cdot,0), u_{xxxt}(\cdot,0)\big), \qquad (4.35)$$

$$u_{tt}(0,0) = g''(0) \qquad and \qquad \lim_{x \to \infty} u_{tt}(x,0) = 0,$$

where

$$\psi\big(u_t(\cdot,0), f(x), u_{xt}(\cdot,0), u_{xxxt}(\cdot,0)\big)$$
$$= -a'(f(x))f(x)u_t(\cdot,0) - a(f(x))u_{xt}(\cdot,0) - u_{xxxt}(\cdot,0).$$

The use of equation (4.35) shows that

$$||u_{tt}(\cdot,0) - \epsilon u_{xxtt}(\cdot,0)||^2 = ||\psi||^2.$$

Applying elementary inequalities, the results in Lemma 4.7 and the first part of this lemma, one concludes

$$||u_{tt}(\cdot,0)||^2 + 2\epsilon||u_{xtt}(\cdot,0)||^2 + \epsilon^2||u_{xxtt}(\cdot,0)||^2$$
$$= 2\epsilon u_{tt}(0,0)u_{xtt}(0,0) + ||\psi||^2 \qquad (4.36)$$
$$\leq a_9(||f||_6, |g|_{3,T}) + \frac{1}{2}(\epsilon||u_{xtt}(\cdot,0)||^2 + \epsilon^2||u_{xxtt}(\cdot,0)||^2).$$

The result now follows.   □

The main results of this section are collected in the following umbrella theorem.

Theorem 4.10. *Let* $T > 0$, $f \in H^\infty(\mathbb{R}^+)$, $g \in H^\infty(0,T)$, *with*

$$f(0) = g(0).$$

Let u *be the solution of the regularized initial- and boundary-value problem (3.1) corresponding to the given data* f *and* g. *Then there is a constant* a_{10}, *depending on* $||f||_3 + \epsilon^{\frac{1}{2}}||f||_4$ *and* $|g|_{2,T}$, *such that*

$$||u(\cdot,t)||_3 + ||u_t(\cdot,t)|| + \epsilon^{\frac{1}{2}}||u(\cdot,t)||_4 \leq a_{10},$$

for all t *in* $[0,T]$ *and* ϵ *in* $(0,\epsilon_2]$. *Here* ϵ_2 *is the positive constant arising in Lemma 4.4. Moreover if*

$$g'(0) + a(f(0))f'(0) + f'''(0) = 0$$

holds, then

$$||u(\cdot,t)||_4 + ||u_t(\cdot,t)||_1 \leq a_{11} \quad and \quad ||u_t(\cdot,t)||_3 \leq a_{12},$$

where $a_{11} = a_{11}(||f||_4, |g|_{2,T})$ *and* $a_{12} = a_{12}(||f||_6, |g|_{3,T})$ *for all* t *in* $[0,T]$ *and* ϵ *in* $(0,\epsilon_2]$.   □

5. Higher-Order Estimates for the Regularized Problem

The derivation of ϵ-independent bounds for solutions of the regularized initial- and boundary-value problem (3.1) is continued in this section. Smoother solutions would be expected to obtain provided the initial and boundary data are smooth enough. A proof of such further regularity, presented in the next section, is based on the additional estimates to be obtained in the present section.

The assumption that $f \in H^\infty(\mathbb{R}^+)$, $g \in H^\infty(0,T)$, and $f(0) = g(0)$ will continue to be enforced throughout this section. This hypothesis will be recalled informally by the stipulation that the data f and g are smooth and compatible. For simplicity, denote

$$u^{(j)} = \partial_t^j u.$$

The next lemma generalizes Lemma 4.6 and Lemma 4.8.

Lemma 5.1. *Let $T > 0$, $f \in H^\infty(\mathbb{R}^+)$, $g \in H^\infty(0,T)$, with $f(0) = g(0)$. There exists a constant b_1 where*

$$b_1 = b_1(\max_{0 \le j \le k}\{||u^{(j)}(\cdot,0)||_1\}, |g|_{k+1,T}),$$

such that the solution of (3.1) corresponding to the data f and g satisfies

$$||u^{(k)}(\cdot,t)||_1^2 + \int_0^t [u_{xx}^{(k)}(0,s) - \epsilon u_{xt}^{(k)}(0,s)]^2 ds \le b_1,$$

provided that $t \in [0,T]$ and $\epsilon \in (0,\epsilon_2]$. Moreover there exists a constant b_2 where

$$b_2 = b_2(\max_{0 \le j \le k}\{||u^{(j)}(\cdot,0)||_3\}, |g|_{k+2,T}),$$

such that

$$||u^{(k)}(\cdot,t)||_3^2 + \int_0^t \{[u_{xxx}^{(k)}(0,s)]^2 + [u_x^{(k+1)}(0,s)]^2 + \epsilon[u_{xx}^{(k+1)}(0,s)]^2\} ds \le b_2.$$

Proof: First note that for $k = 1$, the desired result is implied by Lemma 4.6 and Lemma 4.8. The proof proceeds by induction on k. Let $k > 1$, and suppose that the stated estimates hold for all nonnegative integers less than or equal to $k - 1$. The induction hypothesis implies that

$$||u^{(j)}||_{L_\infty(0,T;H^1(\mathbb{R}^+))}, \quad ||u^{(j)}||_{L_\infty(\mathbb{R}^+\times[0,T])},$$

$$\int_0^t \left(u^{(j)}\right)_x^2(0,s)ds + \int_0^t (u_{xx}^{(j)}(0,s) - \epsilon u_x^{(j+1)}(0,s))^2 ds \le b_1, \tag{5.1a}$$

and

$$||u^{(j)}||_{L_\infty(0,T;H^3(\mathbb{R}^+))}, \quad ||u^{(j)}||_{L_\infty(0,T;W^{2,\infty}(\mathbb{R}^+))},$$

$$\int_0^t \{[u_{xxx}^{(j)}(0,s)]^2 + [u_x^{(j+1)}(0,s)]^2 + \epsilon[u_{xx}^{(j+1)}(0,s)]^2\} ds \le b_2, \tag{5.1b}$$

for $0 \le j \le k - 1$. In the remainder of this proof, C will denote various constants which all depend on the same variables as the constant b_1 or b_2 given in the statement of the lemma, but which will always be independent of ϵ. For any integer $j \ge 1$ the function $u^{(j)}$ satisfies the equation

$$u_t^{(j)} + \left(a(u)u^{(j)} + h_j(u)\right)_x + u_{xxx}^{(j)} - \epsilon u_{xxt}^{(j)} = 0, \text{ for } (x,t) \in \bar{\mathbb{R}}^+ \times [0,T], \tag{5.2}$$

where

$$h_j(u) = \sum_{i=1}^{j-1} \binom{j-1}{i} a(u)^{(i)} u^{(j-i)} \quad \text{and} \quad h_1(u) = 0.$$

The induction hypothesis implies that

$$||h_k(u)||_{L_\infty(0,T;W^{2,\infty}(\mathbb{R}^+))} \le c||h_k(u)||_{L_\infty(0,T;H^3(\mathbb{R}^+))} \le C. \tag{5.3}$$

Let $v = u^{(k)}$, where u is the solution of the regularized initial- and boundary-value problem (3.1) corresponding to the given smooth and compatible data f and g. Then v satisfies the equation

$$v_t + \left(a(u)v + h_k(u)\right)_x + v_{xxx} - \epsilon v_{xxt} = 0. \tag{5.4}$$

For t in $[0,T]$, define

$$A^2(t) = ||v(\cdot,t)||_1^2 + \int_0^t [v_{xx}(0,s) - \epsilon v_{xt}(0,s)]^2 ds.$$

The function A(t) will be estimated via an energy inequality derived from equation (5.4). Multiply this equation by $2v$ and integrate the result over $\mathbb{R}^+ \times (0,t)$ to obtain

$$||v(\cdot,t)||^2 + \epsilon||v_x(\cdot,t)||^2 + \int_0^t v_x^2(0,s)ds$$

$$= ||v(\cdot,0)||^2 + \epsilon||v_x(\cdot,0)||^2 + \int_0^t a(g(s))\left(g^{(k)}(s)\right)^2 ds$$

$$+ \int_0^t 2g^{(k)}(s)[v_{xx}(0,s) - \epsilon v_{xt}(0,s)]ds + \int_0^t \int_0^{+\infty} \left[a'(u)v^2 - 2v\left(h_k\right)_x\right]dxds.$$

The induction hypothesis implies that for any $\delta > 0$, there is a constant C_δ such that

$$||v(\cdot,t)||^2 + \epsilon||v_x(\cdot,t)||^2 + \int_0^t v_x^2(0,s)ds$$

$$\le C_\delta\left(1 + \int_0^t ||v(\cdot,s)||_1^2 ds\right) + \delta \int_0^t [v_{xx}(0,s) - \epsilon v_{xt}(0,s)]^2 ds. \tag{5.5}$$

Multiply (5.4) by $2(\epsilon v_{xt} - a(u)v - v_{xx})$ and integrate the results over $\mathbb{R}^+ \times (0,t)$. After simplifying, one obtains

$$||v_x(\cdot,t)||^2 + \int_0^t \left(v_{xx}(0,s) - \epsilon v_{xt}(0,s) + a(g(s))g^{(k)}(s)\right)^2 ds$$

$$= ||v_x(\cdot,0)||^2 + \int_0^{+\infty} a(u(x,t))v^2(x,t)dx - \int_0^{+\infty} a(f(x))v^2(x,0)dx$$

$$+ \int_0^t \left(\epsilon(g^{(k+1)}(s))^2 - 2g^{(k+1)}(s)v_x(0,s)\right)ds \tag{5.6}$$

$$+ \int_0^t \int_0^{+\infty} \left[2\left(h_k(u)\right)_x(v_{xx} + a(u)v - \epsilon v_{xt}) - a'(u)u_t v^2\right]dxds.$$

Integrations by parts show

$$\int_0^t \int_0^{+\infty} [h_k(u)]_x v_{xx} \, dx \, ds$$

$$= -\int_0^t \left(h_k(u(0,s))\right)_x v_x(0,s) \, ds - \int_0^t \int_0^{+\infty} \left(h_k(u)\right)_{xx} v_x \, dx \, ds,$$

and

$$\epsilon \int_0^t \int_0^{+\infty} \left(h_k(u)\right)_x v_{xt} \, dx \, ds = \epsilon \int_0^{+\infty} \left(h_k(u(x,s))\right)_x v_x(x,s) \, dx \Big|_{s=0}^{s=t}$$

$$- \epsilon \int_0^t \int_0^{+\infty} \left(h_k(u)\right)_{xt} v_x \, dx \, ds.$$

Note that the leading term of the expression $\left(h_k(u)\right)_{xt}$ is $\left(H(u)v\right)_x$, where $H(u)$ is a function depending on $u, u_t, \cdots, u^{(k-1)}$. By the induction hypothesis (5.2) and (5.3), it transpires that for all $t \in [0, T]$,

$$\left| \int_0^t \int_0^{+\infty} [h_k(u)]_x [v_{xx} + a(u)v - \epsilon v_{xt}] \, dx \, ds \right|$$

$$\leq \epsilon^2 C \|v_x(\cdot, t)\|^2 + C \left(1 + \int_0^t \|v(\cdot, s)\|_1^2 \, ds + \int_0^t v_x^2(0, s) \, ds \right).$$

From (5.5) and (5.6), it follows that

$$\|v(\cdot, t)\|_1^2 + \int_0^t (v_{xx}(0,s) - \epsilon v_{xt}(0,s))^2 \, ds \leq C + C \int_0^t \|v(\cdot, s)\|_1^2 \, ds. \qquad (5.7)$$

Gronwall's lemma implies that there is a constant b_1 with

$$b_1 = b_1 \left(\max_{0 \leq j \leq k} \{\|u^{(j)}(\cdot, 0)\|_1\}, |g|_{k+1, T} \right),$$

such that

$$A^2(t) = \|v(\cdot, t)\|_1^2 + \int_0^t (v_{xx}(0,s) - \epsilon v_{xt}(0,s))^2 \, ds \leq b_1,$$

for all $t \in [0, T]$.

To finish the lemma we need to control

$$\|v(\cdot, t)\|_3^2 \quad \text{and} \quad \int_0^t \{v_{xxx}^2(0,s) + v_{xt}^2(0,s) + \epsilon v_{xxt}^2(0,s)\} \, ds.$$

Define the quantity $B(t)$ to be the sum of these two quantities,

$$B^2(t) = \|v(\cdot, t)\|_3^2 + \int_0^t \{v_{xxx}^2(0,s) + v_{xt}^2(0,s) + \epsilon v_{xxt}^2(0,s)\} \, ds.$$

Differentiate (5.4) once with respect to x, multiply by $-2v_{xxx}$ and integrate the resulting expression over $\mathbb{R}^+ \times (0, t)$ to reach the equation

$$||v_{xx}(\cdot, t)||^2 + \epsilon||v_{xxx}(\cdot, t)||^2 + \int_0^t v_{xxx}^2(0, s)ds$$

$$= ||v_{xx}(\cdot, 0)||^2 + \epsilon||v_{xxx}(\cdot, 0)||^2 - 2\int_0^t v_{xt}(0, s)v_{xx}(0, s)ds \qquad (5.8)$$

$$+ 2\int_0^t \int_0^{+\infty} \big(a(u)v + h_k(u)\big)_{xx} v_{xxx} dx ds.$$

Inequalities (5.1) and (5.3) imply that

$$\int_0^t \int_0^{+\infty} \big(a(u)v + h_k(u)\big)_{xx} v_{xxx} dx ds \leq C\left(1 + \int_0^t ||v(\cdot, s)||_3^2 ds\right).$$

Note that

$$- 2\int_0^t v_{xt}(0, s)v_{xx}(0, s)ds$$

$$= -2\int_0^t v_{xt}(0, s)[v_{xx}(0, s) - \epsilon v_{xt}(0, s)]ds - 2\epsilon \int_0^t v_{xt}^2(0, s)ds.$$

Then using the induction hypothesis (5.1), one shows that for any $\delta > 0$, (5.8) can be estimated in the form

$$||v_{xx}(\cdot, t)||^2 + \epsilon||v_{xxx}(\cdot, t)||^2 + \int_0^t v_{xxx}^2(0, s)ds + 2\epsilon \int_0^t v_{xt}^2(0, s)ds$$

$$\leq C_\delta + \delta \int_0^t v_{xt}^2(0, s)ds + C\int_0^t B^2(s)ds. \qquad (5.9)$$

Next, multiply (5.4) by $2v_{xxxt}$ and integrate the results over $\mathbb{R}^+ \times (0, t)$ to come to the equation

$$||v_{xxx}(\cdot, t)||^2 + \int_0^t [v_{xt}^2(0, s) + \epsilon v_{xxt}^2(0, s)]ds$$

$$= ||v_{xxx}(\cdot, 0)||^2 + \int_0^t 2v_{xxt}(0, s)v_t(0, s)ds$$

$$- \int_0^{+\infty} \Big[2v_{xxx}(x, t)\big(a(u(x, t))v(x, t) + h_k(u(x, t))\big)_x$$ $$\qquad (5.10)$$

$$\qquad\qquad + 2v_{xxx}(x, 0)\big(a(f(x))v(x, 0) + h_k(u(x, 0))\big)_x\Big]dx$$

$$- \int_0^t \int_0^{+\infty} 2v_{xxx}\Big[\big(a(u)v_{xt} + a'(u)u_x v_t\big) + \big(a'(u)u_t v + h_{k+1}(u)\big)_x\Big]dx ds.$$

Because

$$||v||_{L_\infty(0,t;W^{2,\infty}(\mathbb{R}^+))} \le C_\delta + \delta||v(\cdot,t)||_3^2, \tag{5.11}$$

and because of the relations $v_{tt}(0,s) = g^{(k+2)}(s)$ and $v_t(0,s) = g^{(k+1)}(s)$, it is adduced that corresponding to any $\delta > 0$ there is a constant C_δ for which

$$\int_0^t 2v_{xxt}(0,s)v_t(0,s)ds = 2v_{xx}(0,s)v_t(0,s)\big|_{s=0}^{s=t} - \int_0^t 2v_{xx}(0,s)v_{tt}(0,s)ds$$

$$\le C_\delta + \delta||v(\cdot,t)||_3^2 + C\int_0^t ||v(\cdot,s)||_3^2 ds.$$

One also easily obtains the inequality

$$\left|\int_0^{+\infty} v_{xxx}(x,t)\Big(a\big(u(x,t)\big)v(x,t) + h_k\big(u(x,t)\big)\Big)_x dx\right| \le C_\delta + \delta||v(\cdot,t)||_3^2.$$

Differentiating equation (5.4) with respect to x yields

$$-v_{xt} = \big(a(u)v + h_k(u)\big)_{xx} + v_{xxxx} - \epsilon v_{xxxt}.$$

Using the above equation leads to the inequality

$$\int_0^t \int_0^{+\infty} a(u)v_{xt}v_{xxx}dxds$$

$$= -\int_0^t \int_0^{+\infty} a(u)v_{xxx}\Big[\big(a(u)v + h_k(u)\big)_{xx} + v_{xxxx} - \epsilon v_{xxxt}\Big]dxds$$

$$\le C + \epsilon||a(u)||_{C_b(\mathbb{R}^+\times[0,T])}||v_{xxx}(\cdot,t)||^2$$

$$+ C\int_0^t v_{xxx}^2(0,s)ds + C\int_0^t (||v_{xxx}(\cdot,s)||^2 + ||v_{xx}(\cdot,s)||^2)ds$$

$$\le C + C\int_0^t v_{xxx}^2(0,s)ds + C\epsilon||v_{xxx}(\cdot,t)||^2 + C\int_0^t B^2(s)ds.$$

By again using equation (5.4), and the induction hypothesis, one sees that

$$||v_t(\cdot,t) - \epsilon v_{xxt}(\cdot,t)||^2 \le ||v_{xxx}(\cdot,t)||^2 + C||v(\cdot,t)||_2^2.$$

Expanding the norm on the left-hand side of this inequality and integrating by parts the mixed term gives

$$||v_t(\cdot,t)||^2 + 2\epsilon||v_{xt}(\cdot,t)||^2 + \epsilon^2||v_{xxt}(\cdot,t)||^2$$
$$\le C||v(\cdot,t)||_3^2 + 2\epsilon|v_t(0,t)v_{xt}(0,t)|$$
$$\le C||v(\cdot,t)||_3^2 + 4\epsilon|g^{(k+1)}(t)|||v_{xt}(\cdot,t)||^{\frac{1}{2}}||v_{xxt}(\cdot,t)||^{\frac{1}{2}}$$
$$\le C||v(\cdot,t)||_3^2 + C\epsilon^{\frac{1}{2}}|g^{(k+1)}(t)|^2 + \epsilon||v_{xt}(\cdot,t)||^2 + \frac{1}{2}\epsilon^2||v_{xxt}(\cdot,t)||^2,$$

from which one obtains

$$||v_t(\cdot,t)||^2 + \epsilon||v_{xt}(\cdot,t)||^2 + \frac{1}{2}\epsilon^2||v_{xxt}(\cdot,t)||^2 \leq C||v(\cdot,t)||_3^2 + C\epsilon^{\frac{1}{2}}|g^{(k+1)}(t)|^2.$$

Using this last information yields

$$\left|\int_0^t\int_0^{+\infty} a'(u)u_xv_tv_{xxx}\,dxds\right| \leq C\int_0^t ||v_t(\cdot,s)||\,||v_{xxx}(\cdot,s)||ds$$

$$\leq C + C\int_0^t B(s)^2ds.$$

Note also that $|h_{k+1}(u)| \leq |H_k(u)v|$, where $H_k(u)$ is a function containing terms in h_k. Using this information and the induction hypothesis shows that

$$\int_0^t\int_0^{+\infty} v_{xxx}\big(a'(u)u_tv + h_{k+1}(u)\big)_x\,dxds \leq \int_0^t C||v_{xxx}(\cdot,s)||||v_x(\cdot,s)||ds$$

$$\leq C\int_0^t B^2(s)ds.$$

If ϵ is chosen small enough, (5.10) can be estimated as

$$||v_{xxx}(\cdot,t)||^2 + \int_0^t [v_{xt}^2(0,s) + \epsilon v_{xxt}^2(0,s)]ds$$

$$\leq C + C\int_0^t v_{xxx}^2(0,s)ds + C\int_0^t B^2(s)ds.$$

(5.12)

Multiply (5.9) by a suitable constant and add the result to (5.12). Then applying Gronwall's lemma shows that $B(t)$ is bounded by a constant b_2, as advertised. This completes the induction argument and hence the proof of Lemma 5.1. □

The bounds established in the last lemma are just what will be needed in Section 6, except that, so far as is known now, not all the arguments of the constant b_1 and b_2 are independent of ϵ. To attain the goal for this section, it will suffice to give conditions on the data f and g which imply that $||u^j(\cdot,0)||_3$ and $||u^{j+1}(\cdot,0)||_1$ are bounded independently of ϵ for $0 \leq j \leq k$.

Let $\phi^{(0)}(x) = f(x)$, and for each integer $j \geq 1$ define functions $\phi^{(j)}$ inductively by the recurrence

$$\phi^{(j+1)} = -\left[\phi_{xxx}^{(j)} + \left(\sum_{i=0}^{j-1} \binom{j-1}{i}(a(\phi))^{(i)}\phi^{(j-i)}\right)_x\right].$$

(5.13)

Here is the result to which allusion was just made.

Lemma 5.2. *Let $f \in H^\infty(\mathbb{R}^+)$, $g \in H^\infty(0,T)$ be given, with $f(0) = g(0)$. Let $k \geq 1$ be a given integer and suppose additionally that*

$$g^{(j)}(0) = \phi^{(j)}(0), \quad for\ j = 1, 2, \cdots, k. \tag{5.14}$$

Then there exists a constant b_3, depending continuously on $\|f\|_{3k+1}$ and $|g|_{k+1,T}$ such that

$$\|u^{(j)}(\cdot,0)\|_1 + \epsilon^{\frac{1}{2}}\|u^{(j)}(\cdot,0)\|_2 \leq b_3,$$

and there exists a constant b_4, depending continuously on $\|f\|_{3(k+1)}$ and $|g|_{k+2,T}$ such that

$$\|u^{(j)}(\cdot,0)\|_3 + \epsilon^{\frac{1}{2}}\|u^{(j)}(\cdot,0)\|_4 \leq b_4,$$

for $0 \leq j \leq k$ and all $\epsilon \in (0,1]$.

Proof: Note that when $k = 1$, the desired result is implied by Lemma 4.7 and Lemma 4.9. The proof proceeds by induction on k. Let $k > 1$, and suppose that the stated estimates hold for all non-negative integers less than or equal to $k - 1$. Let $u^{(k)}(x,0) = v(x,0)$. From equation (5.2), $v(x,0)$ satisfies the boundary-value problem

$$v - \epsilon v_{xx} = \phi^{(k)}, \tag{5.15}$$

with

$$v(0,0) = g^{(k+1)}(0), \quad and \quad \lim_{x \to +\infty} v(x,0) = 0.$$

By the compatibility conditions (5.14), one has $v_{xx}(0,0) = 0$. Following the line of argument introduced in proving Lemma 4.7, but using (5.15), one shows that there is a constant $b_3 = b_3(\|f\|_{3k+1}, |g|_{k+1,T})$ such that

$$\|u^{(j)}(\cdot,0)\|_1 + \epsilon^{\frac{1}{2}}\|u^{(j)}(\cdot,0)\|_2 \leq b_3$$

for $0 \leq j \leq k$ and all $\epsilon \in (0,1]$. As worked out in Lemma 4.9, one then shows that there is a constant $b_4 = b_4(\|f\|_{3(k+1)}, |g|_{k+2,T})$, such that

$$\|u^{(j)}(\cdot,0)\|_3 + \epsilon^{\frac{1}{2}}\|u^{(j)}(\cdot,0)\|_4 \leq b_4,$$

for $0 \leq j \leq k$ and all $\epsilon \in (0,1]$. $\square$

It is worth summarizing the accomplishments of this section in the following theorem. This is a higher-order analogue of Theorem 4.10. In the statement of the theorem, ϵ_2 is the same positive constant that already appeared in Theorem 4.10.

Theorem 5.3. *Let $T > 0$ and a positive integer k be given. Let $f \in H^\infty(\mathbb{R}^+)$ and $g \in H^\infty(0,T)$ be given with $f(0) = g(0)$. Furthermore, suppose f and g satisfy*

$$g^{(j)}(0) = \phi^{(j)}(0), \quad for\ 1 \leq j \leq k,$$

where the functions $\phi^{(j)}$ are related to f as in (5.13). Then there exists a constant $b_5 = b_5(\|f\|_{3k+1}, |g|_{k+1,T})$, such that

$$\|u^{(j)}(\cdot,t)\|_1 \leq b_5,$$

holds for $1 \leq j \leq k$ and all $\epsilon \in (0, \epsilon_2]$.

Moreover there exists a constant $b_6 = b_6(||f||_{3(k+1)}, |g|_{k+2,T})$, depending continuously on its arguments, such that

$$||u^{(j)}(\cdot, t)||_3 + \epsilon^{\frac{1}{2}} ||\partial_x^4 u^{(j)}(\cdot, t)|| \leq b_6,$$

holds for $1 \leq j \leq k$ and all $\epsilon \in (0, \epsilon_2]$.

6. Existence and Uniqueness of Solutions

Using the theory developed in Sections 3, 4 and 5, it is comparatively simple to prove existence of smooth solutions of the quarter-plane problem for the equation

$$u_t + P(u)_x + u_{xxx} = 0, \qquad \text{for } x, \ t > 0, \tag{6.1a}$$

subject to the auxiliary conditions,

$$\begin{aligned} u(x, 0) &= f(x) & \text{for } x \geq 0, \\ u(0, t) &= g(t) & \text{for } t \geq 0, \end{aligned} \tag{6.1b}$$

where f and g are given functions.

It is useful to first settle uniqueness of solutions of this initial- and boundary-value problem.

Theorem 6.1. *Let $T > 0$ and $s > \frac{3}{2}$. Then, corresponding to given auxiliary data f and g, there is at most one solution of (6.1) in the function class $L_\infty(0, T; H^s(\mathbb{R}^+))$.*

Proof: This result is proved as in Theorem 6.1 of [17] for the KdV equation (1.2). $\square$

Theorem 6.2. *Let k be a positive integer. Suppose $f \in H^{3k+1}(\mathbb{R}^+)$ and $g \in H_{loc}^{k+1}(\mathbb{R}^+)$, or $f \in H^{3k+3}(\mathbb{R}^+)$ and $g \in H_{loc}^{k+2}(\mathbb{R}^+)$, and the $k+1$ compatibility conditions*

$$g^{(j)}(0) = \phi^{(j)}(0) \quad \text{for } 0 \leq j \leq k,$$

hold, where $\phi^{(j)}$ is defined in (5.13). Then, corresponding to the given auxiliary data f and g, there exists a unique solution u of (6.1) in the function class $L_\infty(0, T; H^s(\mathbb{R}^+))$, where $s = 3k+1$, or $3k+3$, respectively. Furthermore there exist a constant C_{3k+1} depending on $||f||_{3k+1}$ and $|g|_{k+1,T}$, and a constant C_{3k+3} depending on $||f||_{3k+3}$ and $|g|_{k+2,T}$ such that for $0 \leq j \leq k$,

$$\int_0^T [(u_x^{(j)})^2(0, s) + (u_{xx}^{(j)})^2(0, s)] ds \leq C_{3j+1} \tag{6.2}$$

and

$$\int_0^T (u_x^{(k+1)})^2(0, s) ds \leq C_{3k+3}. \tag{6.3}$$

If $k > 1$ in the first case, or $k \geq 1$ in the second case, u defines a classical solution, up to the boundary, of (6.1) in the quarter-plane $\mathbb{R}^+ \times \mathbb{R}^+$.

The proof is made in two steps. First, existence of a smooth solution of (6.1) corresponding to smooth initial data and smooth boundary data is established. Then a limit is taken through smooth solutions of (6.1) to infer existence of solutions corresponding to initial data in $H^{3k+1}(\mathbb{R}^+)$, or $H^{3(k+1)}(\mathbb{R}^+)$, and boundary data in $H_{loc}^{k+1}(\mathbb{R}^+)$, or $H_{loc}^{k+2}(\mathbb{R}^+)$, respectively. In pursuing this program, the following technical lemma (see [**16**], Lemma 7) is useful.

Lemma 6.3. *Suppose $u_n \to u$ weak-star in $L_\infty(0, T; H^s(\mathbb{R}^+))$ where $s > \frac{1}{2}$ and $\partial_t u_n \to \partial_t u$ weak-star in $L_2(0, T; H^r(\mathbb{R}^+))$ for some real r. Then there exists a subsequence $\{u_l\}$ of $\{u_n\}$ such that $u_l \to u$ pointwise almost everywhere in $[0, T] \times \mathbb{R}^+$ and $a(u_l)\partial_x u_l \to a(u)u_x$ in $\mathcal{D}'([0, T] \times \mathbb{R}^+)$ in the usual sense of distributions ($a(u)u_x$ is interpreted as $\partial_x P(u)$ in case $s < 1$ and similarly for $a(u_l)\partial_x u_l$).*

First, it is established that solutions of (6.1) exist in case f and g happen to be infinitely smooth.

Proposition 6.4. *Let $T > 0$ and k a positive integer. Let $f \in H^\infty(\mathbb{R}^+)$ and $g \in H^\infty(0, T)$ satisfy the $k + 1$ compatibility conditions,*

$$g^{(j)}(0) = \phi^{(j)}(0), \qquad for\ 0 \leq j \leq k.$$

Then there exists a solution u of (6.1) in $L_\infty(0, T; H^{3k+3}(\mathbb{R}^+))$ corresponding to the data f and g. Moreover, there exist constants

$$b_{3k+1} = b_{3k+1}(\|f\|_{3k+1}, |g|_{k+1,T}) \qquad and \qquad b_{3k+3} = b_{3k+3}(\|f\|_{3k+3}, |g|_{k+2,T})$$

such that

$$\|u^{(j)}(\cdot, t)\|_3 \leq b_{3k+1}, \quad for\ 0 \leq j < k, \quad \|u^{(k)}(\cdot, t)\|_1 \leq b_{3k+1}, \qquad (6.4)$$

and

$$\|u^{(j)}(\cdot, t)\|_3 \leq b_{3k+3}, \qquad (6.5)$$

for $0 \leq j \leq k$. The constants b_s for the various values of s can be chosen to depend continuously on their arguments.

Proof: The argument very closely parallels others appearing in many standard works, so it is presented in outline only. Throughout, $T > 0$ will be fixed, but arbitrary.

According to Corollary 3.8, for any $\epsilon > 0$ there is a smooth solution u_ϵ of the regularized initial- and boundary-value problem (3.1) corresponding to the data f and g. And by Theorem 5.3, there is a constant $b = b(\|f\|_{3(k+1)}, |g|_{k+2,T})$ depending continuously on its arguments, but independent of ϵ in $(0, \epsilon_2]$, such that

$$\|u_\epsilon^{(j)}(\cdot, t)\|_3 \leq b,$$

for $0 \le j \le k$. Moreover, for all nonnegative integers i and m,

$$\partial_t^i u_\epsilon \in C(0, T; H^m(\mathbb{R}^+)), \tag{6.6}$$

by Corollary 4.2 and in particular for $0 \le j \le k$, $\{\partial_t^j u_\epsilon\}_{0 < \epsilon \le \epsilon_2}$ is bounded in $L_\infty(0, T; H^3(\mathbb{R}^+))$, independently of ϵ.

These bounds together with standard compactness results due to Aubin and Lions (cf. [**34**], Lemma 6.3), and diagonalization procedures imply that there is a sequence $\{\epsilon_n\}_{n=1}^\infty$ tending to zero as n tends to infinity such that if $u_n = u_{\epsilon_n}$, $n = 1, 2, \cdots$, then

$$\partial_t^j u_n \to \partial_t^j u, \quad \text{weak-star in } L_\infty(0, T; H^3(\mathbb{R}^+)) \tag{6.7}$$

for $0 \le j \le k$ and

$$a(u_n)\partial_x u_n \to a(u)\partial_x u \text{ in } \mathcal{D}'(\mathbb{R}^+ \times [0, T]), \tag{6.8}$$

as $n \to +\infty$. Moreover, the function u satisfies the generalized KdV-equation (6.1a) on $\mathbb{R}^+ \times [0, T]$; if $k = 0$, u satisfies the equation in the $L_2(\mathbb{R}^+)$-sense, while if $k \ge 1$, the solution is classical. The initial and boundary conditions are easily inferred to be taken on in appropriate senses. In particular, by standard interpolation results (cf. [**35**]) it is inferred that

$$\partial_t^j u \in C(0, T; H^{3(k-j)+\frac{3}{2}}(\mathbb{R}^+)) \tag{6.9}$$

for $0 \le j < k$.   □

With a little change in the details, one shows that the argument of Proposition 6.4 also applies to the case when $s = 3$.

Corollary 6.5. *Let* $f \in H^\infty(\mathbb{R}^+)$ *and* $g \in H^\infty_{loc}(\mathbb{R}^+)$ *with* $f(0) = g(0)$. *Then there exists a unique solution* u *of (6.1) in* $L_\infty(0, T; H^3(\mathbb{R}^+))$ *and*

$$\|u(\cdot, t)\|_3 \le b(\|f\|_3, |g|_{2,T}).$$

Proof: By following the line of argument adopted in proving Proposition 6.4, and using Lemma 4.4, one shows that u lies in $L_\infty(0, T; H^3(\mathbb{R}^+))$ and $a(u)u_x$ lies in $L_\infty(0, T; H^2(\mathbb{R}^+))$. Hence, from the differential equation (6.1), $u_t \in L_\infty(0, T; L_2(\mathbb{R}^+))$. Moreover, according to Lemma 4.4, If u_ϵ is the approximating solution of the regularized problem corresponding to the value of the perturbation parameter ϵ, there is a constant a depending continuously on $\|f\|_3 + \epsilon^{\frac{1}{2}}\|f^{(4)}\|$ and $|g|_{2,T}$ such that

$$\|u_\epsilon(\cdot, t)\|_3 \le a,$$

at least for ϵ sufficiently small. Since the solution u of (6.1) pertains to the weak limit as ϵ tends to zero, it follows that

$$\|u(\cdot, t)\|_3 \le \limsup_{\epsilon \to 0} a = b,$$

say, where $b = b(\|f\|_3, |g|_{2,T})$, as advertised.   □

Now we pass to the second stage of the proof of Theorem 6.2, where it is supposed that the initial condition f is constrained only to lie in $H^{3k}(\mathbb{R}^+)$ or $H^{3k+1}(\mathbb{R}^+)$, and the boundary condition g to be a member of $H_{loc}^{k+1}(\mathbb{R}^+)$, for some integer $k \geq 1$. The following result from [17] will prove to be useful in the present context.

Lemma 6.6. *Let f and g satisfy the conditions in Theorem 6.2. Then there exist sequences $\{f_N\}_{N=1}^{\infty} \subset H^{\infty}(\mathbb{R}^+)$ and $\{g_N\}_{N=1}^{\infty} \subset C^{\infty}(\mathbb{R}^+)$ such that*

$$g_N^{(j)}(0) = \phi_N^{(j)}(0) \quad \text{for } 0 \leq j \leq k$$

and for which, as $N \to \infty$,

$$
\begin{aligned}
&(i) \quad f_N \to f \quad \text{in} \quad H^{3k+1}(\mathbb{R}^+), \quad g_N \to g \quad \text{in} \quad H_{loc}^{k+1}(\mathbb{R}^+) \quad \text{or} \\
&(ii) \quad f_N \to f \quad \text{in} \quad H^{3k+3}(\mathbb{R}^+), \quad g_N \to g \quad \text{in} \quad H_{loc}^{k+2}(\mathbb{R}^+)
\end{aligned}
$$

where $\phi_N^{(j)}$ is as defined in (5.13) with f_N replacing f, and $g_N^{(j)} = \partial_t^j g_N$.

Proof of Theorem 6.2: Suppose that $f \in H^{3k+3}(\mathbb{R}^+)$ and $g \in H_{loc}^{k+2}(\mathbb{R}^+)$ are fixed, and that f and g satisfy the first $k+1$ compatibility conditions as in the statement of the theorem. For fixed $T > 0$, Lemma 6.6 implies there exist sequences $\{f_N\}_{N=1}^{\infty} \subset H^{\infty}(\mathbb{R}^+)$ and $\{g_N\}_{N=1}^{\infty} \subset H^{\infty}(\mathbb{R}^+)$ such that

$$f_N \to f \quad \text{in} \quad H^{3k+3}(\mathbb{R}^+) \quad \text{and} \quad g_N \to g \quad \text{in} \quad H^{k+2}(0, T) \tag{6.10}$$

as $N \to +\infty$. For each $N > 0$, f_N and g_N satisfy the same $k+1$ compatibility conditions satisfied by f and g. From Proposition 6.4, corresponding to the auxiliary data f_N and g_N there is a solution u_N of (6.1) defined on $\mathbb{R}^+ \times [0, T]$ such that $\partial_t^j u_N \in L_{\infty}(0, T; H^{(3(k+1)-3j)}(\mathbb{R}^+))$ for $0 \leq j \leq k$, $N = 1, 2, \cdots$. Moreover, there exist constants

$$b_{3k+3}^N = b_{3k+3}(\|f_N\|_{3k+3}, |g_N|_{k+2,T})$$

such that

$$\|\partial_t^j u_N\|_{L_{\infty}(0,T;H^3(\mathbb{R}^+))} \leq b_{3k+3}^N, \quad \text{for } 0 \leq j \leq k. \tag{6.11}$$

Because of (6.10) and the fact that b_{3k+3} is uniformly bounded when its arguments vary over a bounded set, there is a constant B_{3k+3}, independent of N, such that

$$\|\partial_t^j u_N\|_{L_{\infty}(0,T;H^3(\mathbb{R}^+))} \leq B_{3k+3}, \tag{6.12}$$

for $0 \leq j \leq k$. Similarly, one shows that if f lies in $H^{3k+1}(\mathbb{R}^+)$ and g lies in $H_{loc}^{k+1}(\mathbb{R}^+)$, there exists a solution u_N of (6.1) on $\mathbb{R}^+ \times [0, T]$ with initial- and boundary-data f_N and g_N, respectively, for which $\partial_t^j u_N \in L_{\infty}(0, T; H^{(3k+1-3j)}(\mathbb{R}^+))$ for $0 \leq j \leq k$, and a constant B_{3k+1} such that

$$
\begin{aligned}
&\|\partial_t^j u_N\|_{L_{\infty}(0,T;H^3(\mathbb{R}^+))} \leq B_{3k+1}, \quad \text{for } 0 \leq j < k, \text{ and} \\
&\|\partial_t^k u_N\|_{L_{\infty}(0,T;H^1(\mathbb{R}^+))} \leq B_{3k+1}.
\end{aligned}
\tag{6.13}
$$

In consequence of the bounds expressed in (6.12) and (6.13), the arguments of Proposition 6.4 may be repeated without essential change (the extra smoothness available during the proof of the proposition was not used, nor was the regularizing term $-\epsilon u_{xxt}$). It is concluded therefore that $\{u_N\}_1^\infty$ converges to a function u_T, say, in the various ways already detailed in the proof of Proposition 6.4. As in the proposition, u_T provides a solution of (6.1) corresponding to the data f and g, in $H^{3k+1}(\mathbb{R}^+)$ and $H^{k+1}_{loc}(\mathbb{R}^+)$, or in $H^{3(k+1)}(\mathbb{R}^+)$, and $H^{k+2}_{loc}(\mathbb{R}^+)$.

The above argument applies for any fixed $T > 0$. Define a function U on $\mathbb{R}^+ \times \mathbb{R}^+$ by,

$$U(x,t) = u_T(x,t),$$

provided that $t < T$. This is well defined because of the uniqueness result. It is clear that U provides the solution whose existence was contemplated in the statement of Theorem 6.2. The fact that U is a classical solution of the problem (6.1), if $f \in H^{3k+1}(\mathbb{R}^+)$ for $k > 1$, or $f \in H^{3k+3}(\mathbb{R}^+)$ for $k \geq 1$, follows exactly as in the proof of the Proposition 6.4. This finishes the proof of Theorem 6.2. $\square$

The above arguments also apply to the case $s = 3$ because of Corollary 6.5.

Corollary 6.7. *Let $f \in H^3(\mathbb{R}^+)$, and $g \in H^2_{loc}(\mathbb{R}^+)$, with $f(0) = g(0)$. Then, corresponding to given auxiliary data f and g, there exists a unique solution u of (6.1) in the function class $L_\infty(0,T; H^3(\mathbb{R}^+))$. Moreover, for any $T > 0$, there is a constant C depending on $\|f\|_3$ and $|g|_{2,T}$, such that $\int_0^T u_{xt}^2(0,s)ds \leq C$.*

If $f \in H^1(\mathbb{R}^+)$ and $g \in H^1_{loc}(\mathbb{R}^+)$, then Theorem 6.2 also holds because of the ϵ-independent $H^1(\mathbb{R}^+)$-bound established in Lemma 4.1. In this case, the equation will be satisfied in the sense of distributions. However, the uniqueness result does not apply. The proof of existence of these weaker solutions fits more or less directly into the framework exposed in the proof of Proposition 6.4. The outcome is stated here.

Theorem 6.8. *Let $f \in H^1(\mathbb{R}^+)$, and $g \in H^1_{loc}(\mathbb{R}^+)$, with $f(0) = g(0)$. Then for any $T > 0$, there exists a solution of (6.1) in the function class $L_\infty(0,T; H^1(\mathbb{R}^+))$ corresponding to the initial data f and boundary data g.*

7. Solutions in More Restricted Spaces
& Continuous-Dependence Results

In Section 6, we obtained a unique solution of the generalized KdV equation posed in a quarter-plane. Thus if $f \in H^s(\mathbb{R}^+)$ where $s = 3k$, or $3k + 1$, and $g \in H^{k+1}_{loc}(\mathbb{R}^+)$ satisfy the appropriate compatibility conditions at $(x,t) = (0,0)$, then the quarter-plane problem has a solution in $L^\infty_{loc}(\mathbb{R}^+; H^s(\mathbb{R}^+))$. In this section it will be shown that the solutions lie in $C(\mathbb{R}^+; H^s(\mathbb{R}^+))$, and a result of continuous dependence of solutions on the data in spaces that are as restrictive as the solutions allow will be established.

To prove such a result, we follow the line worked out for the KdV equation (1.2) in [18]. Since we deal with a more general nonlinearity and our theory implies that solutions exist corresponding to weaker assumptions on the auxiliary data, details of the proof are provided. We first show that the solution u_n corresponding to smooth initial data f_n and boundary data g_n which approximate f in $H^{3k}(\mathbb{R}^+)$ and g in $H^{k+1}_{loc}(\mathbb{R}^+)$ appropriately are Cauchy in $C(0,T;H^{3k}(\mathbb{R}^+))$ for any fixed $T > 0$. It follows that for any $T > 0$, $u_n \to u$ strongly in $C(0,T;H^{3k}(\mathbb{R}^+))$ for some u. This is accomplished by deriving bounds for the difference between two solutions, say u_1 and u_2, of the initial- and boundary-value problem (6.1). These bounds may be expressed in terms of corresponding differences in the initial and the boundary data for the two solutions. After such bounds are obtained, the result that $\{u_n\}_{n=1}^\infty$ is Cauchy in $C(0,T;H^{3k}(\mathbb{R}^+))$ follows by choosing appropriate approximations to the initial and boundary data.

Throughout this section X_{3k} will be the set

$$X_{3k} = \{(f,g) \in H^{3k}(\mathbb{R}^+) \times H^{k+1}_{loc}(\mathbb{R}^+): \phi^{(j)}(0) = g^{(j)}(0) \text{ for } 0 \le j \le k-1\}, \quad (7.1)$$

and X_{3k+1} will be the set

$$X_{3k+1} = \{(f,g) \in H^{3k+1}(\mathbb{R}^+) \times H^{k+1}_{loc}(\mathbb{R}^+): \phi^{(j)}(0) = g^{(j)}(0) \text{ for } 0 \le j \le k\}, \quad (7.2)$$

where ϕ is expressed in terms of the initial data f as in (5.13) and k is a positive integer. Assume that (f_1,g_1) and (f_2,g_2) are two sets of data for the problem (6.1) which lie in X_{3k}, or X_{3k+1}. By Corollary 6.7 and Theorem 6.2, the corresponding solutions u_1 and u_2 of (6.1) will be elements of $L^\infty_{loc}(\mathbb{R}^+;H^{3k}(\mathbb{R}^+))$, or $L^\infty_{loc}(\mathbb{R}^+;H^{3k+1}(\mathbb{R}^+))$, respectively. Moreover, there exist constants $C^m_{k,T}$ which depend only on T, $\|f_m\|_{3k}$, or $\|f_m\|_{3k+1}$, and $|g|_{k+1,T}$, which bound above

$$\|u_m\|_{L_\infty(0,T;H^{3k}(\mathbb{R}^+))} \quad \text{or} \quad \|u_m\|_{L_\infty(0,T;H^{3k+1}(\mathbb{R}^+))} \quad (7.3)$$

for $m = 1,2$. As a corollary of Theorem 6.2, one also has that

$$|\partial_x^{3k+1} u_m(0,s)|_{0,T} \quad \text{or} \quad |\partial_x^{3k+2} u_m(0,s)|_{0,T} \quad (7.4)$$

is bounded above by $C^m_{k,T}$, for $m = 1,2$. The constants $C^m_{k,T}$ will be shown to depend continuously on T, $\|f_m\|_{3k}$, or $\|f_m\|_{3k+1}$, and $|g|_{k+1,T}$, $m = 1,2$. For convenience in writing the difference between datum, denote

$$f \equiv f_1 - f_2, \quad g \equiv g_1 - g_2, \quad \text{and} \quad w \equiv u_1 - u_2.$$

Then from (6.1), the function w satisfies the initial- and boundary-value problem

$$w_t + a(u_1)w_x + [a(u_1) - a(u_2)](u_2)_x + w_{xxx} = 0 \quad \text{for } x, t \in \mathbb{R}^+ \times \mathbb{R}^+, \quad (7.5\text{a})$$

$$w(x,0) = f(x) \qquad\qquad \text{for } x \ge 0, \quad (7.5\text{b})$$

$$w(0,t) = g(t) \qquad\qquad \text{for } t \ge 0. \quad (7.5\text{c})$$

As before, for non-negative integers j, denote by $w^{(j)}$ the temporal derivative $\partial_t^j w$. Then $w^{(k)}$ satisfies the partial differential equation

$$
\begin{aligned}
w_t^{(k)} + (a(u_1))^{(k)} w_x + a(u_1) w_x^{(k)} + [a(u_1) - a(u_2)]^{(k)} (u_2)_x \\
+ [a(u_1) - a(u_2)](u_2^{(k)})_x + w_{xxx}^{(k)} + F_k = 0,
\end{aligned} \tag{7.6}
$$

where $F_1 = 0$ and F_k is defined by

$$
F_k = \sum_{j=1}^{k-1} \binom{k}{j} \left[(a(u_1))^{(k-j)} w_x^{(j)} + [(a(u_1))^{(k-j)} - a((u_2))^{(k-j)}](u_2)_x^{(j)} \right].
$$

Lemma 7.1. *Let* $(f_m, g_m) \in X_3$ *for* $m = 1, 2$. *Then for any* $T > 0$, *there is a constant* C_T *depending continuously on* T, $\|f_m\|_3$ *and* $|g_m|_{2,T}$, $m = 1, 2$, *such that*

$$
\|w(\cdot, t)\|^2 + \int_0^t w_x^2(0, s) ds \le C_T \{\|f\|^2 + |g|_{1,T}^2\} \tag{7.7}
$$

and

$$
\|w_x(\cdot, t)\|^2 + \int_0^t w_{xx}^2(0, s) ds \le C_T \{\|f\|_1^2 + |g|_{1,T}^2\} \tag{7.8}
$$

for $0 \le t \le T$. *If* $(f_m, g_m) \in X_4$ *for* $m = 1, 2$, *then there is a constant* $C_{1,T}^1$ *depending continuously on* T, $\|f_m\|_3$ *and* $|g_m|_{2,T}$, $m = 1, 2$, *such that*

$$
\|w(\cdot, t)\|_3^2 + \int_0^t w_{xxxx}^2(0, s) ds \tag{7.9}
$$
$$
\le C_{1,T}^1 \{\|f\|_3^2 + |g|_{2,T}^2 + \|w\|_{L_\infty(\mathbb{R}^+ \times (0,T))}^2 \|u_2\|_{L_2(0,T;H^4(\mathbb{R}^+))}^2\}.
$$

Proof: Let $(f_m, g_m) \in X_3$ for $m = 1, 2$. It follows from Corollary 6.7 that u_m lies in $L_\infty(0, T; H^3(\mathbb{R}^+))$ and there is a constant C_T such that

$$
\|u_m\|_{L_\infty(0,T;H^3(\mathbb{R}^+))} \le C_T \tag{7.10}
$$

for $m = 1, 2$. Here, and below, C_T will denote different constants possessing the same properties as the constant C_T specified in the statement of the lemma.

First, it is shown that $\|w(\cdot, t)\|$ is bounded by $C_T\{\|f\|^2 + |g|_{1,T}^2\}$. Let $U(x, t) = g(t)e^{-x}$ and $y = w - U$. Then y satisfies the initial- and boundary-value problem

$$
\begin{aligned}
y_t + [a(y_1 + U_1) - a(y_2 + U_2)](y_2 + U_2)_x & \\
+ a(y_1 + U_1) y_x + y_{xxx} = h, & \quad \text{for } x, t \in \mathbb{R}^+ \times \mathbb{R}^+, \\
y(x, 0) = f(x) - g(0)e^{-x}, & \quad \text{for } x \ge 0, \\
y(0, t) = 0, & \quad \text{for } t \ge 0,
\end{aligned} \tag{7.11}
$$

where $y_m = u_m - g_m(t)e^{-x} \equiv u_m - U_m$, $m = 1, 2$, and $h = U_t + U_{xxx} + a(y_1 + U_1)U_x$. Note that $y = y_1 - y_2$. To establish the advertised bound for $\|w(\cdot, t)\|$, one need only establish a similar estimate for y. Multiply (7.11) by $2y$ and then integrate the results over $\mathbb{R}^+ \times (0, t)$. After integrations by parts, there appears

$$\|y(\cdot, t)\|^2 + \int_0^t y_x^2(0, s)ds$$

$$= \|f - U(x, 0)\|^2 + \int_0^t \int_0^\infty 2y\Big(h - a(y_1 + U_1)y_x \tag{7.12}$$

$$- [a(y_1 + U_1) - a(y_2 + U_2)](y_2 + U_2)_x\Big)dxds.$$

By a further integration by parts, one sees that

$$\int_0^\infty 2a(y_1 + U_1)yy_x dx = -\int_0^\infty [a(y_1 + U_1)]_x y^2 dx.$$

From (7.10) and the definition of U, it therefore follows that

$$\|[a(y_1 + U_1)]_x\|_{L_\infty(\mathbb{R}^+ \times (0,T))} \le C_T, \qquad \|h\|_{L_2(\mathbb{R}^+ \times (0,T))} \le C_T|g|_{1,T}$$

$$\text{and} \quad \|U\|_{L_\infty(\mathbb{R}^+ \times (0,T))} \le C_T|g|_{1,T}.$$

Because of the Lipschitz-condition satisfied by P, one infers that

$$\|(a(y_1 + U_1) - a(y_2 + U_2))\|_{L_2(\mathbb{R}^+ \times (0,T))} \le C_T\gamma(\mathcal{B})\|y + U\|_{L_2(\mathbb{R}^+ \times (0,T))}, \tag{7.13}$$

where $\mathcal{B} = [0, C_T]$. Note that this follows since both arguments $y_1 + U_1 = u_1$ and $y_2 + U_2 = u_2$ lie in the set $\{u : \|u\|_{L_\infty(\mathbb{R}^+ \times (0,T))} \le C_T\}$. With the above estimates in hand, it follows from (7.11) and Gronwall's lemma that for $0 \le t \le T$,

$$\|y(\cdot, t)\|^2 + \int_0^t y_x^2(0, s)ds \le C_T\{\|f\|^2 + |g|_{1,T}^2\}.$$

To control $\|w_x(\cdot, t)\|$, multiply equation (7.5a) by $-2w_{xx}$ and integrate the result over $\mathbb{R}^+ \times (0, t)$ to obtain

$$\|w_x(\cdot, t)\|^2 + \int_0^t w_{xx}^2(0, s)ds = \|f'\|^2 - \int_0^t 2g'(s)w_x(0, s)ds$$

$$+ \int_0^t \int_0^\infty 2w_{xx}[a(u_1)w_x + (a(u_1) - a(u_2))(u_2)_x]dxds. \tag{7.14}$$

By integrations by parts and using (7.7) and (7.10), one shows that

$$\int_0^t \int_0^\infty 2w_{xx}a(u_1)w_x dxds$$

$$= -\int_0^t a(u_1(0, s))w_x^2(0, s)ds - \int_0^t \int_0^\infty [a(u_1)]_x w_x^2 dxds \tag{7.15}$$

$$\le C_T\{\|f\|_1^2 + |g|_{1,T}^2 + \int_0^t \|w_x(\cdot, s)\|^2 ds\}.$$

Also note that

$$\int_0^t \int_0^\infty 2w_{xx}[a(u_1) - a(u_2)](u_2)_x \, dx \, ds$$

$$= -\int_0^t \int_0^\infty \left[2w_x[a(u_1) - a(u_2)]_x(u_2)_x + 2w_x[a(u_1) - a(u_2)](u_2)_{xx} \right] dx \, ds$$

$$- \int_0^t 2w_x(0,s)[a(u_1(0,s)) - a(u_2(0,s))](u_2)_x(0,s)ds \tag{7.16}$$

$$\leq C_T \left(\int_0^t w_x^2(0,s)ds \right)^{\frac{1}{2}} \gamma(\mathcal{B}) \left(\int_0^t g^2(s)ds \right)^{\frac{1}{2}} + C_T \gamma(\mathcal{B}) \int_0^t \|w(\cdot,s)\|_1^2 ds,$$

where $\mathcal{B} = [0, C_T]$ again, by using (7.7), (7.10) and (7.13). Hence using (7.15) and (7.16), (7.14) is reduced to

$$\|w_x(\cdot,t)\|^2 + \int_0^t w_{xx}^2(0,s)ds \leq C_T \{\|f\|_1^2 + |g|_{1,T}^2 + \int_0^t \|w(\cdot,s)\|_1^2 ds\} \tag{7.17}$$

for $0 \leq t \leq T$. Applying the first result and Gronwall's lemma to the above inequality gives the second result.

To obtain the last result, let $(f_m, g_m) \in X_4$, $m = 1, 2$. Differentiate equation (7.5a) once and multiply the result by $2w_{xxx}$. Differentiate equation (7.5a) twice and multiply the result by $2w_{xxxx}$. Add the above results together and then integrate over $\mathbb{R}^+ \times [0, t]$. After integrations by parts, there appears

$$\|w_{xx}(\cdot,t)\|^2 + \|w_{xxx}(\cdot,t)\|^2 + \int_0^t \left[w_{xxx}^2(0,s) + w_{xxxx}^2(0,s) \right] ds$$

$$= \|w_{xx}(\cdot,0)\|^2 + \|w_{xxx}(\cdot,0)\|^2$$

$$- 2\int_0^t \left[w_{xx}(0,s)w_{xs}(0,s) + w_{xxx}(0,s)w_{xxs}(0,s) \right.$$

$$\left. + w_{xxx}(0,s)\left(a(u_1)w_x + (a(u_1) - a(u_2))(u_2)_x \right)_{xx}(0,s) \right] ds$$

$$+ 2\int_0^t \int_0^{+\infty} \left[[a(u_1)w_x + (a(u_1) - a(u_2))(u_2)_x]_x \right. \tag{7.18}$$

$$- \sum_{j=1}^3 \binom{3}{j} \left[\partial_x^j(a(u_1)) \partial_x^{(3-j)}(w_x) \right.$$

$$\left. + \partial_x^j(a(u_1) - a(u_2)) \partial_x^{(3-j)}((u_2)_x) \right]$$

$$\left. - a(u_1)w_{xxxx} - (a(u_1) - a(u_2))(u_2)_{xxxx} \right] w_{xxx} \, dx \, ds.$$

By using (7.7), (7.8) and equation (7.5a), the first boundary term in (7.18) is estimated from above as follows:

$$\left|2\int_0^t w_{xx}(0,s)w_{xs}(0,s)ds\right| \leq \frac{1}{\delta}\int_0^t w_{xx}^2(0,s)ds + \delta\int_0^t w_{xs}^2(0,s)ds$$

$$\leq C_T(\|f\|_1,|g|_{1,T}) + \delta\int_0^t w_{xxxx}^2(0,s)ds, \quad (7.19)$$

for any $\delta > 0$. Similarly, one sees that for any $\delta > 0$,

$$-\int_0^t w_{xxx}(0,s)w_{xxs}(0,s)ds$$

$$= \int_0^t \left(g'(s)+a(g_1(s))w_x(0,s)+[a(g_1(s))-a(g_2(s))](u_2)_x(0,s)\right)w_{xxs}(0,s)ds$$

$$= \left(g'(s) + a(g_1(s))w_x(0,s) + [a(g_1(s)) - a(g_2(s))](u_2)_x(0,s)\right)w_{xx}(0,s)\Big|_0^t$$

$$-\int_0^t \left[g''(s) + a'(g_1(s))g_1'(s)w_x(0,s) + a(g_1(s))w_{xs}(0,s)\right.$$

$$\left. + \left([a(g_1(s)) - a(g_2(s))](u_2)_x(0,s)\right)_s\right]w_{xx}(0,s)ds \quad (7.20)$$

$$\leq C_{1,T}^1(\|f\|_3,|g|_{2,T}) + \delta\int_0^t w_{xs}^2(0,s)ds$$

$$\leq C_{1,T}^1(\|f\|_3,|g|_{2,T}) + \delta\int_0^t w_{xxxx}^2(0,s)ds.$$

Note that the inequality

$$\int_0^t \left((u_m)_{xs}(0,s)\right)^2 ds \leq C(\|f_m\|_3,|g_m|_{2,T})$$

has been used in (7.20) for $m = 1,2$. This fact is obtained via Corollary 6.7. Again by using (7.7) and (7.8), the last boundary integral in (7.18) is bounded above by a constant $C_T = C_T(\|f\|_1,|g|_{1,T})$.

There is a constant $C_{1,T}^1 = C_{1,T}^1(\|f\|_3,|g|_{2,T})$ such that all the terms featuring double integrals in (7.18) except the last two terms can be bounded above by

$$C_{1,T}^1(\|f\|_3,|g|_{2,T})\int_0^t \|w(\cdot,s)\|_3^2 ds.$$

By integrations by parts and using once more the hypothesis on P, one shows that

$$\left| \int_0^t \int_0^{+\infty} 2\big[a(u_1)w_{xxxx} + (a(u_1) - a(u_2))(u_2)_{xxxx}\big] w_{xxx}\,dx\,ds \right|$$

$$\leq \int_0^t a(g_1(s))w_{xxx}^2(0,s)\,ds + C_T \int_0^t ||w_{xxx}(\cdot,s)||^2 ds \tag{7.21}$$

$$+ C_T(||f||_1, |g|_{1,T})\gamma(\mathcal{B})||w||^2_{L_\infty(\mathbb{R}^+ \times (0,T))}||u_2||^2_{L_2(0,T;H^4(\mathbb{R}^+))},$$

where $\mathcal{B}$ is as before. Using the preceding information in (7.18) and choosing δ small enough, one obtains

$$||w(\cdot,t)||_3^2 \leq C_{1,T}^1\{||f||_3 + |g|_{2,T} + \int_0^t ||w(\cdot,s)||_3^2 ds$$

$$+ ||w||^2_{L_\infty(\mathbb{R}^+ \times (0,T))}||u_2||^2_{L_2(0,T;H^4(\mathbb{R}^+))}\}.$$

Applying Gronwall's lemma to the above inequality shows that

$$||w(\cdot,t)||_3^2 \leq C_{1,T}^1\{||f||_3 + |g|_{2,T} + ||w||^2_{L_\infty(\mathbb{R}^+ \times (0,T))}||u_2||^2_{L_2(0,T;H^4(\mathbb{R}^+))}\}.$$

The proof is complete. $\quad\square$

Bounds on higher-order Sobolev norms are needed for proving continuous-dependence results in higher-order Sobolev spaces. The next step is to derive such bounds. In the preliminary results stated and proved next, as with the results in Lemma 7.1, bounds on the H^s-norm of solutions are couched in the terms of H^{s+1}-norms of the initial data f if $s = 3k$ or $s = 3k + 1$. This apparent defect in the theory in which smoothness is lost is remedied at a later stage.

Lemma 7.2. Let $(f_m, g_m) \in X_{3k+1}$ for $m = 1, 2$. Then for any $T > 0$, there exists a constant $C_{k,T}^1$ depending continuously on T, $||f_m||_{3k}$ and $|g_m|_{k+1,T}$, such that

$$||w^{(k)}(\cdot,t)||^2 + \int_0^t (w_x^{(k)}(0,s))^2 ds$$

$$\leq C_{k,T}^1\{||f||_{3k}^2 + |g|_{k+1,T}^2 + ||w||^2_{L_\infty(\mathbb{R}^+ \times (0,T))}||u_2^{(k)}||^2_{L_2(0,T;H^1(\mathbb{R}^+))}\}. \tag{7.22}$$

If $(f_m, g_m) \in X_{3k+3}$, then there exists a constant $C_{k,T}^2$ depending continuously on T, $||f_m||_{3k+1}$ and $|g_m|_{k+1,T}$ such that for $0 \leq t \leq T$,

$$||w_x^{(k)}(\cdot,t)||^2 + \int_0^t (w_{xx}^{(k)}(0,s))^2 ds$$

$$\leq C_{k,T}^2\{||f||_{3k+1}^2 + |g|_{k+1,T}^2 + ||w||^2_{L_\infty(\mathbb{R}^+ \times (0,T))}||u_2^{(k)}||^2_{L_2(0,T;H^2(\mathbb{R}^+))}\}. \tag{7.23}$$

Furthermore if $(f_m, g_m) \in X_{3k+4}$, *then*

$$||w^{(k)}(\cdot, t)||_3^2 + \int_0^t (w_{xxxx}^{(k)}(0, s))^2 ds$$

$$\leq C_{k+1,T}^1 \{||f||_{3(k+1)}^2 + |g|_{k+2,T}^2 + ||w||_{L_\infty(\mathbb{R}^+ \times (0,T))}^2 ||u_2^{(k)}||_{L_2(0,T;H^4(\mathbb{R}^+))}^2 \}. \quad (7.24)$$

Proof: The lemma is proved by induction on k. The estimates (7.22), (7.23) and (7.24) are true when $k = 0$ by virtue of Lemma 7.1, and the constants $C_{0,T}^1$, $C_{0,T}^2$ and $C_{1,T}^1$ depend continuously on T, $||f_m||_3$ and $|g_m|_{2,T}$. Assume that the estimates (7.22), (7.23) and (7.24) regarding $w^{(j)}$ hold for $0 \leq j < k$. To obtain (7.22) for $j = k$, let $(f_m, g_m) \in X_{3k+1}$ for $m = 1, 2$. Then Theorem 6.2 implies that $u_m \in L_\infty(0, T; H^{3k+1}(\mathbb{R}^+))$. Hence the following calculations make sense.

First, make a transformation such that the solution vanishes on the boundary. This can be achieved by setting

$$y = w^{(k)} - g^{(k)}(t)e^{-x} \equiv w^{(k)} - U^{(k)}.$$

Then y satisfies the initial- and boundary-value problem

$$y_t + a(y_1 + U_1)y_x + [a(y_1 + U_1) - a(y_2 + U_2)]^{(k)}(y_2 + U_2)_x$$

$$+ [a(y_1 + U_1) - a(y_2 + U_2)](y_2^{(k)} + U_2^{(k)})_x + y_{xxx} = h_k, \quad (7.25a)$$

$$\text{for } x, t \in \mathbb{R}^+ \times \mathbb{R}^+,$$

$$y(x, 0) = \phi^{(k)}(x) - U^{(k)}, \qquad \text{for } x \geq 0, \quad (7.25b)$$

$$y(0, t) = 0, \qquad \text{for } t \geq 0, \quad (7.25c)$$

where $y_m = u_m^{(k)} - g_m^{(k)}(t)e^{-x} \equiv u_m^{(k)} - U_m^{(k)}$ for $m = 1, 2$,

$$h_k = -F_k + [U_t^{(k)} + U_{xxx}^{(k)} - (a(y_1 + U_1))^{(k)}w_x + (a(y_1 + U_1))U_x^{(k)}],$$

and $\phi^{(k)}(x)$ takes the form presented in (5.13) with $\phi^{(0)}(x) = f(x)$. After multiplying (7.25a) by $2y$ and integrating over $\mathbb{R}^+ \times (0, t)$, there appears

$$||y(\cdot, t)||^2 + \int_0^t y_x^2(0, s)ds = ||y(\cdot, 0)||^2 + \int_0^t \int_0^\infty \Big[2h_k y + (a(y_1 + U_1))_x y^2$$

$$- 2y[a(y_1 + U_1) - a(y_2 + U_2)]^{(k)}(y_2 + U_2)_x$$

$$- 2y[a(y_1 + U_1) - a(y_2 + U_2)](y_2^{(k)} + U_2^{(k)})_x\Big]dxds. \quad (7.26)$$

Taking account of the definition of $U^{(k)}$ and $y(\cdot, 0)$, one sees immediately that

$$||y(\cdot, 0)|| \leq C_{k,T}^1 \{||f||_{3k} + |g|_{k+1,T}\}.$$

By the induction hypothesis, there obtains

$$||h_k||_{L_2(\mathbb{R}^+ \times (0,T))} \leq C_{k,T}^1 \{||f||_{3k} + |g|_{k+1,T}\}.$$

Hence the first term in the double integral on the right-hand side of (7.26) is bounded above as follows:

$$\int_0^t \int_0^\infty 2yh_k\,dx\,ds \leq C_{k,T}^1\{||f||_{3k}^2 + |g|_{k+1,T}^2 + \int_0^t ||y(\cdot,s)||^2 ds\}.$$

The assumption about the nonlinearity P implies

$$||[a(u_1) - a(u_2)]^{(j)}||_{L_2(0,T;L_2(\mathbb{R}^+))} \leq \gamma(\mathcal{B})C_{k,T}^1 \sum_{n=0}^j ||w^{(n)}||_{L_2(0,T;L_2(\mathbb{R}^+))}, \qquad (7.27)$$

where $0 \leq j \leq k$ and $\mathcal{B} = [0, C_T]$ as before. Note that

$$||(a(y_1 + U_1))_x||_{L_\infty(\mathbb{R}^+\times(0,T))} \leq C_{k,T}^1.$$

With this information in hand, it is readily deduced that the second and third terms in the double integral in (7.26) are bounded above; viz.

$$\int_0^t \int_0^\infty \left[(a(y_1 + U_1))_x y^2 - 2y[a(y_1 + U_1) - a(y_2 + U_2)]^{(k)}(y_2 + U_2)_x\right] dx\,ds$$

$$\leq C_{k,T}^1\{||f||_{3k}^2 + ||g||_{k+1,T}^2 + \int_0^t ||y(\cdot,s)||^2 ds\}.$$

Finally, note that

$$\left|2\int_0^t \int_0^\infty y[a(y_1 + U_1) - a(y_2 + U_2)](y_2^{(k)} + U_2^{(k)})_x dx\,ds\right|$$

$$\leq \int_0^t \int_0^\infty \left[y^2 + [a(y_1 + U_1) - a(y_2 + U_2)]^2[(y_2^{(k)} + U_2^{(k)})_x]^2\right] dx\,ds$$

$$\leq C_{k,T}^1\gamma(\mathcal{B})||w||_{L_\infty(\mathbb{R}^+\times(0,T))}^2||u_2^{(k)}||_{L_2(0,T;H^1(\mathbb{R}^+))}^2 + \int_0^t ||y(\cdot,s)||^2 ds.$$

Combining the above estimates, (7.26) is reduced to

$$||y(\cdot,t)||^2 + \int_0^t y_x^2(0,s)ds \leq C_{k,T}^1\{||f||_{3k}^2 + |g|_{k+1,T}^2$$

$$+ ||w||_{L_\infty(\mathbb{R}^+\times(0,T))}^2||u_2^{(k)}||_{L_2(0,T;H^1(\mathbb{R}^+))}^2 + \int_0^t ||y(\cdot,s)||^2 ds\}$$

for $0 \leq t \leq T$. If Gronwall's lemma is applied to this last inequality, one obtains the result (7.22).

To obtain (7.23), let $(f_m, g_m) \in X_{3k+3}$ for $m = 1, 2$. Theorem 6.2 implies that $u_m \in L_\infty(0, T; H^{3k+3}(\mathbb{R}^+))$. Multiply (7.6) by $-2w_{xx}^{(k)}$ and integrate the result over

$\mathbb{R}^+ \times (0, t)$. After integrations by parts, there appears

$$||w_x^{(k)}(\cdot, t)||^2 + \int_0^t (w_{xx}^{(k)}(0, s))^2 ds$$

$$= ||w_x^{(k)}(\cdot, 0)||^2 - \int_0^t 2g^{(k+1)}(s) w_x^{(k)}(0, s) ds$$

$$+ \int_0^t \int_0^\infty 2w_{xx}^{(k)} \Big[a(u_1) w_x^{(k)} + [a(u_1) - a(u_2)]^{(k)} (u_2)_x \tag{7.28}$$

$$+ [a(u_1) - a(u_2)](u_2^{(k)})_x + (a(u_1))^{(k)} w_x + F_k \Big] dx ds.$$

First note that

$$\int_0^\infty 2w_{xx}^{(k)} a(u_1) w_x^{(k)} dx = -a(u_1(0, s))[w_x^{(k)}]^2 (0, s) - \int_0^\infty a(u_1)_x [w_x^{(k)}]^2 dx.$$

Therefore by using (7.3) and (7.22) one obtains

$$\int_0^t \int_0^\infty 2a(u_1) w_{xx}^{(k)} w_x^{(k)} dx ds$$

$$\leq C_{k,T}^1 \{||f||_{3k}^2 + |g|_{k+1,T}^2 + \int_0^t ||w_x^{(k)}(\cdot, s)||^2 ds\}, \tag{7.29}$$

for $0 \leq t \leq T$. By another integration by parts, one also shows that

$$\int_0^\infty 2w_{xx}^{(k)} [(a(u_1))^{(k)} w_x + F_k] dx$$

$$= -2w_x^{(k)}(0, s)[a(u_1(0, s))^{(k)} w_x(0, s) + F_k(0, s)]$$

$$- \int_0^\infty 2w_x^{(k)} \Big[((a(u_1))^{(k)})_x w_x + (a(u_1))^{(k)} w_{xx} + (F_k)_x \Big] dx.$$

Note that (7.3), (7.22) and the induction hypothesis entail the inequality

$$||w^{(j)}||_{L_\infty(0,T;H^3(\mathbb{R}^+))} \leq C_{k,T}^1 \{||f||_{3k} + |g|_{k+1,T}\}$$

for $0 \leq j \leq k - 1$. This in turn implies that

$$||\partial_x(F_k)||_{L_2(\mathbb{R}^+ \times (0,T))} \leq C_{k,T}^1 \{||f||_{3k} + |g|_{k+1,T}\}.$$

By the induction hypothesis and the first result of the lemma, one obtains

$$|F_k(0, s)|_T \leq C_{k,T}^1 \{||f||_{3k} + |g|_{k+1,T}\}.$$

The preceding facts show that

$$\int_0^t \int_0^\infty 2w_{xx}^{(k)}[(a(u_1))^{(k)}w_x + F_k]dxds$$

$$\le \int_0^t ||w_x^{(k)}(\cdot,s)||^2 ds + C_{k,T}^2\{||f||_{3k+1}^2 + |g|_{k+1,T}^2\}.$$

(7.30)

The inequality (7.27) implies

$$\int_0^t \int_0^\infty 2w_{xx}^{(k)}[a(u_1) - a(u_2)]^{(k)}(u_2)_x dxds$$

$$= -\int_0^t \int_0^\infty 2w_x^{(k)}\Big[\partial_x\Big([a(u_1)-a(u_2)]^{(k)}\Big)(u_2)_x + [a(u_1)-a(u_2)]^{(k)}(u_2)_{xx}\Big]dxds$$

$$-\int_0^t \{2w_x^{(k)}[a(u_1) - a(u_2)]^{(k)}(u_2)_x\}(0,s)ds$$

(7.31)

$$\le C_{k,T}^1\{||f||_{3k}^2 + |g|_{k+1,T}^2\}\Big[\int_0^t ||w_x^{(k)}(\cdot,s)||^2 ds + \gamma\Big].$$

Finally, integration by parts gives

$$\int_0^t \int_0^\infty 2w_{xx}^{(k)}[a(u_1) - a(u_2)](u_2^{(k)})_x dxds$$

$$= -\int_0^t \{2w_x^{(k)}[a(u_1) - a(u_2)](u_2^{(k)})_x\}(0,s)ds$$

$$-\int_0^t \int_0^\infty 2w_x^{(k)}\Big[(a(u_1) - a(u_2))_x(u_2^{(k)})_x + [a(u_1) - a(u_2)](u_2^{(k)})_{xx}\Big]dxds.$$

Note that

$$\Big|\int_0^t \int_0^\infty \{2w_x^{(k)}(a(u_1) - a(u_2))_x(u_2^{(k)})_x\}(x,s)dxds\Big|$$

$$\le \int_0^t ||w_x^{(k)}(\cdot,s)||^2 ds + C_{k,T}^2\{||f||_{3k+1}^2 + |g|_{k+1,T}^2\},$$

(7.32)

and that

$$\Big|\int_0^t \int_0^\infty 2w_x^{(k)}[a(u_1) - a(u_2)](u_2^{(k)})_{xx}dxds\Big|$$

$$\le \int_0^t \int_0^\infty \Big[[w_x^{(k)}]^2 + [a(u_1) - a(u_2)]^2[(u_2^{(k)})_{xx}]^2\Big]dxds$$

(7.33)

$$\le \int_0^t ||w_x^{(k)}(\cdot,s)||^2 ds + \gamma||w||_{L_\infty(\mathbb{R}^+\times(0,T))}^2||u_2^{(k)}||_{L_2(0,T;H^2(\mathbb{R}^+))}^2.$$

The definition of $w^{(k)}$ together with equation (7.6) entail that

$$||w_x^{(k)}(\cdot, 0)|| \leq C_{k,T}^2 \{||f||_{3k+1} + |g|_{k+1,T}\}. \tag{7.34}$$

Using the preceding relations (7.29), (7.30), (7.31), (7.32), (7.33) and (7.34) in (7.28) and applying Gronwall's lemma, it is confirmed that (7.19) holds.

To finish the induction, let $(f_m, g_m) \in X_{3k+4}$. Differentiate (7.6) once and multiply the result by $2w_{xxx}^{(k)}$. Differentiate (7.6) twice and multiply the result by $2w_{xxxx}^{(k)}$. Add the results together and then integrate over $\mathbb{R}^+ \times [0,t]$. After integrations by parts, we have

$$||w_{xx}^{(k)}(\cdot, t)||^2 + ||w_{xxx}^{(k)}(\cdot, t)||^2 + \int_0^t \left[(w_{xxx}^{(k)}(0, s))^2 + (w_{xxxx}^{(k)}(0, s))^2 \right] ds$$

$$= ||w_{xx}^{(k)}(\cdot, 0)||^2 + ||w_{xxx}^{(k)}(\cdot, 0)||^2$$

$$- 2 \int_0^t \left[w_{xx}^{(k)}(0, s) w_{xs}^{(k)}(0, s) + w_{xxx}^{(k)}(0, s) w_{xxs}^{(k)}(0, s) \right.$$

$$+ w_{xxx}^{(k)}(0, s) \left(a(u_1) w_x^{(k)} + (a(u_1) - a(u_2))^{(k)} (u_2)_x \right.$$

$$\left. + \left(a(u_1) \right)^{(k)} w_x + (a(u_1) - a(u_2))(u_2^{(k)})_x + F_k \right)_{xx} (0, s) \Big] ds$$

$$+ 2 \int_0^t \int_0^{+\infty} \left[\left(a(u_1) w_x^{(k)} + (a(u_1) - a(u_2))^{(k)} (u_2)_x \right. \right. \tag{7.35}$$

$$\left. + \left(a(u_1) \right)^{(k)} w_x + (a(u_1) - a(u_2))(u_2^{(k)})_x + F_k \right)_x$$

$$- \left((a(u_1) - a(u_2))^{(k)} (u_2)_x + \left(a(u_1) \right)^{(k)} w_x + F_k \right)_{xxx}$$

$$- \sum_{j=1}^3 \binom{3}{j} \left(\partial_x^j (a(u_1)) \partial_x^{(3-j)} (w_x^{(k)}) + \partial_x^j (a(u_1) - a(u_2)) \partial_x^{(3-j)} ((u_2^{(k)})_x) \right)$$

$$\left. - a(u_1) w_{xxxx}^{(k)} - (a(u_1) - a(u_2))(u_2^{(k)})_{xxxx} \right] w_{xxx}^{(k)} dx ds.$$

Using the first two results of the lemma and equation (7.6) shows that the first boundary term in (7.35) is bounded above, viz.

$$\left| 2 \int_0^t w_{xx}^{(k)}(0, s) w_{xs}^{(k)}(0, s) ds \right| \leq \frac{1}{\delta} \int_0^t (w_{xx}^{(k)}(0, s))^2 ds + \delta \int_0^t (w_{xs}^{(k)}(0, s))^2 ds$$

$$\leq C_{k,T}^1 (||f||_{3k+1}, |g|_{k+1,T}) + \delta \int_0^t (w_{xxxx}^{(k)}(0, s))^2 ds \tag{7.36}$$

for any $\delta > 0$. Note that by Theorem 6.2, there is a constant C depending on $||f_m||_{3k+3}$ and $|g_m|_{k+2,T}$ such that for $m = 1, 2$,

$$\int_0^t \left((u_m^{(k+1)})_x(0,s)\right)^2 ds \leq C(||f_m||_{3k+3}, |g_m|_{k+2,T}).$$

These inequalities, when combined with equation (7.6) show that for any $\delta > 0$,

$$-\int_0^t w_{xxx}^{(k)}(0,s)w_{xxs}^{(k)}(0,s)ds$$

$$= \int_0^t \left[g^{(k+1)}(s) + \left(a(g_1(s))\right)^{(k)}w_x(0,s) + [a(g_1(s)) - a(g_2(s))]^{(k)}(u_2)_x(0,s)\right.$$

$$\left. +a(g_1(s))w_x^{(k)}(0,s)+[a(g_1(s))-a(g_2(s))](u_2^{(k)})_x(0,s)+F_k(0,s)\right]w_{xxs}(0,s)ds$$

$$= \left[g^{(k+1)}(s) + \left(a(g_1(s))^{(k)}\right)w_x(0,s) + [a(g_1(s)) - a(g_2(s))]^{(k)}(u_2)_x(0,s)\right.$$

$$\left. +a(g_1(s))w_x^{(k)}(0,s)+[a(g_1(s))-a(g_2(s))](u_2^{(k)})_x(0,s)+F_k(0,s)\right]w_{xx}(0,s)\Big|_0^t$$

$$\tag{7.37}$$

$$-\int_0^t \left[g^{(k+1)}(s)+(a(g_1(s)))^{(k)}w_x(0,s)+[a(g_1(s))-a(g_2(s))]^{(k)}(u_2)_x(0,s)\right.$$

$$\left. +a(g_1(s))u_x^{(k)}(0,s)+[a(g_1(s))-a(g_2(s))](u_2^{(k)})_x(0,s)+F_k(0,s)\right]_s w_{xx}(0,s)ds$$

$$\leq C_{k+1,T}^1(||f||_{3k+3}, |g|_{k+2,T}) + \delta\int_0^t \left(w_{xs}^{(k)}(0,s)\right)^2 ds$$

$$\leq C_{k+1,T}^1(||f||_{3k+3}, |g|_{k+2,T}) + \delta\int_0^t \left(w_{xxxx}^{(k)}(0,s)\right)^2 ds.$$

Again using (7.22) and (7.23), the last temporal integral is bounded above by a constant $C_{k+1,T}^1 = C_{k+1,T}^1(||f||_{3k+3}, |g|_{k+2,T})$. There is a constant depending only on $C_{k+1,T}^1$ such that all terms in the double integral in (7.35), except the last two, can be bounded above by

$$C_{k+1,T}^1(||f||_{3k+3}, |g|_{k+2,T}) \int_0^t ||w^{(k)}(\cdot,s)||_3^2 ds.$$

By integrations by parts and the hypothesis on P, one shows that

$$\left|\int_0^t \int_0^{+\infty} 2\left[a(u_1)w_{xxxx}^{(k)} + (a(u_1) - a(u_2))(u_2^{(k)})_{xxxx}\right]w_{xxx}^{(k)} dxds\right|$$

$$\leq \int_0^t a(g_1(s))(w_{xxx}^{(k)})^2(0,s)ds + C_T \int_0^t ||w_{xxx}^{(k)}(\cdot,s)||^2 ds$$

$$\tag{7.38}$$

$$+ C_T(||f||_1, |g|_{1,T})\gamma(\mathcal{B})||w||_{L_\infty(\mathbb{R}^+\times(0,T))}^2 ||u_2^{(k)}||_{L_2(0,T;H^4(\mathbb{R}^+))}^2.$$

Using this information in (7.35) and choosing δ small enough, one obtains

$$||w^{(k)}(\cdot,t)||_3^2 \le C_{k+1,T}^1 \{||f||_{3k+3} + |g|_{k+2,T} + \int_0^t ||w^{(k)}(\cdot,s)||_3^2 ds$$

$$+ ||w||_{L_\infty(\mathbb{R}^+\times(0,T))}^2 ||u_2^{(k)}||_{L_2(0,T;H^4(\mathbb{R}^+))}^2 \}.$$

Applying Gronwall's lemma to the above inequality shows that

$$||w^{(k)}(\cdot,t)||_3^2 \le C_{k+1,T}^1 \{||f||_{3k+3} + |g|_{k+2,T} + ||w||_{L_\infty(\mathbb{R}^+\times(0,T))}^2 ||u_2^{(k)}||_{L_2(0,T;H^4(\mathbb{R}^+))}^2 \}.$$

The proof is complete. $\square$

An inductive use of (7.5a) and (7.6), combined with the estimates derived in Lemma 7.2, gives immediately the following estimates for $||w(\cdot,t)||_{3k}$ and $||w(\cdot,t)||_{3k+1}$.

Lemma 7.3. *Let k be a positive integer. If $(f_m, g_m) \in X_{3k+1}$ for $m = 1, 2$, then for any $T > 0$, there are constants $C_{k,T}^1$ depending continuously on T, $||f_m||_{3k}$ and $|g_m|_{k+1,T}$, such that*

$$||w(\cdot,t)||_{3k}^2 + \int_0^t (\partial_x^{3k+1}w(0,s))^2 ds$$

$$\le C_{k,T}^1 \{||f||_{3k}^2 + |g|_{k+1,T}^2 + ||w||_{L_\infty(\mathbb{R}^+\times(0,T))}^2 ||u_2||_{L_2(0,T;H^{3k+1}(\mathbb{R}^+))}^2 \}. \quad (7.39)$$

If $(f_m, g_m) \in X_{3k+3}$, there are constants $C_{k,T}^2$ depending continuously on T, $||f_m||_{3k+1}$ and $|g_m|_{k+1,T}$, such that

$$||w(\cdot,t)||_{3k+1}^2 + \int_0^t (\partial_x^{3k+2}w(0,s))^2 ds$$

$$\le C_{k,T}^2 \{||f||_{3k+1}^2 + |g|_{k+1,T}^2 + ||w||_{L_\infty(\mathbb{R}^+\times(0,T))}^2 ||u_2||_{L_2(0,T;H^{3k+2}(\mathbb{R}^+))}^2 \}, \quad (7.40)$$

for $0 \le t \le T$.

Since Lemma 7.3 requires that the corresponding initial- and boundary-data lie in the space X_{3k+1}, or X_{3k+3}, we can not directly obtain that the map $(f,g) \to u$ is continuous from X_s into $C(\mathbb{R}^+; H^s(\mathbb{R}^+))$ where $s = 3k$, or $s = 3k+1$. But by choosing smooth and compatible approximations of the data, this goal can be achieved. Such approximations of the auxiliary data were constructed in [**18**, Proposition 4.1] and used in the study of the initial-boundary-value problem for the KdV-equation. For the reader's convenience, these results are restated in the next Proposition.

Proposition 7.4. *Let $(f,g) \in X_s$ where $s = 3k$, or $s = 3k + 1$. Then for any integer $n \ge 0$ and $\epsilon \in (0,1]$, there exist functions*

$$(f_\epsilon, g_\epsilon) \in X_{s(n)} \cap (H^\infty(\mathbb{R}^+) \times H_{loc}^\infty(\mathbb{R}^+)),$$

where $s(n) = 3(k + n)$, *or* $s(n) = 3(k + n) + 1$ *depending on whether* $s = 3k$ *or* $s = 3k + 1$, *such that for any* $T > 0$,

(i) $||f_\epsilon - f||_{s-3j}$, $|g_\epsilon - g|_{k+1-j,T} = o(\epsilon^j)$ *for* $k \geq j \geq 0$, *and*

(ii) $||f_\epsilon||_{s+3j}$, $|g_\epsilon|_{k+j+1,T} \leq c\epsilon^{-j}$ *for* $j \geq 0$

as $\epsilon \downarrow 0$, *where the constant* c *depends only on* $||f||_s$, $|g|_{k+1,T+1}$, j, n *and* T. *Furthermore, the convergence in* (i) *depends upon* j, n *and* T, *but is uniform on compact subsets of* $H^s(\mathbb{R}^+) \times H^{k+1}(0, T+1)$. *Finally, for any fixed* $\epsilon \in (0, 1]$, *the map* $(f, g) \to (f_\epsilon, g_\epsilon)$ *is continuous from* X_s *into* $X_{s(n)} \cap (H^\infty(\mathbb{R}^+) \times H^\infty_{loc}(\mathbb{R}^+))$.

Lemma 7.5. *Assume that* $(f, g) \in X_s$ *where* $s = 3k$ *or* $3k + 1$, k *a positive integer, and let* $(f_\epsilon, g_\epsilon) \in X_{s(1)}$ *for* $\epsilon \in (0, 1]$ *be approximations to* (f, g) *whose existence is guaranteed by Proposition 7.4 with* $n = 1$. *If* u_ϵ *denotes the solution of* (6.1) *with data* (f_ϵ, g_ϵ), *then for any* $T > 0$,

$$||u_\epsilon||_{L_\infty(0,T;H^{3k+1}(\mathbb{R}^+))} \leq C^1_{k,T}\epsilon^{-\frac{1}{3}} \quad and \quad ||u_\epsilon||_{L_\infty(0,T;H^{3k+2}(\mathbb{R}^+))} \leq C^2_{k,T}\epsilon^{-\frac{1}{2}},$$

where the constants $C^i_{k,T}$, $i = 1, 2$, *depend only on* T, $||f||_s$ *and* $|g|_{k+1,T+1}$.

Proof: Suppose $(f, g) \in X_{3k}$ so that $(f_\epsilon, g_\epsilon) \in X_{3k+3}$. Note that from (7.3), (7.4) and Proposition 7.4, it follows that

$$||u_\epsilon||_{L_\infty(0,T;H^{3k}(\mathbb{R}^+))} + \int_0^t (\partial_x^{(3k+1)}u_\epsilon(0,s))^2 ds \leq C^1_{k,T}. \tag{7.41}$$

The constants $C^i_{k,T}$ will have the same dependence on parameters and data as those specified in the statement of the lemma.

For the moment, denote u_ϵ by simply u. Then for $k \geq 1$ the function $u^{(k)}$ satisfies the equation

$$u_t^{(k)} + a(u)u_x^{(k)} + u_{xxx}^{(k)} = h_k(u), \tag{7.42}$$

where $h_1 = -(a(u))_x u_t$ and for $k > 1$,

$$h_k(u) = -(a(u))_x u^{(k)} - \partial_x \sum_{i=1}^{k-1} \binom{k-1}{i} (a(u))^{(i)} u^{(k-i)}.$$

Multiply (7.42) by $-2u_{xx}^{(k)}$ and integrate the results over $\mathbb{R}^+ \times (0, t)$. After suitable integrations by parts, there appears

$$||u_x^{(k)}(\cdot, t)||^2 + 2\int_0^t g_\epsilon^{(k+1)}(s)u_x^{(k)}(0, s)ds + \int_0^t (u_{xx}^{(k)}(0, s))^2 ds$$

$$= ||u_x^{(k)}(\cdot, 0)||^2 + 2\int_0^t \int_0^\infty u_{xx}^{(k)}[(a(u))u_x^{(k)} - h_k(u)]dxds. \tag{7.43}$$

Note that

$$\int_0^\infty 2u_{xx}^{(k)} a(u) u_x^{(k)} dx = -a(u(0,s))[u_x^{(k)}(0,s)]^2 - \int_0^\infty a(u)_x [u_x^{(k)}]^2 dx.$$

Therefore by (7.41), it follows that

$$\int_0^t \int_0^\infty 2a(u) u_{xx}^{(k)} u_x^{(k)} dx ds \le C_{k,T}^1 \{1 + \int_0^t ||u_x^{(k)}(\cdot, s)||^2 ds\} \tag{7.44}$$

for $0 \le t \le T$. Note also that (7.41) entails as before the inequality

$$||u^{(j)}||_{L_\infty(0,T;H^3(\mathbb{R}^+))} \le C_{k,T}^1$$

for $0 \le j \le k-1$. This in turn implies that

$$||\partial_x(h_k)||_{L_2(\mathbb{R}^+ \times (0,T))} \le C_{k,T}^1 \left(\int_0^t ||u_x^{(k)}(\cdot, s)||^2 ds \right)^{\frac{1}{2}}.$$

Then since

$$\int_0^\infty 2u_{xx}^{(k)} h_k(u) dx = -2h_k(u(0,s)) u_x^{(k)}(0,s) - \int_0^\infty 2u_x^{(k)} (h_k)_x dx,$$

$$|h_k(u(0,s))|_T \le C_{k,T}^1 \quad \text{and} \quad ||[h_k(u(\cdot,s))]_x|| \le C_{k,T}^1(1 + ||u_x^{(k)}(\cdot, s)||),$$

one obtains

$$\left| \int_0^t \int_0^\infty 2u_{xx}^{(k)} h_k dx ds \right| \le \int_0^t ||u_x^{(k)}(\cdot, s)||^2 ds + C_{k,T}^1. \tag{7.45}$$

Taking recourse again to (7.41) together with Proposition 7.4 gives

$$\left| \int_0^t g_\epsilon^{(k+1)} u_x^{(k)}(0,s) ds \right| \le C_{k,T}^1. \tag{7.46}$$

Finally, from the equation (7.42) and Proposition 7.4, one shows that

$$||u_x^{(k)}(\cdot, 0)|| \le C_{k,T}^1 ||u(\cdot, 0)||_{3k+1} \le C_{k,T}^1 ||f||_{3k+1} \le C_{k,T}^1 \epsilon^{-\frac{1}{3}}. \tag{7.47}$$

The above estimates and Gronwall's lemma imply the first-stated inequality in the lemma.

The second inequality will follow if

$$||u_\epsilon||_{L_\infty(0,T; H^{3k+3}(\mathbb{R}^+))} \le C_{k,T}^2 \epsilon^{-1}. \tag{7.48}$$

Let $(f,g) \in X_{3k+1}$, so that $(f_\epsilon, g_\epsilon) \in X_{3k+4}$. Then from (7.3), (7.4) and Proposition 7.4, one has

$$||u_\epsilon||_{L_\infty(0,T; H^{3k+1}(\mathbb{R}^+))} + \int_0^t (\partial_x^{(3k+2)} u_\epsilon(0,s))^2 ds \le C_{k,T}^2. \tag{7.49}$$

Replace k by $k+1$ in (7.42) and let

$$y = u_\epsilon^{(k+1)} - g_\epsilon^{(k+1)}(t)e^{-x} \equiv u^{(k+1)} - V^{(k+1)}.$$

Then y satisfies the equation

$$y_t + a(u)y_x + y_{xxx} = h \tag{7.50}$$

with initial- and boundary-values given by

$$y(0,t) = 0, \quad y(x,0) = u^{(k+1)}(x,0) - V^{(k+1)}(x,0),$$

where

$$h = -\left(V_t^{(k+1)} + a(u)V_x^{(k+1)} + V_{xxx}^{(k+1)} + h_{k+1}(u)\right).$$

Multiply (7.50) by $2y$ and integrate the results over $\mathbb{R}^+ \times (0,t)$. Suitable integrations by parts lead to

$$||y(\cdot,t)||^2 + \int_0^t y_x^2(0,s)ds + 2\int_0^t \int_0^\infty \{a(u)y_x y\}(x,s)dxds$$

$$= ||y(\cdot,0)||^2 + 2\int_0^t \int_0^\infty hy\,dxds. \tag{7.51}$$

By using (7.49), one has

$$2\left|\int_0^t \int_0^\infty a(u)y_x y\,dxds\right| = \left|-\int_0^t \int_0^\infty a'(u)u_x y^2\,dxds\right|$$

$$\leq C_{k,T}^2 \int_0^t ||y(\cdot,s)||^2 ds$$

for $0 \leq t \leq T$. The properties of the net (f_ϵ, g_ϵ) mentioned in Proposition 7.4 imply

$$||y(\cdot,0)|| \leq C_{k,T}^2 \epsilon^{-\frac{2}{3}}$$

and

$$||h||_{L_2(\mathbb{R}^+ \times (0,T))} \leq C_{k,T}^2 \epsilon^{-1}.$$

Hence (7.51) and Gronwall's lemma yield

$$||y||_{L_\infty(0,T;L_2(\mathbb{R}^+))} \leq C_{k,T}^2 \epsilon^{-1}.$$

By the definition of y and referring again to Proposition 7.4, it is adduced that

$$||u_\epsilon||_{L_\infty(0,T;H^{3k+2}(\mathbb{R}^+))} \leq C_{k,T}^2 \epsilon^{-\frac{1}{2}}.$$

This completes the proof.   □

The information derived from Lemmas 7.1 to 7.5 allows us to prove the following proposition.

Proposition 7.6. *Let $(f, g) \in X_s$ where $s = 3k$, or $s = 3k + 1$ for some integer $k \geq 1$. Then there exists a unique solution u of (6.1) in $C(\mathbb{R}^+; H^s(\mathbb{R}^+))$ corresponding to the given data f and g.*

Proof: Denote by s_i the quantity $3k + i - 1$ for $i = 1, 2$. Fix a positive value of T, let a net $\{(f_\epsilon, g_\epsilon)\}_{\epsilon \in (0,1]} \subset X_{s_i+3}$ of approximations (f_ϵ, g_ϵ) to the data (f, g) be constructed for which the properties delineated in Proposition 7.4 hold and let $\{u_\epsilon\}_{\epsilon \in (0,1]}$ denote the corresponding family of solutions of (6.1). From Theorem 6.2, we have

$$u_\epsilon \in L_\infty(0, T; H^{s_i+3}(\mathbb{R}^+)) \quad \text{and} \quad \partial_t u_\epsilon \in L_\infty(0, T; H^{s_i}(\mathbb{R}^+)).$$

Hence for all $\epsilon \in (0, 1]$, u_ϵ certainly lies in $C(0, T; H^{s_i}(\mathbb{R}^+))$. It will now be shown that $\{u_\epsilon\}$ is Cauchy in $C(0, T; H^{s_i}(\mathbb{R}^+))$. Suppose that $0 < \delta < \epsilon \leq 1$. From Lemma 7.3 and Proposition 7.4, there follows the existence of constants $C^i_{k,T}$ depending continuously on $\|f\|_{s_i}$, $|g|_{k+1,T+1}$ and T such that for $0 \leq t \leq T$,

$$
\begin{aligned}
\|u_\epsilon(\cdot, t) - u_\delta(\cdot, t)\|_{s_i} &\leq C^i_{k,T} \{ \|f_\epsilon - f_\delta\|_{s_i} + |g_\epsilon - g_\delta|_{k+1,T} \\
&\quad + \|u_\epsilon - u_\delta\|_{L_\infty(\mathbb{R}^+ \times (0,T))} \|u_\epsilon\|_{L_2(0,T;H^{3k+i}(\mathbb{R}^+))} \}.
\end{aligned}
\tag{7.52}
$$

From Lemma 7.1 and Proposition 7.4 follows the inequality

$$\|u_\epsilon - u_\delta\|_{L_\infty(\mathbb{R}^+ \times (0,T))} \leq \|u_\epsilon - u_\delta\|_{L_\infty(0,T;H^1(\mathbb{R}^+))} \leq C^i_{k,T} \epsilon^k,$$

and from Lemma 7.5 it is seen that

$$\|u_\epsilon\|_{L_\infty(0,T;H^{3k+1}(\mathbb{R}^+))} \leq C^1_{k,T} \epsilon^{-\frac{1}{3}} \quad \text{and} \quad \|u_\epsilon\|_{L_\infty(0,T;H^{3k+2}(\mathbb{R}^+))} \leq C^2_{k,T} \epsilon^{-\frac{1}{2}}.$$

Moreover, the construction of the regularized data (f_ϵ, g_ϵ) (see again Proposition 7.4) entails

$$\|f_\epsilon - f_\delta\|_{s_i} \to 0 \quad \text{and} \quad |g_\epsilon - g_\delta|_{k+1,T} \to 0 \quad \text{as } \epsilon \to 0.$$

The last three inequalities imply that $\{u_\epsilon\}_{\epsilon \in (0,1]}$ is Cauchy in $C(0, T; H^{s_i}(\mathbb{R}^+))$. Hence, as $\epsilon \to 0$, $\{u_\epsilon\}_{\epsilon \in (0,1]}$ converges to a function $\bar{u} \in C(0, T; H^{s_i}(\mathbb{R}^+))$. By continuity, it certainly follows that $\bar{u}$ satisfies the differential equation (6.1) in the sense of distributions on $\mathbb{R}^+ \times (0, T)$. Furthermore,

$$\|\bar{u}(\cdot, 0) - f\|_{s_i} \leq \|\bar{u}(\cdot, 0) - u_\epsilon(\cdot, 0)\|_{s_i} + \|f_\epsilon - f\|_{s_i} \to 0$$

as $\epsilon \downarrow 0$, and

$$|\bar{u}(0, \cdot) - g|_{k+1,T} \leq |\bar{u}(0, \cdot) - u_\epsilon(0, \cdot)|_{k+1,T} + |g_\epsilon - g|_{k+1,T} \to 0$$

as $\epsilon \downarrow 0$. Hence $\bar{u}$ is a solution of (6.1) with initial and boundary data f and g, respectively. From the uniqueness result of Theorem 6.2 it is therefore implied that $u \equiv \bar{u} \in C(0, T; H^{s_i}(\mathbb{R}^+))$. $\square$

Theorem 7.7. *Let $(f, g) \in X_s$, where $s = 3k$, or $s = 3k + 1$ for some integer $k \geq 1$. Then the map $(f, g) \to u$ is continuous from X_s into $C(\mathbb{R}^+; H^s(\mathbb{R}^+))$.*

Proof: Let $\{(f_n, g_n)\}_{n=1}^{\infty}$ be a sequence in X_s that converges to $(f, g) \in X_s$. Thus for any $T > 0$,

$$\|f_n - f\|_s + |g_n - g|_{k+1, T} \to 0 \quad \text{as} \quad n \to \infty.$$

Let u_n and u be the solutions of (6.1) corresponding to the data (f_n, g_n) and (f, g), respectively, $n = 1, 2, \cdots$, and let $T > 0$ be fixed but arbitrary. By Proposition 7.6 it is known that u_n and u lie in $C(0, T; H^s(\mathbb{R}^+))$ for all $n \geq 1$. For any $n \geq 1$ and $\epsilon \in (0, 1]$ define approximations $(f_{n,\epsilon}, g_{n,\epsilon}) \in X_{s+3}$ of (f_n, g_n) for which the properties in Proposition 7.4 hold. Let also (f_ϵ, g_ϵ) be similar approximations of (f, g) and let $u_{n,\epsilon}$ and u_ϵ be the solutions of (6.1) corresponding to the data $(f_{n,\epsilon}, g_{n,\epsilon})$ and (f_ϵ, g_ϵ), respectively, $n = 1, 2, \cdots$. From the proof of Proposition 7.6, one has

$$\|u_\epsilon - u\|_{C(0,T; H^s(\mathbb{R}^+))} \to 0 \quad \text{as} \quad \epsilon \to 0. \tag{7.53}$$

Next consider the difference $u_{n,\epsilon} - u_n$. Again from Proposition 7.6, one also has

$$\|u_{n,\epsilon} - u_n\|_{C(0,T; H^s(\mathbb{R}^+))} \to 0, \quad \text{as} \quad \epsilon \to 0, \tag{7.54}$$

for each fixed $n \geq 1$. Furthermore, by Proposition 7.4, we know that

$$\|f_{n,\epsilon} - f_\epsilon\|_s \to 0 \quad \text{and} \quad |g_{n,\epsilon} - g_\epsilon|_{k+1, T} \to 0 \tag{7.55}$$

as $\epsilon \to 0$, uniformly in n. From Lemma 7.1, Lemma 7.3, Lemma 7.5 and Proposition 7.4, one shows that when $s = 3k$,

$$\|u_{n,\epsilon} - u_n\|_{L_\infty(\mathbb{R}^+ \times (0,T))} \|u_{n,\epsilon}\|_{L_\infty(0,T; H^{3k+1}(\mathbb{R}^+))} \leq C_{k,T}^1 \epsilon^{k - \frac{1}{3}},$$

and when $s = 3k + 1$,

$$\|u_{n,\epsilon} - u_n\|_{L_\infty(\mathbb{R}^+ \times (0,T))} \|u_{n,\epsilon}\|_{L_\infty(0,T; H^{3k+2}(\mathbb{R}^+))} \leq C_{k,T}^2 \epsilon^{k - \frac{1}{2}},$$

where $C_{k,T}^i$ is independent of n for $i = 1, 2$. Because of these inequalities and the relations (7.55) and (7.52), it is seen that the convergence in (7.54) is uniform in n. Let $\gamma > 0$ be arbitrary. Because of the convergence in (7.53) and the uniform convergence in (7.54), there exists an $\epsilon_1 \in (0, 1]$ such that

$$\|u_{n,\epsilon} - u_n\|_{C(0,T; H^s(\mathbb{R}^+))} + \|u_\epsilon - u\|_{C(0,T; H^s(\mathbb{R}^+))} \leq \gamma, \tag{7.56}$$

for all $\epsilon \in (0, \epsilon_1]$ and all $n \geq 1$. Fix a value of ϵ in the interval $(0, \epsilon_1)$. From Lemma 7.3, one has, when $s = 3k$, that

$$\|u_{n,\epsilon} - u_\epsilon\|_{C(0,T; H^{3k}(\mathbb{R}^+))} \leq C_{k,T}^1 \{\|f_{n,\epsilon} - f_\epsilon\|_{3k} + |g_{n,\epsilon} - g_\epsilon|_{k+1,T}$$
$$+ \|u_{n,\epsilon} - u_\epsilon\|_{L_\infty(\mathbb{R}^+ \times (0,T))} \|u_\epsilon\|_{L_\infty(0,T; H^{3k+1}(\mathbb{R}^+))}\}$$

(see (7.39)), and when $s = 3k + 1$,

$$\|u_{n,\epsilon} - u_\epsilon\|_{C(0,T;\,H^{3k+1}(\mathbb{R}^+))} \leq C^2_{k,T}\{\|f_{n,\epsilon} - f_\epsilon\|_{3k+1} + |g_{n,\epsilon} - g_\epsilon|_{k+1,T}$$
$$+ \|u_{n,\epsilon} - u_\epsilon\|_{L_\infty(\mathbb{R}^+ \times (0,T))}\|u_\epsilon\|_{L_\infty(0,T;\,H^{3k+2}(\mathbb{R}^+))}\}$$

(see (7.40)), where the constants $C^i_{k,T}$ are independent of n for $i = 1, 2$. The conti-
nuity of the map $(f, g) \to (f_\epsilon, g_\epsilon)$ in $H^s(\mathbb{R}^+) \times H^{k+1}_{loc}(\mathbb{R}^+)$, implies that

$$\|f_{n,\epsilon} - f_\epsilon\|_s + |g_{n,\epsilon} - g_\epsilon|_{k+1,T} \to 0,$$

as $n \to \infty$. Also by Lemma 7.1 and Proposition 7.4, one has

$$\|u_{n,\epsilon} - u_\epsilon\|_{L_\infty(\mathbb{R}^+ \times (0,T))} \to 0,$$

as $n \to \infty$. It follows that, for fixed ϵ,

$$\lim_{n \to \infty} \|u_{n,\epsilon} - u_\epsilon\|_{C(0,T;H^s(\mathbb{R}^+))} = 0 \tag{7.57}$$

if $(f, g) \in X_s$, for $s = 3k$ or $s = 3k + 1$. Thus if we write

$$u_n - u = u_n - u_{n,\epsilon} + u_{n,\epsilon} - u_\epsilon + u_\epsilon - u,$$

then since $\epsilon \in (0, \epsilon_1)$, (7.56) and (7.57) imply that

$$\limsup_{n \to \infty} \|u_n - u\|_{C(0,T;H^s(\mathbb{R}^+))} \leq \gamma$$

if $(f, g) \in X_s$. Since $\gamma > 0$ and $T > 0$ were arbitrary, the result follows and the
proof is complete. $\square$

Remark. A careful perusal of the preceding arguments indicates that the solution map $(f, g) \to u$ for the initial-boundary-value problem (6.1) is in fact Lipschitz
continuous.

8. Conclusion

The well-posedness of the initial- and boundary-value problem (1.1) for the
generalized KdV equation has been studied here. Well-posedness locally in time
requires only suitable smoothness of the nonlinearity P, while our theory of global
well-posedness uses more restrictive assumptions. Precisely stated, if the initial
value $f \in H^{3k}(\mathbb{R}^+)$ or $f \in H^{3k+1}(\mathbb{R}^+)$ and the boundary value $g \in H^{k+1}_{loc}(\mathbb{R}^+)$
satisfy the appropriate compatibility conditions at $(x, t) = (0, 0)$ (see Lemma 4.7
and Formula (5.14)) and the growth of the nonlinearity P satisfies the one-sided
condition (**) put forward in the beginning of §4, then there corresponds a unique
global solution u of the initial-boundary-value problem (1.1) which, for each $T > 0$,
lies in $C(0, T; H^{3k}(\mathbb{R}^+))$ or $C(0, T; H^{3k+1}(\mathbb{R}^+))$, respectively. Moreover, u depends
continuously in the relevant function classes on the pair (f, g). It is worth particular
note that the $H^1(\mathbb{R}^+)$-bound obtained in our theory grows roughly linearly with

the energy $|g|_{1,T}$ supplied by the wavemaker (see Lemma 4.1). This is a satisfactory aspect of the theory as it corresponds well with what is observed in experiments (see [11]). It is also worth note that if $|g|_{1,T}$ is sufficiently small, then the well-posedness theory can proceed under weaker growth conditions on the nonlinearity, namely that $\limsup_{s \to +\infty} \Lambda(s)/s^5 \leq 0$. Thus for small boundary forcing, the theory comes in line to some extent with that available for the pure initial-value problem (see Kato [29], Schechter [40] and more recent work of Kenig et al. [30, 31]).

Despite the complexity of the developments presented here, there are many obvious issues left open. Perhaps the foremost is that to which allusion was just made, namely whether or not problem (1.1) is globally well posed for nonlinearities P whose growth at infinity is less than quintic. There is also the slightly unsatisfactory aspect that certain regularity classes for initial data are missing in the results (e.g. $H^{3k+2}(\mathbb{R}^+)$, $k = 0, 1, \cdots$). This is a technical point with little impact on the assessment of (1.1) as a model of real phenomena. However, it presents an interesting analytical challenge. Indeed, using additional techniques, the authors have been able to fill in the gap just mentioned, but we eschew detailed discussion of this point here.

Other mathematical aspects deserve further attention. The question of smoothing and an associated well-posedness theory set in weak function classes is an interesting and topical issue. As mentioned briefly above, the smoothing established by Kato [29] (see also Faminskii [21]) holds in the situation envisaged here. In particular, if the initial data lies in a suitable weighted Sobolev class, then it gains smoothness for positive time in relation to the rate of its decay at infinity. However, the more subtle results of Ginibre & Velo [25], Kenig et al. [30, 31] and Bourgain [19] have not been considered in the context of initial-boundary-values problems other than with periodic boundary conditions.

The issue mentioned parenthetically at the end of Section 7 of smoothness of the mapping that associates the solution to given initial- and boundary-data also deserves further study. For the pure initial-value problem, this map is known to be analytic and we expect the same is true for the present initial-boundary-value problem provided the nonlinearity P is entire, say, or analytic in an appropriate neighborhood of the origin in any case.

Finally, since the initial-boundary-value problem is well posed, say for the KdV equation, it would be worthwhile to develop a numerical scheme for this problem along the lines of that put forward for the quarter-plane problem for the regularized long-wave equation in [11] and test the model (1.1) quantitatively against experimental data. As in [11], damping will need to be incorporated into the model. This in itself presents an interesting challenge, both as regards modelling and from the view of analysis since dissipation may well be a non-local effect at the level of approximation corresponding to that already in effect for nonlinearity and dispersion.

Acknowledgements. The authors thank Professors Thanasis Fokas, Diane Henderson, Julian Maynard, Eugene Wayne, Ragnar Winther and Jinchao Xu for their interest and advice. LL thanks Professor Rumei Luo for his encouragement.

References

1. J.P. Albert and J.L. Bona, Comparisons between model equations for long waves, *J. Nonlinear Sci.,* **1** (1991), 345-374.
2. T.B. Benjamin, Lectures on nonlinear wave motion, in *Lectures in Applied Mathematics,* **15** (A. Newell, ed.) pp. 3-47, American Mathematical Society; Providence, RI (1974).
3. T.B. Benjamin, J.L. Bona and J.J. Mahony, Model equations for long waves in nonlinear dispersive systems, *Philos. Trans. Royal. Soc. London Ser. A,* **272** (1972), 47-78.
4. J.L. Bona and P.J. Bryant, A mathematical model for long waves generated by wavemakers in nonlinear dispersive systems, *Proc. Cambridge Philos. Soc.,* **73** (1973), 391-405.
5. J.L. Bona and V.A. Dougalis, An initial- and boundary-value problem for a model equation for propagation of long waves, *J. Math. Anal. Appl.,* **75** (1980), 503-522.
6. J.L. Bona and V.A. Dougalis and O.A. Karakashian, Fully discrete numerical schemes for the Korteweg-de Vries equation, *Computat. Math. Appl.,* **12A** (1986), 859-884.
7. J.L. Bona and V.A. Dougalis, O.A. Karakashian and W. McKinney, Computations of blow up and decay for periodic solutions of the generalized Korteweg-de Vries equation, *Applied Numerical Math.,* **10** (1992), 335-355.
8. J.L. Bona, V.A. Dougalis, O.A. Karakashian and W. McKinney, Conservative, high-order numerical schemes for the generalized Korteweg-de Vries equation, *Philos. Trans. Royal Soc. London Ser. A,* **351** (1995), 107-164.
9. J.L. Bona and L. Luo, Initial-boundary-value problems for model equations for the propagation of long waves, in *Evolution Equations,* Lecture Notes in Pure and Applied Mathematics, Vol. **168** (G. Ferreyra, G. Goldstein & F. Neubrander, ed.) pp. 65-93, American Mathematical Society; Providence, RI (1994).
10. ———, Nonhomogeneous initial- and boundary-value problems for the KdV equation. (preprint).
11. J.L. Bona, W.G. Pritchard and L.R. Scott, An evaluation of a model equation for water waves, *Philos. Trans. Royal Soc. London Ser. A,* **302** (1981), 457-510.
12. ———, A comparison of solutions of model equations for long waves, in *Lectures in Applied Mathematics,* Vol. **20** (N. Lebovitz, ed.) pp. 235-267, American Mathematical Society; Providence, RI (1983).
13. J.L. Bona and J.-C. Saut, Singularités dispersives de solutions d'équations de type Korteweg-de Vries, *C. R. Acad. Sci. Paris,* **303** (1986), 101-103.
14. ———, Dispersive blow-up of solutions of generalized Korteweg-de Vries equations, *J. Differential Equations,* **103** (1993), 3-57.
15. J.L. Bona and R. Scott, Solutions for the Korteweg-de Vries equation in fractional order Sobolev spaces, *Duke Math. J.,* **43** (1976), 87-99.
16. J.L. Bona and R. Smith, The initial-value problem for the Korteweg-de Vries equation, *Philos. Trans. Royal. Soc. London Ser. A,* **278** (1975), 555-601.
17. J.L. Bona and R. Winther, The Korteweg-de Vries equation, posed in a quarter plane, *SIAM J. Math. Anal.,* **14** (1983), 1056-1106.

18. ———, Korteweg-de Vries equation in a quarter plane, continuous dependence results, *Differential and Integral Equations,* **2** (1989), 228-250.

19. J. Bourgain, Fourier transform restriction phenomena for certain lattice subsets and applications to nonlinear evolution equation, *II* the Korteweg-de Vries equation, *Geom. Funct. Anal.,* **3** (1993), 209-262.

20. A.V. Faminskii, The Cauchy problem and the mixed problem in the half strip for equations of Korteweg-de Vries type, (Russian) *Dinamika Sploshn. Sredy.,* No. 63, **162** (1983), 152-158.

21. ———, The Cauchy problem for the Korteweg-de Vries equation and its generalizations, (Russian) *Trudy Sem. Petrovsk.,* **256-257** (1988), 56-105.

22. ———, A mixed problem in a semistrip for the Korteweg-de Vries equation and its generalizations, (Russian) *Trudy Moskov. Mat. Obshch.,* No. 51, **258** (1988), 54-94.

23. B. Fornberg and G. Whitham, A numerical and theoretical study of certain nonlinear wave phenomena, *Philos. Trans. Royal. Soc. London Ser. A,* **289** (1978), 373-404.

24. J. Ginibre and Y. Tsutsumi, Uniqueness of solutions for the generalized Korteweg-de Vries equation, *SIAM J. Math. Anal.,* **6** (1989), 1388-1425.

25. J. Ginibre and G. Velo, Smoothing properties and retarded estimates for some dispersive evolution equations, *Comm. Math. Phys.,* **144** (1992), 163-188.

26. J.L. Hammack, A note on tsunamis: their generation and propagation in an ocean of uniform depth, *J. Fluid Mech.,* **60** (1973), 769-799.

27. J.L. Hammack and H. Segur, The Korteweg-de Vries equation and water waves. 2. Comparison with experiments, *J. Fluid Mech.,* **65** (1974), 237-246.

28. T. Kato, On the Korteweg-de Vries equation *Manuscripta Math.,* **29** (1979), 89-99.

29. ———, On the Cauchy problem for the (generalized) Korteweg-de Vries equation, *Studies in Applied Math., Advances in Math. Suppl. Studies,* **8** (1983), 93-128.

30. C.E. Kenig, G. Ponce and L. Vega, Well-posedness and scattering results for the generalized Korteweg-de Vries equation via the contraction principle, *Comm. Pure Appl. Math.,* **46** (1993), 527-620.

31. C.E. Kenig, G. Ponce and L. Vega, The Cauchy problem for the Korteweg-de Vries equation in Sobolev spaces of negative indices, *Duke. Math. J.,* **71** (1993), 1-21.

32. D.J. Korteweg and G. de Vries, On the change of form of long wave advancing in a rectangular canal, and on a new type of long stationary waves, *Philos. Mag.,* **39** (1895), 422-443.

33. S.N. Kruzhkov and A.V. Faminskii, Generalized solutions of the Cauchy problem for the Korteweg-de Vries equation, *Math. USSR Sb.,* **48** (1984), 391-421.

34. J.L. Lions, Quelques méthodes de resolution des problèmes aux limites non linéaires, Dunod, Paris (1969).

35. J.L. Lions and E. Magenes, Non-Homogeneous Boundary Value Problems and Applications, Springer-Verlag, New York (1972).

36. R.M. Miura, The Korteweg-de Vries equation: A survey of results, *SIAM Review,* **18** (1976), 412-459.

37. T. Maxworthy, A note on the internal solitary waves produced by tidal flow over a three-dimensional ridge, *Journal of Geophysical Research,* **84** (1979), 338-346.

38. L.G. Redekopp, Nonlinear waves in geophysics: Long internal waves, in *Lectures in Applied Mathematics,* Vol. **20** (N. Lebovitz, ed) pp. 59-78, American Mathematical Society; Providence, RI (1983).

39. J.-C. Saut, Sur quelque généralisations de l'equation de Korteweg-de Vries, *J. Math. Pure Appl.,* **58** (1979), 21-61.

40. E. Schechter, *Well-behaved evolutions and the Trotter product formulas,* Ph.D. Thesis, University of Chicago (1978).

41. J. Tasi, A 2nd-order Korteweg-de Vries equation for a lattice, *J. Appl. Phys.,* **51** (1969), 5816-5827.

42. F. Treves, *Topological Vector Spaces, Distributions and Kernels,* Academic Press: New York (1967).

43. M. Tsutsumi, On global solutions of the generalized Korteweg-de Vries equation, *Publ. Res. Inst. Math. Sci.*, **7** (1972), 329-344.

44. N.J. Zabusky and C.J. Galvin, Shallow-water waves, the Korteweg-de Vries equation and solitons, *J. Fluid Mech.*, **47** (1971), 811-824.

DEPARTMENT OF MATHEMATICS AND TEXAS INSTITUTE FOR COMPUTATIONAL AND APPLIED MATHEMATICS, THE UNIVERSITY OF TEXAS, AUSTIN, TX 78712 USA AND

CENTRE DE MATHÉMATIQUES ET DE LEURS APPLICATIONS, ECOLE NORMAL SUPERIEURE DE CACHAN, 94235 CACHAN CEDEX, FRANCE

E-mail address: `bona@math.utexas.edu`

DEPARTMENT OF MATHEMATICS, IMPERIAL COLLEGE, 180 QUEEN'S GATE, LONDON, SW7 2BZ UK

Current address: Department of Mathematics, Arizona State University, Tempe, AZ 85287 USA

E-mail address: `l.luo@ic.ac.uk`

Contemporary Mathematics
Volume **221**, 1999

AN INVERSE FUNCTION THEOREM

Alfonso Castro and J.W. Neuberger

Abstract

In this note we present a local surjectivity result which is applicable to differential equations for which full boundary conditions may not be known. Our method uses continuous steepest descent and Sobolev gradients.

The main purpose of this note is the proof of the following.

Main Theorem. *Suppose that each of H and K is a Hilbert space, $r > 0$, $Q > 0$, F is a C^1 function from H to K which has a locally Lipschitzian derivative and $F(0) = 0$. Finally suppose that there is $c > 0$ such that if $u \in H$, $\|u\| \leq r$, and $g \in K, \|g\|_K = 1$, then*

$$\langle F'(u)v, g \rangle_K \geq c, \text{ for some } v \in H \text{ with } \|v\|_H \leq Q. \tag{1}$$

If $y \in K$ and $\|y\|_K < \frac{rc}{Q}$), then

$$u = \lim_{t \to \infty} z(t) \text{ satisfies } F(u) = y \text{ and } \|u\|_H \leq r$$

where z is the unique function from $[0, \infty)$ to H so that

$$z(0) = 0, \ z'(t) = -F'(z(t))^*(F(z(t)) - y), \ t \geq 0. \tag{2}$$

From condition (1) we have that if $u \in H, \|u\| \leq r, g \in K$, then

$$\|F'(u)^* g\|_H \geq \frac{c}{Q} \|g\|_K \tag{3}$$

where $F'(u)^*$ denotes the member of $L(K, H)$ so that

$$\langle F'(u)v, g \rangle = \langle v, F'(u)^* g \rangle, v \in H, g \in K \tag{4}$$

---

[1]The first author was supported by NSF Grant DMS-9215027
[2]Keywords: Implicit Function, Sobolev Gradient, Steepest Descent
[3]Subject classification: Primary 34B15, Secondary 35J65

Although some of the ingredients needed to prove the Theorem can be found in [3] and [4] we have elected to give a more nearly self-contained argument. References [3], [4], [5] mainly concern global results and details on numerical tracking of solutions z to equations such as (2).

Proof. Let $y \in K$. If $\phi : H \to R$ is defined by

$$\phi(x) = \|F(x) - y\|_K^2/2, \quad x \in H,$$

then

$$\phi'(x) = \langle F'(x)h, F(x) - y\rangle_K = \langle h, F'(x)^*(F(x) - y)\rangle_H, \quad x, h \in H.$$

We denote $F'(x)^*(F(x) - y)$ by $(\nabla \phi)(x)$ and call this element the gradient of ϕ at the point x, $x \in H$.

Assertion 1. For each $y \in K$ the equation (2) has a unique solution defined on $[0, \infty)$. In fact, since F' is locally Lipschitzian so is $\nabla \phi$. From basic theory of ordinary differential equations there is $d_0 > 0$ so that the equation in (2) has a solution on $[0, d_0)$. Suppose now that the set of all such numbers d_0 is bounded and denote by d its least upper bound. Denote by z the unique solution to the equation to (2) on $[0, d)$. We will show that $\lim_{t \to d-} z(t)$ exists. Note that if $0 \leq a < b < d$, then

$$\|z(b) - z(a)\|_H^2 = \|\int_a^b z'\|_H^2 \leq (\int_a^b \|z'\|_H)^2 \leq (b - a) \int_a^b \|z'\|_H^2 \qquad (5)$$

Note also that if $0 \leq t < d$, then

$$\phi(z)'(t) = \phi'(z(t))z'(t) = < z'(t), (\nabla\phi)(z(t)) > = -\|(\nabla\phi)(z(t))\|_H^2 \qquad (6)$$

and so

$$\phi(z(b)) - \phi(z(a)) = \int_a^b \phi(z)' = -\int_a^b \|z'\|_H^2.$$

Hence

$$\int_a^b \|z'\|_H^2 \leq \phi(z(a)), a \leq b < d. \qquad (7)$$

By the Cauchy-Schwartz inequality and (7) we see that

$$(\int_a^b \|z'\|_H)^2 \leq (b - a) \int_a^b \|z'\|_H^2.$$

Therefore

$$\int_a^b \|z'\|_H \leq (d\phi(z(a)))^{1/2}, a \leq b < d.$$

But this implies that $\int_a^{d-} \|z'\|_H$ exists and so $q \equiv \lim_{t \to d-} z(t)$ exists. But again from basic theory, there is $d_1 > d$ for which there is a function f defined on $[c, d_1)$ such that

$$f(d) = q, f'(t) = -(\nabla\phi)(f(t)), \ t \in [d, d_1).$$

But the function w on $[0, d_1)$ so that $w(t) = z(t)$, $t \in [0, d)$, $w(d) = q$, $w(t) = f(t)$, $t \in (d, d_1)$, satisfies

$$w(0) = 0, w'(t) = -(\nabla\phi)(w(t)), \ t \in [0, d_1),$$

contradicting the nature of d since $d < d_1$. Hence there is a solution to (2) defined on $[0, \infty)$. Uniqueness follows from the basic theory. This completes an argument for Assertion 1, a known result ([1] ,[5]).

Assertion 2. Let $y \in K$ and z satisfy (2). If for some $C > 0$

$$\|(\nabla\phi)(z(t))\|_H \geq C\|F(z(t)) - y\|_K, \ t \geq 0, \tag{8}$$

then

$$u = \lim_{t \to \infty} z(t) \text{ exists and } F(u) = y.$$

To prove this assertion, note first that if $(\nabla\phi)(0) = 0$, then the conclusion holds (with $u = 0$). Suppose then that $(\nabla\phi)(0) \neq 0$. This last supposition has the consequence that $(\nabla\phi)(z(t)) \neq 0, t \geq 0$. [If $(\nabla\phi)(z(t_0)) = 0$ for some $t_0 \geq 0$, then the function w on $[0, \infty)$ defined by $w(t) = z(t_0)$, $t \geq 0$, would satisfy

$$w' = (\nabla\phi)(w) \text{ and } w(t_0) = z(t_0) \tag{9}$$

and hence $(\nabla\phi)(w(t)) = 0, t \geq 0$. But if w is replaced in (9) by z, then (9) still holds. This violates the fact that (9) does not have two solutions.

As in (6), $\phi(z)' = -\|(\nabla\phi)(z)\|_H^2$ and so using (8)

$$\phi(z)'(t) \leq -C^2\|F(z(t)) - y\|_H^2 = -2C^2\phi(z(t)).$$

Therefore

$$\phi(z)'(t)/\phi(z(t)) \leq -2C^2, \ t \geq 0.$$

Accordingly,
$$\ln(\phi(z(t))/\phi(0)) \leq -2C^2 t$$
and
$$\phi(z(t)) \leq \phi(0) \exp(-2C^2 t), \ t \geq 0.$$
Therefore
$$\lim_{t \to \infty} \phi(z(t)) = 0.$$
Moreover if n is a positive integer,
$$\left(\int_n^{n+1} \|z'\|_H\right)^2 \leq \int_n^{n+1} \|z'\|_H^2 = \phi(z(n)) - \phi(z(n+1)) \leq \phi(0)\exp(-2C^2 n)$$
and so
$$\int_0^\infty \|z'\|_H \leq \phi(0)^{1/2} \sum_{n=0}^\infty \exp(-C^2 n) = \phi(0)^{1/2}/(1 - \exp(-C^2)).$$

Consequently, $\|z'\|_H \in L_1([0,\infty)]$ and so $u = \lim_{t\to\infty} z(t)$ exists. Moreover $\phi(u) = 0$ since $0 = \lim_{t\to\infty} \phi(z(t))$. This completes an argument Assertion 2. This is essentially Theorem 1 of [3] (see also Theorem 2 of [4]).

Assertion 3. Suppose $y \in K$ and $\|y\|_K < \frac{rc}{Q}$ then $\|z(t)\|_H < r$ for all $t \in [0,\infty)$. By the Cauchy-Schwartz inequality and (1) we have
$$c\|g\|_K \leq \langle F'(u)v, g\rangle_K = \langle v, F'(u)^* g\rangle_H \leq Q\|F'(u)^* g\|_H$$
if $\|u\|_H \leq r$ and $g \in K$. Consequently
$$\|(\nabla\phi)(u)\|_H = \|F'(u)^*(F(u) - y)\|_H \geq C\|F(u) - y\|_K, u \in H, \|u\|_H \leq r,$$
where $C = c/Q$. Hence
$$\|(\nabla\phi)(z(t))\|_H = \|F'(z(t))^*(F(z(t)) - y)\|_H \geq C\|F(z(t)) - y\|_K \qquad (10)$$
provided that
$$t \geq 0 \text{ and } \|z(t)\|_H \leq r.$$

Finally we show that if $\|y\| < Cr$ then $\|z(t)\| \leq r$ for all $t \in [0,\infty)$. To prove this assertion first note that if $(\nabla\phi)(0) = 0$, then $z(t) = 0$, $t \geq 0$, and so the conclusion holds with $u = 0$.

Suppose now that $(\nabla\phi)(0) \neq 0$. Thus $(\nabla\phi)(z(t)) \neq 0$ for all $t \geq 0$ as noted in the argument for the previous assertion.

Define α as the function which satisfies

$$\alpha(0) = 0, \quad \alpha'(t) = 1/\|(\nabla\phi)(z(\alpha(t)))\|_H, \quad t \in [0, d), \tag{11}$$

where d is as large as possible, possibly $d = \infty$.

Define v by $v(t) = z(\alpha(t))$, $t \in D(\alpha)$. Thus

$$\begin{aligned} v'(t) &= \alpha'(t)z'(\alpha(t)) = -\alpha'(t)(\nabla\phi)(z(\alpha(t))) \\ &= -(1/\|(\nabla\phi)(z(\alpha(t)))\|_H)(\nabla\phi)(z(\alpha(t))) \end{aligned} \tag{12}$$

and so

$$v(0) = 0, \ v'(t) = -(1/\|(\nabla\phi)(v(t))\|_H)(\nabla\phi)(v(t)), \ t \in D(v) = D(\alpha). \tag{13}$$

Note also that

$$\|v(t)\|_H = \|v(t) - v(0)\|_H = \|\int_0^t v'\|_H \leq \int_o^t \|v'\|_H = t, t \in D(v).$$

Hence

$$\phi(v)' = \phi'(v)v' = -\langle(\nabla\phi)(v), (\nabla\phi)(v)\rangle_H/\|(\nabla\phi)(v)\|_H = -\|(\nabla\phi)(v)\|_H,$$

and so if $t \in D(v)$ and $t \leq r$, then

$$\phi(v)'(t) = -\|(\nabla\phi)(v(t))\|_H \leq -C\|F(v(t)) - y\|_H = -C(2\phi(v(t)))^{1/2},$$

and thus

$$\phi(v)'(t)/\phi(v(t))^{1/2} \leq -C2^{1/2}.$$

This differential inequality is solved to yield

$$2\phi(v(t))^{1/2} - 2\phi(v(0))^{1/2} \leq -2^{1/2}Ct, \ t \in D(v), \ t \leq r.$$

But this is equivalent to

$$\|F(v(t)) - y\|_K \leq \|F(0) - y\|_K - Ct \leq C(r - t) \tag{14}$$

since $F(0) = 0$ and $\|y\|_K < \frac{rc}{Q}$. It follows from (14) that $d \leq r$ since if not, there would be $t \in [0, r]$ such that $F(v(t_0)) = y$ and hence $(\nabla\phi)(v(t_0)) = 0$

and consequently, $(\nabla\phi)(z(\alpha^{-1}(t_0))) = 0$, a contradiction. From this it follows that $\|v(t)\|_H \leq r$, $t \in D(v)$. Since $D(\alpha)$ is a bounded set, and α is increasing, it must be that $\lim_{t \to d^-} \alpha(t) = \infty$ (if $\lim_{t \to d^-} \alpha(t) = q < \infty$, it would follow that $(\nabla\phi)(z(q)) = 0$, a contradiction). Therefore $\|z(t)\|_H \leq r$, $t \geq 0$. This and Assertion 2 prove the Theorem.

That the pairing between

$$g \in K, \|g\| < \frac{rc}{Q}$$

and the corresponding

$$u = \lim_{t \to \infty} z(t)$$

yields a function is intended to justify the use of the term 'inverse function theorem' in the title. We hope to eventually make an extension of the present process in the direction of [2].

References

[1] M. Berger, *Nonlinearity and Functional Analysis*, Academic Press (1977).

[2] R. Hamilton, *The Inverse Function Theorem of Nash and Moser*, Bulletin of the Amer. Math. Soc., 7 (1982), 65-222.

[3] J. W. Neuberger, *Steepest Descent and Differential Equations*, J. Math. Soc. Japan, 37 (1985), 187-195.

[4] J. W. Neuberger, *Constructive Variational Methods for Differential Equations*, Nonlinear Analysis, Theory, Meth. & Appl., 13 (1988), 413-426.

[5] J. W. Neuberger, *Sobolev Gradients and Differential Equations*, Springer Verlag Lecture Notes in Mathematics, No. 1670, 1997.

Department of Mathematics
University of North Texas
Denton, TX 76205
email: acastro@unt.edu, jwn@unt.edu

Contemporary Mathematics
Volume **221**, 1999

Toward a Characterization of Elliptic Solutions of Hierarchies of Soliton Equations

F. Gesztesy and R. Weikard

ABSTRACT. The current status of an explicit characterization of all elliptic algebro-geometric solutions of hierarchies of soliton equations is discussed and the case of the KdV hierarchy is considered in detail. More precisely, we review our recent result that an elliptic function q is a solution of some equation of the stationary KdV hierarchy, if and only if the associated differential equation $\psi''(E, z) + q(z)\psi(E, z) = E\psi(E, z)$ has a meromorphic fundamental system for every complex value of the spectral parameter E.

This result also provides an explicit condition under which a classical theorem of Picard holds. This theorem guarantees the existence of solutions which are elliptic of the second kind for second-order ordinary differential equations with elliptic coefficients associated with a common period lattice. The fundamental link between Picard's theorem and elliptic algebro-geometric solutions of completely integrable hierarchies of nonlinear evolution equation is the principal new aspect of our approach.

In addition, we describe most recent attempts to extend this circle of ideas to n-th-order scalar differential equations and first-order $n \times n$ systems of differential equations with elliptic functions as coefficients associated with Gelfand-Dickey and matrix-valued hierarchies of soliton equations.

1. Introduction

The principal purpose of this review is to describe the basic ideas underlying an efficient characterization of elliptic algebro-geometric solutions of general hierarchies of soliton equations. Since at this time the only case worked out in all details is that of the KdV hierarchy, we will focus to a large extent on this case and turn in our final two sections to possible extensions to the Gelfand-Dickey and matrix-valued hierarchies.

Before describing our approach in some detail, we shall give a brief account of the history of the problem involved. This theme dates back to a 1940 paper of Ince [65] who studied what is presently called the Lamé–Ince potential

$$(1.1) \qquad q(x) = -g(g + 1)\wp(x + \omega_3), \; g \in \mathbb{N}, \; x \in \mathbb{R}$$

1991 *Mathematics Subject Classification.* Primary 35Q53, 34L05; Secondary 58F07.

Based upon work supported by the National Science Foundation under Grants No. DMS-9623121 and DMS-9401816.

in connection with the second-order ordinary differential equation

(1.2) $\psi''(E,x) + q(x)\psi(E,x) = E\psi(E,x),\ E \in \mathbb{C}.$

Here $\wp(x) = \wp(x; \omega_1, \omega_3)$ denotes the elliptic Weierstrass function with fundamental periods $2\omega_1$ and $2\omega_3$ ($\text{Im}(\omega_3/\omega_1) \neq 0$). In the special case where ω_1 is real and ω_3 is purely imaginary, the potential $q(x)$ in (1.1) is real-valued and Ince's striking result [65], in modern spectral theoretic terminology, yields that the spectrum of the unique self-adjoint operator associated with the differential expression $L_2 = d^2/dx^2 + q(x)$ in $L^2(\mathbb{R})$ exhibits finitely many bands (respectively gaps), that is,

(1.3) $\sigma(L_2) = (-\infty, E_{2g}] \cup \bigcup_{m=1}^{g} [E_{2m-1}, E_{2m-2}],\ E_{2g} < E_{2g-1} < \ldots < E_0.$

What we call the Lamé–Ince potential has, in fact, a long history and many investigations of it precede Ince's work [65]. Without possibly trying to be complete we refer the interested reader, for instance, to [3], [4], Sect. 59, [6], Ch. IX, [9], Sect. 3.6.4, [18], Sects. 135–138, [19], [20], [22], [54], [63], p. 494–498, [64], p. 118–122, 266–418, 475–478, [66], p. 378–380, [69], [71], p. 265–275, [87], [88], [118], [120], [124], Ch. XXIII as pertinent publications before and after Ince's fundamental paper.

Following the traditional terminology, any real-valued potential q that gives rise to a spectrum of the type (1.3) is called an algebro-geometric potential. The proper extension of this notion to general complex-valued meromorphic potentials q and its connection with stationary solutions of the KdV hierarchy on the basis of elementary algebro-geometric concepts is then obtained as follows. Let $L_2(t)$ be the second-order differential expression

(1.4) $L_2(t) = \dfrac{d^2}{dx^2} + q(x,t),\ (x,t) \in \mathbb{R}^2,$

where q depends on the additional (deformation) parameter t. It is well known (see, e.g., Ohmiya [87], Schimming [100], Wilson [125], [126]) that one can find coefficients $p_j(x,t)$ in

(1.5) $P_{2g+1}(t) = \dfrac{d^{2g+1}}{dx^{2g+1}} + p_{2g}(x,t)\dfrac{d^{2g}}{dx^{2g}} + \cdots + p_0(x,t),$

in such a way that $[P_{2g+1}, L_2]$ is a multiplication operator. The coefficients p_j are then certain differential polynomials in q, that is, polynomials in q and its x-derivatives. The pair (P_{2g+1}, L_2) is called a Lax pair, and the equation

(1.6) $\dfrac{d}{dt}L_2 = [P_{2g+1}, L_2],\quad \text{that is,}\quad q_t = [P_{2g+1}, L_2]$

is a nonlinear evolution equation for q. The collection of all such equations for all possible choices of P_{2g+1}, $g \in \mathbb{N}_0$ is then called the KdV hierarchy (see Section 2 for more details). Due to the commutator structure in (1.6), solutions $q(.,t)$ of the nonlinear evolution equations of the KdV hierarchy represent isospectral deformations of $L_2(0)$. In this context, $q(x,t)$ is called an algebro-geometric solution of the KdV equation if it satisfies one of the stationary higher-order equations $[P_{2g+1}, L_2] = 0$ for some $g \geq 0$ for some (and hence for all) $t \in \mathbb{R}$.

Novikov [85], Dubrovin [30], Its and Matveev [68], and McKean and van Moerbeke [79] then showed that a real-valued smooth potential q is an algebro-geometric potential if and only if it satisfies one of the higher-order stationary (i.e.,

t-independent) KdV equations. Because of these facts it is common to call any complex-valued meromorphic function q an algebro-geometric potential if q satisfies one (and hence infinitely many) of the equations of the stationary KdV hierarchy. Therefore, without loss of generality, we mostly focus on stationary solutions in the remainder of this review.

The stationary KdV hierarchy, characterized by $q_t = 0$ or $[P_{2g+1}, L_2] = 0$, is intimately connected with the question of commutativity of ordinary differential expressions. Thus, if $[P_{2g+1}, L] = 0$, a celebrated theorem of Burchnall and Chaundy [16], [17] implies that P_{2g+1} and L_2 satisfy an algebraic relation of the form

$$(1.7) \qquad P_{2g+1}^2 = \prod_{m=0}^{2g} (L_2 - E_m), \ \{E_m\}_{m=0}^{2g} \subset \mathbb{C}.$$

The locations E_m of the (finite) branch points and singular points of the associated hyperelliptic curve

$$(1.8) \qquad F^2 = \prod_{m=0}^{2g} (E - E_m)$$

are precisely the band (gap) edges of the spectral bands of L_2 (see (1.3)) whenever $q(x)$ is real-valued and smooth for $x \in \mathbb{R}$ (with appropriate generalizations to the complex-valued case, see Section 2). It is the (possibly singular) hyperelliptic compact Riemann surface K_g of (arithmetic) genus g, obtained upon one-point compactification of the curve (1.8), which signifies that q in $L_2 = d^2/dx^2 + q(x)$ represents an algebro-geometric potential.

While these considerations pertain to general solutions of the stationary KdV hierarchy, we now concentrate on the additional restriction that q be an elliptic function (i.e., meromorphic and doubly periodic) and hence return to our main subject, elliptic algebro-geometric potentials q for $L_2 = d^2/dx^2 + q(x)$, or, equivalently, elliptic solutions of the stationary KdV hierarchy. Ince's remarkable algebro-geometric result (1.3) remained the only explicit elliptic algebro-geometric example until the KdV flow $q_t = \frac{1}{4} q_{xxx} + \frac{3}{2} q q_x$ with the initial condition $q(x,0) = -6\wp(x)$ was explicitly integrated by Dubrovin and Novikov [34] in 1975 (see also [36], [37], [38], [67]), and found to be of the type

$$(1.9) \qquad q(x,t) = -2 \sum_{j=1}^{3} \wp(x - x_j(t))$$

for appropriate $\{x_j(t)\}_{1 \le j \le 3}$. As observed above, all potentials $q(\cdot,t)$ in (1.9) are isospectral to $q(\cdot,0) = -6\wp(\cdot)$. Given these results it was natural to ask for a systematic account of all elliptic solutions of the KdV hierarchy, a problem posed, for instance, in [86], p. 152.

In 1977, Airault, McKean and Moser, in their seminal paper [2], presented the first systematic study of the isospectral torus $I_\mathbb{R}(q_0)$ of real-valued smooth potentials $q_0(x)$ of the type

$$(1.10) \qquad q_0(x) = -2 \sum_{j=1}^{M} \wp(x - x_j)$$

with an algebro-geometric spectrum of the form (1.3). Among a variety of results they proved that any element q of $I_\mathbb{R}(q_0)$ is an elliptic function of the type (1.10) (with different x_j) with M constant throughout $I_{\mathbb{R}}(q_0)$ and that dim $I_{\mathbb{R}}(q_0) \leq M$. In particular, if q_0 evolves according to any equation of the KdV hierarchy it remains an elliptic algebro-geometric potential. The potential (1.10) is intimately connected with completely integrable many-body systems of the Calogero-Moser-type [19], [84] (see also [20], [22]). This connection with integrable particle systems was subsequently exploited by Krichever [74] in his fundamental construction of elliptic algebro-geometric solutions of the Kadomtsev-Petviashvili equation. In particular, he explicitly determined the underlying algebraic curve Γ and characterized the Baker-Akhiezer function associated with it in terms of elliptic functions as well as the corresponding theta function of Γ. The next breakthrough occurred in 1988 when Verdier [119] published new explicit examples of elliptic algebro-geometric potentials. Verdier's examples spurred a flurry of activities and inspired Belokolos and Enol'skii [11], Smirnov [105], and subsequently Taimanov [110] and Kostov and Enol'skii [70] to find further such examples by combining the reduction process of abelian integrals to elliptic integrals (see [7], [8], [9], Ch. 7, [10]) with the afore-mentioned techniques of Krichever [74], [75]. This development finally culminated in a series of recent results of Treibich and Verdier [115], [116], [117] where it was shown that a general complex-valued potential of the form

$$(1.11) \qquad q(x) = -\sum_{j=1}^{4} d_j \, \wp(x - \omega_j)$$

($\omega_2 = \omega_1 + \omega_3$, $\omega_4 = 0$) is an algebro-geometric potential if and only if $d_j/2$ are triangular numbers, that is, if and only if

$$(1.12) \qquad d_j = g_j(g_j + 1) \text{ for some } g_j \in \mathbb{Z}, \ 1 \leq j \leq 4.$$

We shall from now on refer to potentials of the type

$$(1.13) \qquad q(x) = -\sum_{j=1}^{4} g_j(g_j + 1)\wp(x - \omega_j), \ g_j \in \mathbb{Z}, \ 1 \leq j \leq 4$$

as Treibich-Verdier potentials. The methods of Treibich and Verdier are based on hyperelliptic tangent covers of the torus $\mathbb{C}/\Lambda$ (Λ being the period lattice generated by $2\omega_1$ and $2\omega_3$). The state of the art of elliptic algebro-geometric solutions up to 1993 was recently reviewed in issues 1 and 2 of volume 36 of Acta Applicandae Math., see, for instance, [12], [40], [76], [107], [111], [114] and also in [13], [24], [27], [39], [61], [98], [108], [113]. In addition to these investigations on elliptic solutions of the KdV hierarchy, the study of other soliton hierarchies, such as the modified KdV hierarchy, nonlinear Schrödinger hierarchy, and Boussinesq hierarchy has also begun. We refer, for instance, to [21], [35], [54], [55], [77], [80], [82], [96], [97], [103], [104], [106], [109].

Despite the efforts described thus far, an efficient characterization of all elliptic solutions of the KdV hierarchy remained an open problem until 1994. Around 1992 we became aware of this problem and started to develop our own approach toward its solution. In contrast to all existing basically algebro-geometric approaches in this area, we realized early on that the most powerful analytic tool in this context, a theorem of Picard (Theorem 4.1) concerning the existence of solutions which are elliptic of the second kind of ordinary differential equations with elliptic coefficients, had not been applied at all. As we have recently shown in [59] (see also [58]), Picard's theorem combined with Floquet theoretic results indeed provides a very simple and efficient characterization of all elliptic algebro-geometric solutions of the KdV hierarchy. Moreover, for reflection symmetric elliptic algebro-geometric potentials q (i.e., $q(z) = q(2z_0 - z)$ for some $z_0 \in \mathbb{C}$) including Lamé-Ince and Treibich-Verdier potentials, our approach reduces the computation of the branch points and singular points of the underlying hyperelliptic curve K_g to certain (constrained) linear algebraic eigenvalue problems as shown in [**54**], [**55**], and [**56**].

Since the main hypothesis in Picard's theorem for a second-order differential equation of the form

$$(1.14) \qquad \psi''(z) + q(z)\psi(z) = E\psi(z), \ E \in \mathbb{C}$$

with an elliptic potential q assumes the existence of a fundamental system of solutions meromorphic in z for each value of the spectral parameter $E \in \mathbb{C}$, we call any such elliptic function q which gives rise to this property a **Picard potential**. The principal result, a characterization of all elliptic algebro-geometric solutions of the stationary KdV hierarchy, then reads as follows:

THEOREM 1.1. *([**59**]) q is an elliptic algebro-geometric potential if and only if q is a Picard potential (i.e., if and only if for each $E \in \mathbb{C}$ every solution of $\psi''(z) + q(z)\psi(z) = E\psi(z)$ is meromorphic with respect to z).*

In particular, Theorem 1.1 sheds new light on Picard's theorem since it identifies the elliptic coefficients q for which there exists a meromorphic fundamental system of solutions of (1.14) precisely as the elliptic algebro-geometric solutions of the stationary KdV hierarchy. Moreover, we stress its straightforward applicability based on an elementary Frobenius-type analysis which decides whether or not (1.14) has a meromorphic fundamental system for each $E \in \mathbb{C}$. In addition, we might mention the obvious connections between this result and the Weierstrass theory of reduction of abelian to elliptic integrals.

The proof of Theorem 1.1 in Section 4 (Theorem 4.7) relies on two main ingredients: A purely Floquet theoretic part to be discussed in Section 3 and an elliptic function part sketched in Section 4.

The result embodied by Theorem 1.1 in the special context of the KdV hierarchy, uncovers a new general principle in connection with elliptic algebro-geometric solutions of completely integrable systems: The existence of such solutions appears to be in a one-to-one correspondence with the existence of a meromorphic (with respect to z) fundamental system of solutions for the underlying linear Lax differential expression (for all values of the corresponding spectral parameter E).

Having dealt with the second-order Lax differential expression $L_2 = d^2/dx^2 + q$ underlying the KdV hierarchy, it is natural to seek extensions to n-th-order Lax differential expressions L_n associated with the Gelfand-Dickey hierarchy and more generally, to matrix-valued hierarchies of soliton equations. At present our results in these directions are promising but far from being complete. We provide a recent generalization of Picard's theorem to first-order $n \times n$ systems of differential equations in Section 5 and devote Section 6 to partial progress in the context of n-th-order scalar differential expressions with elliptic coefficients.

2. The KdV Hierarchy and Hyperelliptic Curves

In this section we review basic facts on the stationary KdV hierarchy. Since this material is well-known (see, e.g., [**5**], [**23**], [**26**], Ch. 12, [**28**], [**29**], [**49**], [**53**], [**87**], [**100**], [**101**], [**126**]), we confine ourselves to a brief account. Assuming $q \in$

$C^\infty(\mathbb{R})$ or q meromorphic in $\mathbb{C}$ (depending on the particular context in which one is interested) and hence either $x \in \mathbb{R}$ or $x \in \mathbb{C}$, consider the recursion relation

$$(2.1) \qquad \hat{f}'_{j+1}(x) = \frac{1}{4}\hat{f}'''_j(x) + q(x)\hat{f}'_j(x) + \frac{1}{2}q'(x)\hat{f}_j(x),\ 0 \le j \le g,\ \hat{f}_0(x) = 1$$

and the associated differential expressions (Lax pair)

$$(2.2) \qquad\qquad\qquad L_2 = \frac{d^2}{dx^2} + q(x),$$

$$(2.3) \qquad \hat{P}_{2g+1} = \sum_{j=0}^{g}\left[-\frac{1}{2}\hat{f}'_j(x) + \hat{f}_j(x)\frac{d}{dx}\right]L_2^{g-j},\ g \in \mathbb{N}_0$$

(here $\mathbb{N}_0 := \mathbb{N} \cup \{0\}$). One can show that

$$(2.4) \qquad \left[\hat{P}_{2g+1}, L_2\right] = 2\hat{f}'_{g+1} = \frac{1}{2}\hat{f}'''_g(x) + 2q(x)\hat{f}'_g(x) + q'(x)\hat{f}_g(x)$$

($[\cdot, \cdot]$ the commutator symbol) and explicitly computes from (2.1),

$$(2.5) \qquad \hat{f}_0 = 1,\ \hat{f}_1 = \frac{1}{2}q + c_1,\ \hat{f}_2 = \frac{1}{8}q'' + \frac{3}{8}q^2 + \frac{c_1}{2}q + c_2,\quad \text{etc.,}$$

where the c_j are integration constants. Using the convention that the corresponding homogeneous quantities obtained by setting $c_\ell = 0$ for $\ell = 1, 2, \dots$ are denoted by f_j, that is, $f_j = \hat{f}_j(c_\ell \equiv 0)$, the (homogeneous) stationary KdV hierarchy is then defined as the sequence of equations

$$(2.6) \qquad\qquad\qquad \text{KdV}_g(q) = 2f'_{g+1} = 0,\ g \in \mathbb{N}_0.$$

Explicitly, this yields

$$(2.7) \qquad \text{KdV}_0(q) = q_x = 0,\quad \text{KdV}_1(q) = \frac{1}{4}q''' + \frac{3}{2}qq' = 0,\quad \text{etc.}$$

The corresponding non-homogeneous version of $\text{KdV}_g(q) = 0$ is then defined by

$$(2.8) \qquad\qquad\qquad \hat{f}'_{g+1} = \sum_{j=0}^{g}c_{g-j}f'_{j+1} = 0,$$

where $c_0 = 1$ and $c_1, \dots, c_g$ are arbitrary complex constants.

If one assigns to $q^{(\ell)} = d^\ell q/dx^\ell$ the degree $\deg(q^{(\ell)}) = \ell + 2$, $\ell \in \mathbb{N}_0$, then the homogeneous differential polynomial f_j with respect to q turns out to have degree $2j$, that is,

$$(2.9) \qquad\qquad\qquad \deg(f_j) = 2j,\ j \in \mathbb{N}_0.$$

Next, introduce the polynomial $\hat{F}_g(E, x)$ in $E \in \mathbb{C}$,

$$(2.10) \qquad\qquad\qquad \hat{F}_g(E, x) = \sum_{j=0}^{g}\hat{f}_{g-j}(x)E^j.$$

Since $\hat{f}_0(x) = 1$,

$$(2.11) \quad \hat{R}_{2g+1}(E, x) = (E - q(x))\hat{F}_g(E, x)^2 - \frac{1}{2}\hat{F}''_g(E, x)\hat{F}_g(E, x) + \frac{1}{4}\hat{F}'_g(E, x)^2$$

is a monic polynomial in E of degree $2g + 1$. However, equations (2.1) and (2.8) imply that

$$(2.12) \qquad \frac{1}{2}\hat{F}_g''' - 2(E - q)\hat{F}_g' + q'\hat{F}_g = 0$$

and this shows that $\hat{R}_{2g+1}(E, x)$ is in fact independent of x. Hence it can be written as

$$(2.13) \qquad \hat{R}_{2g+1}(E) = \prod_{m=0}^{2g} (E - \hat{E}_m), \quad \{\hat{E}_m\}_{m=0}^{2g} \subset \mathbb{C}.$$

By (2.4) the non-homogeneous KdV equation (2.8) is equivalent to the commutativity of L_2 and $\hat{P}_{2g+1}$. This shows that

$$(2.14) \qquad [\hat{P}_{2g+1}, L_2] = 0,$$

and therefore, if $L_2\psi = E\psi$, this implies that $\hat{P}_{2g+1}^2\psi = \hat{R}_{2g+1}(E)\psi$. Thus $[\hat{P}_{2g+1}, L_2] = 0$ implies

$$(2.15) \qquad \hat{P}_{2g+1}^2 = \hat{R}_{2g+1}(L_2) = \prod_{m=0}^{2g} (L_2 - \hat{E}_m),$$

a celebrated theorem by Burchnall and Chaundy [16], [17] (see, e.g., [27], [47], [61], [98], [126] for more recent accounts).

In the second part of Section 3 we will need the converse of the above procedure. It is given by

LEMMA 2.1. ([59]) *Assume that $\hat{F}_g(E, x)$, given by (2.10) with $\hat{f}_0(x) = 1$, is twice differentiable with respect to x, and that*

$$(2.16) \qquad (E - q(x))\hat{F}_g(E, x)^2 - \frac{1}{2}\hat{F}_g''(E, x)\hat{F}_g(E, x) + \frac{1}{4}\hat{F}_g'(E, x)^2$$

is independent of x. Then $q \in C^\infty(\mathbb{R})$. Also the functions $\hat{f}_j(x)$ are infinitely often differentiable and satisfy the recursion relations (2.1) for $j = 0, ..., g-1$. Moreover, $\hat{f}_g$ satisfies

$$(2.17) \qquad \frac{1}{4}\hat{f}_g'''(x) + q(x)\hat{f}_g'(x) + \frac{1}{2}q'(x)\hat{f}_g(x) = 0,$$

that is, the differential expression $\hat{P}_{2g+1}$ given in (2.3) commutes with the expression $L_2 = d^2/dx^2 + q$.

Equation (2.15) illustrates the intimate connection between the stationary KdV equation $\hat{f}_{g+1}' = 0$ in (2.8) and the compact (possibly singular) hyperelliptic curve K_g of (arithmetic) genus g obtained upon one-point compactification of the curve

$$(2.18) \qquad F^2 = \hat{R}_{2g+1}(E) = \prod_{m=0}^{2g} (E - \hat{E}_m).$$

The above formalism leads to the following standard definition.

DEFINITION 2.2. Any solution q of one of the stationary KdV equations (2.8) is called an **algebro-geometric potential** associated with the KdV hierarchy.

Algebro-geometric potentials q can be expressed in terms of the Riemann theta function or through τ-functions associated with the curve K_g (see, e.g., [68], [102]).

3. Floquet Theory and Algebro-Geometric Potentials

In the first part of this section we discuss Floquet theory in connection with a complex-valued non-constant periodic potential q. In the second part we recall a criterion for q to be an algebro-geometric potential in terms of Floquet solutions.

Suppose

$$(3.1) \qquad q \in L^1_{\text{loc}}(\mathbb{R}),\ q(x + \Omega) = q(x),\ x \in \mathbb{R}$$

for some $\Omega > 0$ and let $\mathcal{L}(E)$ be the (two-dimensional) space of solutions of $L_2 y = E y$. Then $T(E)$, the restriction of the operator defined by $y \mapsto y(\cdot + \Omega)$ to $\mathcal{L}(E)$, commutes with the corresponding restriction of L_2 and hence maps $\mathcal{L}(E)$ to itself. The eigenvalues and eigenfunctions of $T(E)$ are called Floquet multipliers and Floquet solutions of $L_2 y = E y$. On $\mathcal{L}(E)$ we introduce the basis $c(E, x, x_0)$ and $s(E, x, x_0)$ defined by

$$(3.2) \qquad c(E, x_0, x_0) = s'(E, x_0, x_0) = 1,\ c'(E, x_0, x_0) = s(E, x_0, x_0) = 0.$$

Using this basis the operator $T(E)$ is represented by the so called monodromy matrix

$$(3.3) \qquad \begin{pmatrix} c(E, x_0 + \Omega, x_0) & s(E, x_0 + \Omega, x_0) \\ c'(E, x_0 + \Omega, x_0) & s'(E, x_0 + \Omega, x_0) \end{pmatrix}.$$

Since $\det(T(E)) = 1$ the Floquet multipliers $\rho_\pm(E)$ are given by

$$(3.4) \qquad \rho_\pm(E) = \Delta(E) \pm \sqrt{\Delta(E)^2 - 1},$$

where $\Delta(E)$ denotes the Floquet discriminant,

$$(3.5) \qquad \Delta(E) = \frac{1}{2} \operatorname{tr}(T(E)) = [c(E, x_0 + \Omega, x_0) + s'(E, x_0 + \Omega, x_0)]/2.$$

For each $E \in \mathbb{C}$ there exists at least one nontrivial Floquet solution. In fact, since together with $\rho(E)$, $1/\rho(E)$ is also a Floquet multiplier, there are two linearly independent Floquet solutions for a given E provided $\rho(E)^2 \neq 1$. Floquet solutions can be expressed in terms of the fundamental system $c(E, x, x_0)$ and $s(E, x, x_0)$ by

$$(3.6) \qquad \psi_\pm(E, x, x_0) = c(E, x, x_0) + \frac{\rho_\pm(E) - c(E, x_0 + \Omega, x_0)}{s(E, x_0 + \Omega, x_0)} s(E, x, x_0),$$

if $s(E, x_0 + \Omega, x_0) \neq 0$, or by

$$(3.7) \qquad \tilde{\psi}_\pm(E, x, x_0) = s(E, x, x_0) + \frac{\rho_\pm(E) - s'(E, x_0 + \Omega, x_0)}{c'(E, x_0 + \Omega, x_0)} c(E, x, x_0),$$

if $c'(E, x_0 + \Omega, x_0) \neq 0$. If both $s(E, x_0 + \Omega, x_0)$ and $c'(E, x_0 + \Omega, x_0)$ are equal to zero, then $s(E, x, x_0)$ and $c(E, x, x_0)$ are linearly independent Floquet solutions.

Associated with the second-order differential expression $L_2 = d^2/dx^2 + q(x)$ we consider the densely defined closed linear operators H, $H_D(x_0)$, $H(\beta, x_0)$, $\beta \in \mathbb{C}$, and $H(\theta)$, $\theta \in \mathbb{C}$. While H will be an operator in $L^2(\mathbb{R})$, the others will be defined in $L^2(I(x_0))$, where $I(x_0) = (x_0, x_0 + \Omega)$ for some $x_0 \in \mathbb{R}$. Specifically, the operators are given as restrictions of the expression L_2 to the following domains:

$$(3.8) \qquad \mathcal{D}(H) = \{g \in L^2(\mathbb{R}) : g, g' \in AC_{\text{loc}}(\mathbb{R}), (g'' + qg) \in L^2(\mathbb{R})\},$$

$$(3.9) \qquad \mathcal{D}(H_D(x_0)) = \{g \in \mathcal{D}_{x_0} : g(x_0) = g(x_0 + \Omega) = 0\},$$

$$(3.10) \qquad \mathcal{D}(H(\beta, x_0)) = \{g \in \mathcal{D}_{x_0} : U_1(\beta, g)(x_0) = U_1(\beta, g)(x_0 + \Omega) = 0\},$$

$$(3.11) \qquad \mathcal{D}(H(\theta)) = \{g \in \mathcal{D}_{x_0} : U_2(\theta, g)(x_0) = U_2(\theta, g)'(x_0) = 0\},$$

where

(3.12) $\quad \mathcal{D}_{x_0} = \{g \in L^2(I(x_0)) : g, g' \in AC([x_0, x_0 + \Omega]), (g'' + qg) \in L^2(I(x_0))\}$

and $U_1(\beta, y) = y' + \beta y$ and $U_2(\theta, y) = y(\cdot + \Omega) - e^{i\theta} y(\cdot)$. Here $AC(\cdot)$ $(AC_{\text{loc}}(\cdot))$ denotes the set of (locally) absolutely continuous functions. Next we denote the purely discrete spectra of $H_D(x_0)$, $H(\beta, x_0)$, and $H(\theta)$ by $\sigma(H_D(x_0)) = \{\mu_n(x_0)\}_{n \in \mathbb{N}}$, $\sigma(H(\beta, x_0)) = \{\lambda_n(\beta, x_0)\}_{n \in \mathbb{N}_0}$ and $\sigma(H(\theta)) = \{E_n(\theta)\}_{n \in \mathbb{N}_0}$, respectively. While $H(\theta)$ depends on x_0 its spectrum does not. We agree that here, as well as in the rest of the paper, all point spectra (i.e., sets of eigenvalues) are recorded in such a way that all eigenvalues are consistently repeated according to their algebraic multiplicity unless explicitly stated otherwise.

The eigenvalues of $H_D(x_0)$ are called Dirichlet eigenvalues with respect to the interval $[x_0, x_0 + \Omega]$. The eigenvalues of $H(\theta)$ are precisely those values E where $T(E)$ has eigenvalues $\rho = e^{\pm i\theta}$. The eigenvalues $E_n(0)$ $(E_n(\pi))$ of $H(0)$ $(H(\pi))$ are called the periodic (antiperiodic) eigenvalues associated with q. Note that the (anti)periodic eigenvalues $E_n(0)$ $(E_n(\pi))$ are the zeros of $\Delta(\cdot) - 1$ $(\Delta(\cdot) + 1)$ and that their algebraic multiplicities coincide with the orders of the respective zeros (see, e.g., [57]). In the following we denote the zeros of $\Delta(E)^2 - 1$ by E_n, $n \in \mathbb{N}_0$. They are repeated according to their multiplicity and are related to the (anti)periodic eigenvalues via

(3.13)
$$E_{4n} = E_{2n}(0), \quad E_{4n+1} = E_{2n}(\pi), \quad E_{4n+2} = E_{2n+1}(\pi), \quad E_{4n+3} = E_{2n+1}(0)$$

for $n \in \mathbb{N}_0$. We also introduce

(3.14) $$p(E) = \text{ord}_E(\Delta(\cdot)^2 - 1),$$

the order of E as a zero of $\Delta(\cdot)^2 - 1$ $(p(E) = 0$ if $\Delta(E)^2 - 1 \neq 0)$.

Similarly, the eigenvalues of $H_D(x_0)$ and $H(\beta, x_0)$ are the zeros of the functions $s(\cdot, x_0 + \Omega, x_0)$ and $h(\cdot, \beta, x_0) = (\beta^2 s + \beta(s' - c) - c')(\cdot, x_0 + \Omega, x_0)$, respectively. Again their algebraic multiplicities coincide precisely with the multiplicities of the respective zeros (see, e.g., [57]). These multiplicities depend in general on x_0. We introduce the notation

(3.15) $$d(E, x_0) = \text{ord}_E(s(\cdot, x_0 + \Omega, x_0)),$$
(3.16) $$r(E, \beta, x_0) = \text{ord}_E(h(\cdot, \beta, x_0)),$$

and remark that $d(E, x_0)$ and $r(E, \beta, x_0)$ are combinations of movable and immovable parts. Specifically, define $d_i(E) = \min\{d(E, x_0) : x_0 \in \mathbb{R}\}$, $r_i(E, \beta) = \min\{r(E, \beta, x_0) : x_0 \in \mathbb{R}\}$ and $d_m(E, x_0)$ and $r_m(E, x_0)$ by

(3.17) $$d(E, x_0) = d_i(E) + d_m(E, x_0),$$
(3.18) $$r(E, \beta, x_0) = r_i(E, \beta) + r_m(E, \beta, x_0).$$

If $d_i(E) > 0$ then E is a Dirichlet eigenvalue irrespective of the value of x_0 and we will call E an immovable Dirichlet eigenvalue. Otherwise, if $d_i(E) = 0$ but $d(E, x_0) > 0$ we call E a movable Dirichlet eigenvalue. (Note that here we use a notation different from the one in [59], in particular, the multiplicities d, d_i, and d_m now refer to Dirichlet eigenvalues while the multiplicities p refer to periodic or antiperiodic eigenvalues).

The functions $c(\cdot, x, x_0)$ and $s(\cdot, x, x_0)$ and their x-derivatives are entire functions of order $1/2$ for every choice of x and x_0. This and their asymptotic behavior as

$|E|$ tends to infinity is obtained via Volterra integral equations. Invoking Rouché's theorem then yields the following facts:

1. The zeros $\mu_n(x_0)$ of $s(E, x_0 + \Omega, x_0)$ and the zeros $\lambda_n(\beta, x_0)$ of $h(\cdot, \beta, x_0)$ are simple for $n \in \mathbb{N}$ sufficiently large.
2. The zeros E_n of $\Delta(E)^2 - 1$ are at most double for $n \in \mathbb{N}$ large enough.
3. $\mu_n(x_0)$, $\lambda_n(\beta, x_0)$, and E_n can be arranged (and will be subsequently) such that they have the following asymptotic behavior as n tends to infinity:

$$(3.19) \qquad \mu_n(x_0) = -\frac{n^2\pi^2}{\Omega^2} + O(1),$$

$$(3.20) \qquad \lambda_n(\beta, x_0) = -\frac{n^2\pi^2}{\Omega^2} + O(1),$$

$$(3.21) \qquad E_{2n-1}, E_{2n} = -\frac{n^2\pi^2}{\Omega^2} + O(1).$$

The Hadamard factorization of $s(E, x_0 + \Omega, x_0)$ therefore reads

$$(3.22) \qquad s(E, x_0 + \Omega, x_0) = c_1(x_0) \prod_{n=1}^{\infty} \left(1 - \frac{E}{\mu_n(x_0)}\right) = F_D(E, x_0)D(E),$$

where all those factors which do not depend on x_0 are collected in $D(E)$. Here we assume that none of the eigenvalues is equal to zero; otherwise, obvious modifications have to be used.

For more details on algebraic versus geometric multiplicities of eigenvalue problems of the type of $H_D(x_0)$ and $H(\theta)$ see, for instance, [57].

It was shown by Rofe-Beketov [99] that the spectrum of H is equal to the conditional stability set of L_2, that is, the set of all spectral parameters E for which a nontrivial bounded solution of $L_2\psi = E\psi$ exists. Hence

$$(3.23) \qquad \sigma(H) = \bigcup_{\theta \in [0,2\pi]} \sigma(T(\theta)) = \bigcup_{n \in \mathbb{N}_0} \sigma_n, \text{ where } \sigma_n = \bigcup_{\theta \in [0,\pi]} E_n(\theta).$$

We note that in the general case where q is complex-valued some of the spectral arcs σ_n may cross each other, see, for instance, [57] and [91] for explicit examples.

The Green's function $G(E, x, x')$ of H, that is, the integral kernel of the resolvent of H,

$$(3.24) \qquad G(E, x, x') = (H - E)^{-1}(x, x'), \ E \in \mathbb{C}\backslash\sigma(H), \ x, x' \in \mathbb{R},$$

is explicitly given by

$$(3.25) \qquad G(E, x, x') = W(f_-(E, x), f_+(E, x))^{-1} \begin{cases} f_+(E, x)f_-(E, x'), \ x \geq x' \\ f_-(E, x)f_+(E, x'), \ x \leq x' \end{cases}.$$

Here $f_\pm(E, \cdot)$ solve $L_2 f = Ef$ and are chosen such that

$$(3.26) \qquad f_\pm(E, \cdot) \in L^2((R, \pm\infty)), \ E \in \mathbb{C}\backslash\sigma(H), \ R \in \mathbb{R},$$

with $W(f, g) = fg' - f'g$ the Wronskian of f and g.

Equation (3.25) implies that the diagonal Green's function is twice differentiable and satisfies the nonlinear second-order differential equation (see, e.g., [48], [83])

$$(3.27) \qquad 4(E - q(x))G(E, x, x)^2 - 2G(E, x, x)G''(E, x, x) + G'(E, x, x)^2 = 1$$

(the primes denoting derivatives with respect to x).

It follows from (3.23) that $|\rho(E)| \neq 1$ unless $E \in \sigma(H)$. Therefore, if $E \notin \sigma(H)$ there is precisely one Floquet solution in $L^2((-\infty, R))$ and one in $L^2((R, \infty))$. Letting $\rho_\pm(E) = e^{\pm i\theta}$ with $\text{Im}(\theta) > 0$ we obtain $|\rho_+(E)| < 1 < |\rho_-(E)|$. Hence $f_+(E, x) = \psi_+(E, x, x_0)$ and $f_-(E, x) = \psi_-(E, x, x_0)$. Since $\psi_\pm(E, x_0, x_0) = 1$, equations (3.4) and (3.6) imply

$$(3.28) \qquad W(f_-(E, \cdot), f_+(E, \cdot)) = \frac{e^{i\theta} - e^{-i\theta}}{s(E, x_0 + \Omega, x_0)} = -2\frac{[\Delta(E)^2 - 1]^{1/2}}{s(E, x_0 + \Omega, x_0)}.$$

The sign of the square root was chosen such that $[\Delta(E)^2 - 1]^{1/2}$ is asymptotically equal to $\rho_-(E)/2$ for large positive E. Equation (3.28) implies (see also [**50**])

$$(3.29) \qquad G(E, x_0, x_0) = -\frac{s(E, x_0 + \Omega, x_0)}{2[\Delta(E)^2 - 1]^{1/2}}.$$

THEOREM 3.1. ([**59**], [**123**]) *If q is a locally integrable periodic function on $\mathbb{R}$ then $p(E) - 2d_i(E) \geq 0$ for all $E \in \mathbb{C}$.*

PROOF. Equations (3.22), (3.27), and (3.29) show that

(3.30)
$$4(E - q(x))F_D(E, x)^2 - 2F_D(E, x)F_D''(E, x) + F_D'(E, x)^2 = \frac{4(\Delta(E)^2 - 1)}{D(E)^2}.$$

Since the left hand side is entire the claim follows immediately from the definitions of the numbers $p(E)$ and $d_i(E)$. $\qquad\square$

A somewhat bigger effort allows one to prove also

THEOREM 3.2. ([**59**], [**123**]) *If q is a locally integrable periodic function on $\mathbb{R}$ then $d_i(E) = r_i(E, \beta)$ unless q is a constant and $E = q + \beta^2$. Moreover, if $d_i(E) > 0$ then there exist two linearly independent Floquet solutions of $L_2 y = Ey$. Finally, $p(E) - 2d_i(E) > 0$ if and only if there exists an $x_0 \in \mathbb{R}$ such that $W(z, x_0)$, the Wronskian of the Floquet solutions $\psi_\pm$ given by (3.6), tends to zero as z tends to E.*

Hence, if there are not two linearly independent Floquet solutions for $L_2 y = Ey$ then $\rho^2 = 1$ and $p(E) > 0$ but $d_i(E) = 0$ and thus $p(E) - 2d_i(E) > 0$ at all such points.

Nowhere in this section did we use thus far that q is an algebro-geometric potential. Next we give necessary and sufficient conditions for this in terms of properties of multiplicities of eigenvalues of (anti)periodic boundary value problems on one hand and the Dirichlet problem on the other hand. We begin with

DEFINITION 3.3. The number $\text{def}(L_2) = \sum_{E \in \mathbb{C}}(p(E) - 2d_i(E))$ is called the Floquet defect. The number $\sum_{E \in \mathbb{C}} d_m(E, x_0)$ will be called the number of movable Dirichlet eigenvalues; similarly, $\sum_{E \in \mathbb{C}} r_m(E, \beta, x_0)$ denotes the number of movable eigenvalues of $H(\beta, x_0)$.

Note that by Theorem 3.1, $\text{def}(L_2)$ is either infinite or else a nonnegative integer. If it is finite then $\text{def}(L_2) = \deg(4(\Delta^2 - 1)/D^2)$. Both, $\text{def}(L_2)$ and the number of movable Dirichlet eigenvalues are in general infinite.

THEOREM 3.4. ([**59**], [**123**]) *Assume that q is a locally integrable, periodic function of period $\Omega > 0$ on $\mathbb{R}$. Then the following statements are equivalent:*
1. The Floquet defect $\text{def}(L_2)$ equals $2g + 1$.

2. The number of movable Dirichlet eigenvalues equals g.

3. There exists a monic differential expression $\hat{P}_{2g+1}$ of odd order $2g+1$ which commutes with L_2 but none of smaller odd order, i.e., q is an algebro-geometric potential.

In particular, $\mathrm{def}(L_2)$ is either odd or infinite.

SKETCH OF PROOF. If $\mathrm{def}(L_2)$ is finite, asymptotic considerations show that only finitely many Dirichlet eigenvalues can be movable. Hence $F_D(\cdot, x_0)$ is a polynomial, say of degree $\hat{g}$. By equation (3.30) $4(\Delta^2 - 1)/D^2$ is a polynomial of degree $2\hat{g} + 1$. Hence $\hat{g} = g$. This shows the equivalence of the first two statements.

Next one shows that the leading coefficent of $F_D(\cdot, x_0)$ is independent of x_0. The third statement follows then from the second using Lemma 2.1. To prove that the third statement implies the other two one has to show that the zeros of the function $\hat{F}_g(\cdot, x_0)$ in (2.10) are precisely the movable Dirichlet eigenvalues. This follows from applying $\hat{P}_{2g+1}$ as given in (2.3) succesively to the generalized Dirichlet eigenfunctions. □

THEOREM 3.5. ([**59**], [**123**]) *Assume that q is a non-constant, locally integrable, periodic function of period $\Omega > 0$ on $\mathbb{R}$ and that any (and hence all) of the three statements in Theorem 3.4 is satisfied. Then the following statements hold.*

1. The number of movable eigenvalues of $H(\beta, x_0)$ equals $g + 1$, i.e.,

$$(3.31) \qquad \sum_{E \in \mathbb{C}} r_m(E, \beta, x_0) = g + 1.$$

2. $q \in C^\infty(\mathbb{R})$.

3. The differential expression $\hat{P}_{2g+1}$ satisfies the Burchnall-Chaundy relation

$$(3.32) \qquad \hat{P}_{2g+1}^2 = \prod_{z \in \mathbb{C}} (L - z)^{p(z) - 2d_i(z)}.$$

4. The diagonal Green's function $G(\cdot, x, x)$ of H is continuous on $\mathbb{C} \setminus \{z : p(z) - 2d_i(z) > 0\}$ and is of the type

$$(3.33) \qquad G(E, x, x) = -\frac{1}{2} \frac{\prod_{z \in \mathbb{C}} (E - z)^{d_m(z,x)}}{\prod_{z \in \mathbb{C}} (E - z)^{p(z) - 2d_i(z)}}.$$

5. The spectrum of H consists of finitely many bounded spectral arcs $\tilde{\sigma}_n$, $1 \leq n \leq \tilde{g}$ for some $\tilde{g} \leq g$ and one unbounded (semi-infinite) arc $\tilde{\sigma}_\infty$ which tends to $-\infty + < q >$, with $< q >= \Omega^{-1} \int_{x_0}^{x_0 + \Omega} q(x)dx$, that is,

$$(3.34) \qquad \sigma(H) = \left(\bigcup_{n=1}^{\tilde{g}} \tilde{\sigma}_n \right) \cup \tilde{\sigma}_\infty,$$

where each $\tilde{\sigma}_n$ and $\tilde{\sigma}_\infty$ is a union of some of the spectral arcs σ_n in (3.23).

Note that the set B of values of E where $p(E) - 2d_i(E) > 0$ contains B_1, the set of all those points where less than two linearly independent Floquet solutions exist. For $B \setminus B_1$ to be nonempty, it is necessary that $p(z) \geq 3$ for some (anti)periodic eigenvalue z. While it seems difficult to construct an explicit example where $B \setminus B_1 \neq \emptyset$, the very existence of this phenomenon has first been noted in [**59**]. References [**45**], [**46**], [**62**], [**89**], [**90**] treat potentials with $p(E) \leq 2$

and references [14], [15] require that algebraic and geometric multiplicities of all (anti)periodic eigenvalues coincide and hence also that $p(E) \leq 2$. Generically one has $p(E) - 2d_i(E) = 1$ if this is positive at all and $B = B_1$ (cf. [112]).

REMARK 3.6 (Singularity structure of the Green's function). As Theorem 3.5 shows, it is precisely the multiplicity $p - 2d_i$ of the branch and singular points in the Burchnall-Chaundy polynomial (3.32) which determines the singularity structure of the diagonal Green's function $G(E, x, x)$ of H. Moreover, since (see, e.g., [83])

(3.35)

$$G(E, x, x') = [G(E, x, x)G(E, x', x')]^{1/2} \exp[-\frac{1}{2} \int_{\min(x,x')}^{\max(x,x')} G(E, s, s)^{-1} ds],$$

this observation extends to the off-diagonal Green's function $G(E, x, x')$ of H as well.

REMARK 3.7 (Inverse square singularities). The case of the Lamé-Ince potential, where q has singularities of the form $-g(g + 1)/x^2$, indicates the necessity to consider also potentials with such singularities. This is possible by modifying the usual approach via Volterra integral equations which are used to obtain the asymptotic properties (3.19)–(3.21) of the corresponding eigenvalue distributions. One obtains essentially the same results as in the present review, the only difference being that the conditional stability set cannot be interpreted as the spectrum of an operator in $L^2(\mathbb{R})$. This approach has been worked out in detail in [122] and [123].

REMARK 3.8 (Finite-band potentials). For real-valued potentials Novikov [85] and Dubrovin [30] showed that q is an algebro-geometric potential if and only if the spectrum of the operator H consists of only finitely many bands. This is no longer true for complex-valued potentials. In fact, for $q = e^{ix}$ one infers $\sigma(H) = (-\infty, 0]$ but every Dirichlet eigenvalue is movable (see [123]).

4. A Characterization of Elliptic Solutions of the KdV Hierarchy

In this section we discuss the principal result in [59], an explicit characterization of all elliptic algebro-geometric solutions of the KdV hierarchy. One of the two key ingredients in our main Theorem 4.7 (the other being Theorem 3.4) is a systematic use of a powerful theorem of Picard (see Theorem 4.1 below) concerning the existence of solutions which are elliptic functions of the second kind of ordinary differential equations with elliptic coefficients.

We start with Picard's theorem.

THEOREM 4.1. ([93] – [95], see also [4], p. 182–187, [66], p. 375–376) Let q_m, $0 \leq m \leq n$ be elliptic functions with a common period lattice spanned by the fundamental periods $2\omega_1$ and $2\omega_3$. Consider the differential equation

(4.1)
$$\sum_{m=0}^{n} q_m(z)\psi^{(m)}(z) = 0, \quad q_n(z) = 1, \quad z \in \mathbb{C}$$

and assume that (4.1) has a meromorphic fundamental system of solutions. Then there exists at least one solution ψ_0 which is elliptic of the second kind, that is, ψ_0 is meromorphic and

(4.2)
$$\psi_0(z + 2\omega_j) = \rho_j \psi_0(z), \quad j = 1, 3$$

for some constants $\rho_1, \rho_3 \in \mathbb{C}$. If in addition, the characteristic equation corresponding to the substitution $z \to z+2\omega_1$ or $z \to z+2\omega_3$ (see [66], p. 358, 376) has distinct roots then there exists a fundamental system of solutions of (4.1) which are elliptic functions of the second kind.

The characteristic equation associated with the substitution $z \to z+2\omega_j$ alluded to in Theorem 4.1 is given by

$$(4.3) \qquad \det[A - \rho I] = 0,$$

where

$$(4.4) \qquad \phi_\ell(z + 2\omega_j) = \sum_{m=1}^{n} a_{\ell,m}\phi_m(z), \ A = (a_{\ell,m})_{1 \le \ell, m \le n}$$

and $\phi_1, ..., \phi_n$ is any fundamental system of solutions of (4.1).

What we call Picard's theorem following the usual convention in [4], p. 182–185, [18], p. 338–343, [63], p. 536–539, [71], p. 181–189, appears, however, to have a longer history. In fact, Picard's investigations [93], [94], [95] were inspired by earlier work of Hermite in the special case of Lamé's equation [64], p. 118–122, 266–418, 475–478 (see also [9], Sect. 3.6.4 and [124], p. 570–576). Further contributions were made by Mittag-Leffler [81], and Floquet [42], [43], [44]. Detailed accounts on Picard's differential equation can be found in [63], p. 532–574, [71], p. 198–288.

In this context it seems appropriate to recall the well-known fact (see, e.g., [4], p. 185–186) that ψ_0 is elliptic of the second kind if and only if it is of the form

$$(4.5) \qquad \psi_0(z) = Ce^{\lambda z} \prod_{j=1}^{m} [\sigma(z - a_j)/\sigma(z - b_j)]$$

for suitable $m \in \mathbb{N}$ and $C, \lambda, a_j, b_j \in \mathbb{C}$, $1 \le j \le m$. Here $\sigma(z)$ is the Weierstrass sigma function associated with the period lattice Λ spanned by $2\omega_1, 2\omega_3$ (see [1], Ch. 18).

Picard's Theorem 4.1, restricted to the second-order case

$$(4.6) \qquad \psi''(z) + q(z)\psi(z) = E\psi(z),$$

motivates the following definition.

DEFINITION 4.2. *Let q be an elliptic function. Then q is called a **Picard potential** if and only if the differential equation (4.6) has a meromorphic fundamental system of solutions (with respect to z) for each value of the spectral parameter $E \in \mathbb{C}$.*

For completeness we recall the following result.

THEOREM 4.3. *([56]) (i) Any non-constant Picard potential q has a representation of the form*

$$(4.7) \qquad q(z) = C - \sum_{j=1}^{m} s_j(s_j + 1)\wp(z - b_j)$$

for suitable $m, s_j \in \mathbb{N}$ and $C, b_j \in \mathbb{C}$, $1 \le j \le m$, where the b_j are pairwise distinct $\mod(\Lambda)$ and $\wp(z)$ denotes the Weierstrass $\wp$-function associated with the period lattice Λ ([1], Ch. 18).

(ii) Let $q(z)$ be given as in (4.7). If $\psi'' + q\psi = E\psi$ has a meromorphic fundamental system of solutions for a number of distinct values of E which exceeds $\max\{s_1, \dots, s_m\}$, then q is a Picard potential.

We emphasize that while any Picard potential is necessarily of the form (4.7), a potential q of the type (4.7) is a Picard potential only if the constants b_j satisfy a series of additional intricate constraints, see, for instance, Section 3.2 in [**56**].

The following result indicates the connection between Picard potentials and elliptic algebro-geometric potentials.

THEOREM 4.4. *(Its and Matveev [**68**], Krichever [**72**], [**73**], Segal and Wilson [**102**]) Every elliptic algebro-geometric potential q is a Picard potential.*

SKETCH OF PROOF. For nonsingular curves $K_g : F^2 = \prod_{j=0}^{2g}(E - \hat{E}_j)$ associated with q (see (2.18)), where $\hat{E}_\ell \neq \hat{E}_{\ell'}$ for $\ell \neq \ell'$, Theorem 4.4 is obvious from the standard representation of the Baker-Akhiezer function in terms of the Riemann theta function of K_g ([**32**], [**68**], [**72**], [**73**]). For singular curves K_g the result follows from the τ-function representation of the Floquet solutions $\psi_\pm(E, x)$ associated with q

$$(4.8) \qquad \psi_\pm(E, x) = e^{\pm k(E)x} \tau_\pm(E, x)/\tau(x),$$

where

$$(4.9) \qquad q(x) = C + 2\{\ln[\tau(x)]\}''$$

and from the fact that $\tau(x)$ and $\tau_\pm(E, x)$ are entire with respect to x (cf. [**102**]). □

Naturally, one is tempted to conjecture that the converse of Theorem 4.4 is true as well. The rest of this section will explain our proof of this conjecture in [**59**].

We start with a bit of notation. Let $q(z)$ be an elliptic function with fundamental periods $2\omega_1, 2\omega_3$ and assume, without loss of generality, that $\text{Re}(\omega_1) > 0$, $\text{Re}(\omega_3) \geq 0$, $\text{Im}(\omega_3/\omega_1) > 0$. The fundamental period parallelogram then consists of the points $z = 2\omega_1 s + 2\omega_3 t$, where $0 \leq s, t < 1$.

We introduce

$$(4.10) \qquad e^{i\phi} = \frac{\omega_3}{\omega_1} \left| \frac{\omega_1}{\omega_3} \right|, \ \phi \in (0, \pi),$$

and

$$(4.11) \qquad t_j = \omega_j/|\omega_j|, \ j = 1, 3$$

and define

$$(4.12) \qquad q_j(x) := t_j^2 q(t_j x + z_0), \ j = 1, 3$$

for a $z_0 \in \mathbb{C}$ which we choose in such a way that no pole of $q_j, j = 1, 3$ lies on the real axis. (This is equivalent to the requirement that no pole of q lies on the line through the points z_0 and $z_0 + 2\omega_1$ or on the line through z_0 and $z_0 + 2\omega_3$. Since q has only finitely many poles in the fundamental period parallelogram this can always be achieved.) For such a choice of z_0 we infer that $q_j(x)$ are real-analytic and periodic of period $\Omega_j = 2|\omega_j|, \ j = 1, 3$. Comparing the differential equations

$$(4.13) \qquad \psi''(z) + q(z)\psi(z) = E\psi(z)$$

and

(4.14) $$w''(x) + q_j(x)w(x) = \lambda w(x), \; j = 1, 3,$$

connected by the variable transformation

(4.15) $$z = t_j x + z_0, \; \psi(z) = w(x),$$

one concludes that w is a solution of (4.14) if and only if ψ is a solution of (4.13) with

(4.16) $$\lambda = t_j^2 E, \; j = 1, 3.$$

Next, consider $\tilde{q} \in C^0(\mathbb{R})$ of period $\tilde{\Omega} > 0$ and let $\tilde{c}(\lambda, x), \tilde{s}(\lambda, x)$ be the corresponding fundamental system of solutions of $\tilde{w}'' + \tilde{q}\tilde{w} = \lambda\tilde{w}$ defined by

(4.17) $$\tilde{c}(\lambda, 0) = \tilde{s}'(\lambda, 0) = 1, \quad \tilde{c}'(\lambda, 0) = \tilde{s}(\lambda, 0) = 0.$$

The corresponding Floquet discriminant is now given by

(4.18) $$\tilde{\Delta}(\lambda) = [\tilde{c}(\lambda, \tilde{\Omega}) + \tilde{s}(\lambda, \Omega)]/2$$

and Rouché's theorem then yields

(4.19) $$\tilde{\Delta}(\lambda) = \cos[i\tilde{\Omega}\lambda^{1/2}(1 + O(\lambda^{-1}))]$$

as $|\lambda|$ tends to infinity.

LEMMA 4.5. *Let $\tilde{\lambda}_n$ be a periodic or antiperiodic eigenvalue of $\tilde{q}$. Then there exists an $m \in \mathbb{Z}$ such that*

(4.20) $$\left|\tilde{\lambda}_n + m^2\pi^2\tilde{\Omega}^{-2}\right| \leq \tilde{C}$$

for some $\tilde{C} > 0$ independent of $n \in \mathbb{N}_0$. In particular, all periodic and antiperiodic eigenvalues $\tilde{\lambda}_n$, $n \in \mathbb{N}_0$ of $\tilde{q}$ are contained in a half-strip $\tilde{S}$ given by

(4.21) $$\tilde{S} = \{\lambda \in \mathbb{C}| \, |\mathrm{Im}(\lambda)| \leq \tilde{C}, \; \mathrm{Re}(\lambda) \leq \tilde{M}\}$$

for some $\tilde{M} \in \mathbb{R}$.

In order to apply Lemma 4.5 to q_1 and q_3 we note that according to (4.19),

(4.22) $$\Delta_j(\lambda) = \cos[i\Omega_j\lambda^{1/2}(1 + O(\lambda^{-1}))], \; j = 1, 3$$

as $|\lambda|$ tends to infinity, where, in obvious notation, $\Delta_j(\lambda)$ denotes the discriminant of $q_j(x)$, $j = 1, 3$. Next, denote by $\lambda_{j,n}$ an Ω_j-(anti)periodic eigenvalue of $w'' + q_j w = \lambda w$. Then $E_{j,n} = t_j^{-2}\lambda_{j,n}$ is a $2\omega_j$-(anti)periodic eigenvalue of $\psi'' + q\psi = E\psi$ and vice versa. Hence Lemma 4.5 immediately yields the following result.

LEMMA 4.6. *Let $j = 1$ or 3. Then all $2\omega_j$-(anti)periodic eigenvalues $E_{j,n}$, $n \in \mathbb{N}_0$ associated with q lie in the half-strip S_j given by*

(4.23) $$S_j = \{E \in \mathbb{C} : |\mathrm{Im}(t_j^2 E)| \leq C_j, \; \mathrm{Re}(t_j^2 E) \leq M_j\}$$

for suitable constants $C_j > 0, M_j \in \mathbb{R}$. The angle between the axes of the strips S_1 and S_3 equals $2\phi \in (0, 2\pi)$.

Lemmas 4.5 and 4.6 apply to any elliptic potential whether or not they are algebro-geometric. In our final step we shall now invoke Picard's Theorem 4.1 to obtain our characterization of elliptic algebro-geometric potentials.

THEOREM 4.7. *q is an elliptic algebro-geometric potential if and only if q is a Picard potential (i.e., if and only if for each $E \in \mathbb{C}$ every solution of $\psi''(z) + q(z)\psi(z) = E\psi(z)$ is meromorphic with respect to z).*

PROOF. By Theorem 4.4 it remains to prove that a Picard potential is algebro-geometric. Hence we assume in the following that q is a Picard potential. Since all $2\omega_j$-(anti)periodic eigenvalues $E_{j,n}$ of q yield zeros $\lambda_{j,n} = t_j^2 E_{j,n}$ of the entire functions $\Delta_j(\lambda)^2 - 1$, the $E_{j,n}$ have no finite limit point. Next we choose $R > 0$ large enough such that the exterior of the closed disk $\overline{D(0,R)}$ centered at the origin of radius $R > 0$ contains no intersection of S_1 and S_3 (defined in (4.23)), that is,

$$(4.24) \qquad (\mathbb{C} \backslash \overline{D(0,R)}) \cap (S_1 \cap S_3) = \emptyset.$$

Let $\rho_{j,\pm}(\lambda)$ be the Floquet multipliers of $q_j(x)$, that is, the solutions of

$$(4.25) \qquad \rho_j^2 - 2\Delta_j \rho_j + 1 = 0, \; j = 1, 3.$$

Then (4.24) implies that for $E \in \mathbb{C} \backslash \overline{D(0,R)}$, at most one of the numbers $\rho_1(t_1 E)$ and $\rho_3(t_3 E)$ can be in $\{-1, 1\}$. In particular, at least one of the characteristic equations corresponding to the substitution $z \to z + 2\omega_1$ or $z \to 2\omega_3$ (cf. (4.3) and (4.4)) has two distinct roots. Since by hypothesis q is a Picard potential, Picard's Theorem 4.1 applies and guarantees for all $E \in \mathbb{C} \backslash \overline{D(0,R)}$ the existence of two linearly independent solutions $\psi_1(E, z)$ and $\psi_2(E, z)$ of $\psi'' + q\psi = E\psi$ which are elliptic of the second kind. Then $w_{j,k}(x) = \psi_k(t_j x + z_0)$, $k = 1, 2$ are linearly independent Floquet solutions associated with q_j. Therefore the points λ for which $w'' + q_j w = \lambda w$ has only one Floquet solution are necessarily contained in $\overline{D(0,R)}$ and hence finite in number. This is true for both $j = 1$ and $j = 3$. Applying Theorem 3.4 then proves that both q_1 and q_3 are algebro-geometric potentials.

By (2.8) (in slight abuse of notation)

$$(4.26) \qquad \sum_{k=0}^{g} c_{g-k} \frac{df_{k+1}(q_1(x))}{dx} = 0,$$

where $g \in \mathbb{N}_0$, f_{k+1}, $k = 0, ..., g$, are differential polynomials in q_1 homogeneous of degree $2k + 2$ (cf. (2.9)), and c_k, $k = 0, ..., g$ are complex constants. Since

$$(4.27) \qquad q_1^{(\ell)}(x) = t_1^{\ell+2} q^{(\ell)}(z),$$

(where $z = t_1 x + z_0$) we obtain

$$(4.28) \qquad \sum_{k=0}^{g} c_{g-k} t_1^{2k+3} \frac{df_{k+1}(q(z))}{dz} = 0,$$

that is, q is an algebro-geometric potential as well. A similar argument would have worked using the relationship between q_3 and q. In particular, the order of the operators commuting with $d^2/dz^2 + q(z)$, $d^2/dx^2 + q_1(x)$, and $d^2/dx^2 + q_3(x)$, respectively, is the same in all cases, namely $2g + 1$.   □

We add a series of remarks further illustrating the significance of Theorem 4.7.

REMARK 4.8 (Complementing Picard's theorem). First we note that Theorem 4.7 extends and complements Picard's Theorem 4.1 in the sense that it determines the elliptic functions which satisfy the hypothesis of the theorem precisely as (elliptic) algebro-geometric solutions of the stationary KdV hierarchy.

REMARK 4.9 (Characterizing elliptic algebro-geometric potentials). While an explicit proof of the algebro-geometric property of q is, in general, highly nontrivial (see, e.g., the references cited in connection with special cases such as the Lamé-Ince and Treibich-Verdier potentials in Remark 4.11 below), the fact of whether or not $\psi''(z) + q(z)\psi(z) = E\psi(z)$ has a fundamental system of solutions meromorphic in z for a finite (but sufficiently large) number of values of the spectral parameter $E \in \mathbb{C}$ can be decided by means of an elementary Frobenius-type analysis (see, e.g., [54] and [55]). Theorem 4.7 appears to be the only effective tool to identify general elliptic algebro-geometric solutions of the KdV hierarchy.

REMARK 4.10 (Reduction of abelian integrals). Theorem 4.7 is also relevant in the context of the Weierstrass theory of reduction of abelian to elliptic integrals, a subject that attracted considerable interest, see, for instance, [7], [8], [9], Ch. 7, [10], [11], [21], [36], [37], [38], [67], [70], [74], [77], [104], [105], [110]. In particular, the theta functions corresponding to the hyperelliptic curves derived from the Burchnall-Chaundy polynomials (2.15), associated with Picard potentials, reduce to one-dimensional theta functions.

REMARK 4.11 (Computing genus and branch points). Even though Theorem 4.7 characterizes all elliptic algebro-geometric potentials as Picard potentials, it does not yield an effective way to compute the underlying hyperelliptic curve K_g; in particular, its proof provides no means to compute the branch and singular points nor the (arithmetic) genus g of K_g. To the best of our knowledge K_g has been computed only for Lamé-Ince potentials and certain Treibich-Verdier potentials (see, e.g., [6], [11], [70], [86], [105], [110], [118], [120], [124]). Even the far simpler task of computing g previously had only been achieved in the case of Lamé-Ince potentials (see [65] and [115] for the real and complex-valued case, respectively). In [54], [55], and [56] we have treated these problems for Lamé-Ince, Treibich-Verdier, and reflection symmetric elliptic algebro-geometric potentials, respectively. In particular, in [55] we computed g for all Treibich-Verdier potentials and in [56] we reduced the computation of the branch and singular points of K_g for any reflection symmetric elliptic algebro-geometric potential to the solution of constraint linear algebraic eigenvalue problems. We refrain from reproducing a detailed discussion of this matter here, instead we just recall an example taken from [55] which indicates some of the subtleties involved: Consider the potentials

$$(4.29) \qquad q_4(z) = -20\wp(z - \omega_j) - 12\wp(z - \omega_k),$$

$$(4.30) \qquad \hat{q}_4(z) = -20\wp(z - \omega_j) - 6\wp(z - \omega_k) - 6\wp(z - \omega_\ell),$$

$$(4.31) \qquad q_5(z) = -30\wp(z - \omega_j) - 2\wp(z - \omega_k),$$

$$(4.32) \qquad \hat{q}_5(z) = -12\wp(z - \omega_j) - 12\wp(a - \omega_k) - 6\wp(z - \omega_\ell) - 2\wp(z - \omega_m),$$

where $j, k, \ell, m \in \{1, 2, 3, 4\}$ ($\omega_2 = \omega_1 + \omega_3$, $\omega_4 = 0$) are mutually distinct. Then q_4 and $\hat{q}_4$ correspond to (arithmetic) genus $g = 4$ while q_5 and $\hat{q}_5$ correspond to $g = 5$. However, we emphasize that all four potentials contain precisely 16 summands of the type $-2\wp(x - b_n)$ (cf. the discussion following (1.10)). q_5 and $\hat{q}_5$ are isospectral (i.e., correspond to the same curve K_5) while q_4 and $\hat{q}_4$ are not.

5. Picard's Theorem for First-Order Systems

Having characterized all elliptic algebro-geometric solutions of the KdV hierarchy which are related to the second-order expression $L_2 = d^2/dz^2 + q$, it is natural to try to extend Theorem 4.7 to n-th order expressions L_n (connected with the Gel'fand-Dickey hierarchy). Actually, a more general extension to integrable systems related to general first-order $n \times n$ matrix-valued differential expressions seems very desirable in order to include AKNS systems (see, e.g., [51]) and the matrix hierarchies of integrable equations described in detail, for instance, in [26], Sects. 9, 13–16, [31], [33]. Picard's Theorem 4.1 generalizes in a straightforward manner to first-order systems, that is, pairwise distinct Floquet multipliers in one of the fundamental directions and a meromorphic fundamental system of solutions guarantee the existence of a fundamental system of solutions which are elliptic of the second kind (see (5.3)). Moreover, it is possible to obtain the explicit Floquet-type structure of these solutions (cf. Theorem 5.2).

Denote by $M(n)$ the set of $n \times n$ matrices with entries in $\mathbb{C}$ and consider the linear homogeneous system

$$(5.1) \qquad \Psi'(z) = Q(z)\Psi(z), \quad z \in \mathbb{C},$$

where $Q(z) \in M(n)$ and where the entries of $Q(z)$ are elliptic functions with a common period lattice Λ spanned by $2\omega_1$ and $2\omega_3$ which satisfy the same conditions as before.

Assuming without loss of generality that no pole of $Q(z)$ lies on the line containing the segments $[0, 2\omega_j]$, Floquet theory with respect to these directions yields the existence of fundamental matrices $\Phi_j(z)$ of the type

$$(5.2) \qquad \Phi_j(z) = P_j(z)\exp(zK_j),$$

where $P_j(z)$ is a periodic matrix with period $2\omega_j$ and K_j is a constant matrix. The monodromy matrix is given by $M_j = \exp(2\omega_j K_j)$. We want to establish the existence of a Floquet representation *simultaneously* for both directions $2\omega_1$ and $2\omega_3$. More precisely, we intend to find solutions $\underline{\phi}$ of $\Psi'(z) = Q(z)\Psi(z)$ satisfying

$$(5.3) \qquad \underline{\phi}(z + 2\omega_j) = \rho_j \underline{\phi}, \quad j = 1, 3,$$

where $\rho_j \in \mathbb{C}\backslash\{0\}$. Solutions $\underline{\phi}(z)$ of (5.1) satisfying (5.3) are again called elliptic of the second kind.

Even though Picard did mention certain extensions of his result to first-order systems (see, e.g., [60], p. 248–249), apparently he did not seek a Floquet representation for systems in the elliptic case. The first to study such a representation seems to have been Fedoryuk who proved the following result.

THEOREM 5.1. *([41]) Let $Q(z)$ be an $n \times n$ matrix whose entries are elliptic functions with fundamental periods $2\omega_1$ and $2\omega_3$ and suppose that (5.1) has a single-valued fundamental matrix of solutions. Then (5.1) admits a fundamental matrix $\Phi(z)$ of the type*

$$(5.4) \qquad \Phi(z) = D(z)\exp(zS + \zeta(z)T), \quad z \in \mathbb{C},$$

where $S, T \in M(n)$, $D(z)$ is invertible and doubly periodic and

$$(5.5) \qquad S = \frac{1}{\pi i}\left[2\omega_3\zeta(\omega_1)K_3 - 2\omega_1\zeta(\omega_3)K_1\right], \quad T = -\frac{2\omega_1\omega_3}{\pi i}(K_3 - K_1).$$

Moreover, K_1 and K_3, and hence S and T commute.

Fedoryuk's representation (5.4) has the peculiar feature that it seems to stress an apparent essential singularity structure of solutions at $z = 0$. Indeed, since $\zeta(z)$

has a first-order pole at $z = 0$, the term $\exp\left(\zeta(z)T\right)$ in (5.4) exhibits an essential singularity unless T is nilpotent. Hence, the doubly periodic matrix $D(z)$, in general, will cancel the essential singularity of $\exp\left(\zeta(z)T\right)$ and therefore cannot be meromorphic and hence not elliptic. Thus Fedoryuk's result cannot be considered the natural extension of Picard's Theorem 4.1.

In the remainder of this section we shall focus on an alternative to Theorem 5.1 and describe a generalization of Picard's theorem in the context of first-order systems with elliptic coefficients.

THEOREM 5.2. ([52]) *Let $Q(z)$ be an elliptic $n \times n$ matrix with fundamental periods $2\omega_1$ and $2\omega_3$ and suppose that (5.1) has a meromorphic fundamental matrix $\Psi(z)$ of solutions. Then (5.1) admits a fundamental matrix of the type*

$$(5.6) \qquad \Phi(z) = E(z)\sigma(z)^{-1}\sigma\left(zI_n - \frac{2\omega_1\omega_3}{\pi i}\left(K_3 - K_1\right)\right)$$

$$\times \exp\left\{\frac{z}{\pi i}\left[2\omega_3\zeta\left(\omega_1\right)K_3 - 2\omega_1\zeta\left(\omega_3\right)K_1\right]\right\}, \quad z \in \mathbb{C},$$

where $E(z)$ is an elliptic matrix with periods $2\omega_j$ and K_1, K_3 (and hence $M_j = \exp\left(2\omega_j K_j\right)$, $j = 1,3$) are commuting matrices. Moreover, linearly independent solutions $\underline{\phi}_m(z) \in \mathbb{C}^n$, $1 \leq m \leq n$ of (5.1), that is, column vectors of (5.6), are of the type

$$(5.7) \qquad \underline{\phi}_m(z) = \sum_{k_1=0}^{n_1}\sum_{k_2=0}^{n_2}\underline{e}_{m,k_1,k_2}(z)\exp\left(z\mu_{m,k_1,k_2}\right)z^{k_1}\zeta(z)^{k_2},$$

where the vectors $\underline{e}_{m,k_1,k_2}(z)$ are elliptic, the numbers μ_{m,k_1,k_2} denote the (not necessarily distinct) eigenvalues of

$$(5.8) \qquad (1/\pi i)\left[2\omega_3\zeta\left(2\omega_1/2\right)K_3 - 2\omega_1\zeta\left(2\omega_3/2\right)K_1\right],$$

and, most notably, the upper limits of the sums in (5.7) satisfy

$$(5.9) \qquad n_1 + n_2 \leq n - 1.$$

In particular, there exists at least one solution $\underline{\phi}_{m_0}(z)$ of (5.1) which is elliptic of the second kind, that is, $\underline{\phi}_{m_0}(z)$ is meromorphic on $\mathbb{C}$ and

$$(5.10) \qquad \underline{\phi}_{m_0}(z + 2\omega_j) = \rho_{m_0,j}\underline{\phi}_{m_0}(z), \quad j = 1,3, \quad z \in \mathbb{C}$$

for some $\rho_{m_0,j} = \exp\left(2\omega_j\mu_{m_0,0,0}\right) \in \mathbb{C}\backslash\{0\}$, $j = 1,3$. In addition, if all eigenvalues of M_1 or M_3 are distinct, then there exists a fundamental system of solutions $\{\underline{\phi}_m(z)\}_{1 \leq m \leq n}$ of (5.1) with all $\underline{\phi}_m(z)$ elliptic of the second kind.

For the proof one considers a meromorphic fundamental matrix $\tilde{\Psi}(z)$ of 5.1 and defines

$$(5.11) \qquad E(z) = \tilde{\Psi}(z)\exp\left(-\frac{z}{\pi i}\left[2\omega_3\zeta\left(\omega_1\right)K_3 - 2\omega_1\zeta\left(\omega_3\right)K_1\right]\right)$$

$$\times \sigma\left(zI_n - \frac{2\omega_1\omega_3}{\pi i}\left(K_3 - K_1\right)\right)^{-1}\sigma(z).$$

By hypothesis, $E(z)$ is meromorphic and, applying the addition theorem

$$(5.12) \qquad \sigma(z + 2\omega_j) = -\sigma(z)\exp\left\{2\zeta\left(\omega_j\right)\left[z + \left(\omega_j\right)\right]\right\}, \quad 1 \leq j \leq 3$$

(more precisely, a matrix-valued generalization thereof), one verifies that

(5.13) $$E(z + 2\omega_j) = E(z), \quad j = 1, 3,$$

that is, $E(z)$ is elliptic. The remaining assertions in Theorem 5.2 follow by transforming $(K_3 - K_1)$ and, say, K_1 (separately) into their Jordan normal forms (cf. [**52**]).

REMARK 5.3. If M_1 or M_3 has distinct eigenvalues, one can show that equation (5.1) has a fundamental system of solutions $\underline{\phi}_m(z)$ of the form,

(5.14) $$\underline{\phi}_m(z) = \underline{e}_m(z)\sigma(z)^{-1}\sigma\left(z - \frac{2\omega_1\omega_3}{\pi i}k_{3,1,m}\right)$$
$$\times \exp\left(\frac{2\omega_3 z}{\pi i}\zeta(\omega_1)k_{3,1,m}\right)\exp(zk_{1,m}), \quad 1 \leq m \leq n, \quad z \in \mathbb{C},$$

where $\underline{e}_m(z)$ are elliptic with period lattice Λ and $\{k_{3,1,m}\}_{1 \leq m \leq n}$ and $\{k_{1,m}\}_{1 \leq m \leq n}$ are the eigenvalues of $(K_3 - K_1)$ and K_1, respectively.

REMARK 5.4. In the special scalar case (4.1), the bound (5.9) was proved in [**66**], p. 377–378 for $n = 2$ and stated (without proof) for $n \geq 3$.

6. The Higher-Order Scalar Case

In our final section we consider a differential expression L_n of the form

(6.1) $$L_n y = y^{(n)} + q_{n-2}y^{(n-2)} + \ldots + q_0 y,$$

where initially the coefficients $q_0, \ldots, q_{n-2}$ are continuous complex-valued functions of a real variable periodic with period $\Omega > 0$. Again we denote the (n-dimensional) vector space of solutions of the differential equation $L_n y = Ey$ by $\mathcal{L}(E)$ and the operator which shifts the argument of a function in $\mathcal{L}(E)$ by a period Ω by $T(E)$. As before $T(E)$ and L_n commute which implies that $T(E)$ maps $\mathcal{L}(E)$ to itself. Floquet multipliers, that is, eigenvalues of $T(E)$ are given as zeros of the polynomial

(6.2) $$\mathcal{F}(E, \rho) = (-1)^n\rho^n + (-1)^{n-1}a_1(E)\rho^{n-1} + \ldots - a_{n-1}(E)\rho + 1 = 0,$$

where the functions $a_1, \ldots, a_{n-1}$ are entire. This is obvious after choosing the basis $\phi_1(E, x), \ldots, \phi_n(E, x)$ of $\mathcal{L}(E)$ satisfying the initial conditions $\phi_j^{k-1}(E, x_0) = \delta_{j,k}$.

Note that $\mathcal{F}(E, \cdot)$ has n distinct zeros unless the discriminant of $\mathcal{F}(E, \cdot)$, which is an entire function of E, is equal to zero. Thus, this happens at most at countably many points. Denote by $m_g(E, \rho)$ and $m_f(E, \rho)$ the geometric and algebraic multiplicity, respectively, of the eigenvalue ρ of $T(E)$. Then the number $m_f(E, \rho) - m_g(E, \rho) \in \{0, 1, \ldots, n-1\}$ counts the "missing" Floquet solutions of $L_n y = Ey$ with multiplier ρ. We will be interested in the case where this number is positive only for finitely many points E.

For $\theta \in \mathbb{C}$, consider the operator $H(\theta)$ associated with the differential expression L_n in $L^2([x_0, x_0 + \Omega])$ with domain

(6.3)
$$D(H(\theta)) = \{g \in H^{2,n}([x_0, x_0 + \Omega]) : g^{(k)}(x_0 + \Omega) = e^{i\theta}g^{(k)}(x_0), \, k = 0, \ldots, n-1\},$$

where $H^{p,r}(\cdot)$ denote the usual Sobolev spaces with r distributional derivatives in $L^p(\cdot)$.

$H(\theta)$ has discrete spectrum. In fact, its eigenvalues, which will be called Floquet eigenvalues, are given as the zeros of $\mathcal{F}(\cdot, \rho)$. Moreover, the algebraic multiplicity

$m_a(E, \rho)$ of E as an eigenvalue of $H(\theta)$ is given as the order of E as a zero of $\mathcal{F}(\cdot, \rho)$ (see, e.g., [**57**]).

With any differential expression L_n given by (6.1) with continuous complex-valued periodic coefficients $q_0, ..., q_{n-2}$ of a real variable we associate the corresponding closed operator $H : H^{2,n} \to L^2(\mathbb{R})$, $Hy = L_n y$.

Rofe-Beketov's result [**99**], referred to in Section 3, originally was proved for an n-th order operator. Hence the spectrum $\sigma(H)$ of H equals the conditional stability set $\mathcal{S}(L_n)$ of L_n, that is, the set of all complex numbers E for which the differential equation $L_n y = Ey$ has a nontrivial bounded solution. For E to be in $\mathcal{S}(L_n)$ it is necessary and sufficient that $L_n y = Ey$ has a Floquet multiplier of modulus one. Hence

(6.4) $$\mathcal{S}(L_n) = \{E \in \mathbb{C} : \mathcal{F}(E, e^{it}) = 0 \text{ for some } t \in \mathbb{R}\},$$

where $\mathcal{F}$ is given by (6.2). Since $\mathcal{F}$ is entire in both its variables, it follows that $\sigma(H) = \mathcal{S}(L_n)$ consists of (generally) infinitely many regular analytic arcs (i.e., spectral bands). They end at a point where the arc fails to be regular analytic or extends to infinity. Finite endpoints of the spectral bands are called band edges.

DEFINITION 6.1. (i) H is called a **finite-band operator** if and only if $\sigma(H)$ consists of a finite number of regular analytic arcs.
(ii) L_n is called a **Picard differential expression** if and only if all its coefficients are elliptic functions associated with a common period lattice and if $L_n y = Ey$ has a meromorphic fundamental system (with respect to the independent variable) for any value of the spectral parameter $E \in \mathbb{C}$.

Next, let $\Phi(E, x)$ be the fundamental matrix of $L_n y = Ey$ satisfying the initial condition $\Phi(E, x_0) = I_n$ where I_n is the $n \times n$ identity matrix. The Floquet multipliers of the differential equation $L_n y = Ey$ are then the eigenvalues of the monodromy matrix $\Phi(E, x_0 + \Omega)$.

Our aim is to determine multiplicities of Floquet eigenvalues and multipliers for large values of the spectral parameter E. For large values of E the equation $L_n y = Ey$ can be treated as a perturbation of $y^{(n)} = Ey$. In this case there exist n linearly independent Floquet solutions $\exp(\lambda \sigma_k x)$ with associated Floquet multipliers $\exp(\lambda \sigma_k)$, where λ is such that $\lambda^n = -E$ and the σ_k are the different n-th roots of -1. The characteristic polynomial of the associated monodromy matrix is therefore given by

(6.5) $$\mathcal{F}_0(E, \rho) = (-1)^n \rho^n + (-1)^{n-1} a_{F,1}(E) \rho^{n-1} + ... - a_{F,n-1}(E) \rho + 1,$$

where the $a_{F,j}$ are the elementary symmetric polynomials in the variables $\exp(\lambda \sigma_1)$, ..., $\exp(\lambda \sigma_n)$. Perturbation theory now yields that the coefficients a_j in (6.2) are related to the coefficients $a_{F,j}$ in (6.5) by

(6.6) $$a_j(E) = a_{F,j}(E) + b_j(E), \quad j = 1, ...n - 1,$$

where, for some suitable positive constant M,

(6.7) $$|b_j| \le \frac{M}{|\lambda|} |\exp(\lambda \sigma_{n+1-j}) ... \exp(\lambda \sigma_n)|, \quad j = 1, ..., n - 1$$

having ordered the roots in such a way that $|\exp(\lambda \sigma_j)| \le |\exp(\lambda \sigma_{j+1})|$ for $j = 1, ..., n-1$. This allows one to show that asymptotically, in a certain small disk about

$\exp(\lambda\sigma_k)$, there are either one or two Floquet multpliers of $L_n y = Ey$ depending on whether or not the inequality

(6.8) $|\exp(\lambda\sigma_k) - \exp(\lambda\sigma_j)| \geq \gamma \max\{|\exp(\lambda\sigma_k)|, |\exp(\lambda\sigma_j)|\}$ for all $j \neq k$

holds. From this one may prove the following theorem concerning algebraic mult-plicities of Floquet multipliers.

THEOREM 6.2. ([121]) Let L_n be defined as in (6.1). For every $\varepsilon > 0$ there exists a disk $B(\varepsilon) \subset \mathbb{C}$ with the following two properties.
1. All values of E, where at least two Floquet multipliers of the differential equation $L_n y = Ey$ coincide, lie in $B(\varepsilon)$ or in the cone $\{E : |\mathrm{Im}(E)|/|\mathrm{Re}(E)| \leq \varepsilon\}$.
2. Every degenerate Floquet multiplier outside $B(\varepsilon)$ has multiplicity two.

This result has been obtained earlier by McKean [78] for $n = 3$ and by da Silva Menezes [25] for general n.

Moreover the above observations may be used to obtain information concerning algebraic multiplicities of Floquet eigenvalues.

THEOREM 6.3. ([121]) Let $\theta_0 \in \mathbb{C}$. Then there exists an $R > 0$ such that every eigenvalue E of the Floquet operator $H(\theta_0)$, which satisfies $|E| > R$, has at most algebraic multiplicity two.

Now suppose that $\mathcal{F}(E, \rho) = 0$ and that $E \in \mathcal{S}(L_n)$ is suitably large. Since $1 \leq m_f(E, \rho), m_a(E, \rho) \leq 2$ we have to distinguish four cases and Weierstrass's preparation theorem provides us with the following information:
1. If $m_f(E, \rho) = m_a(E, \rho) = 1$ then one spectral band passes through E.
2. If $m_f(E, \rho) = 2$ and $m_a(E, \rho) = 1$ then two (possibly coinciding) spectral bands end in E.
3. If $m_f(E, \rho) = 1$ and $m_a(E, \rho) = 2$ then two spectral bands intersect in E forming a right angle.
4. If $m_f(E, \rho) = m_a(E, \rho) = 2$ then two (possibly coinciding) spectral bands pass through E.

This shows that a necessary condition for a suitably large E to be a band edge is that $m_a(E, \rho) = 1$ and $m_f(E, \rho) = 2$ for some ρ. In particular, such an E is necessarily a point where strictly less than n linearly independent Floquet solutions exist. In addition, when n is odd then $\sigma(H)$ is ultimately in a cone with the imaginary axis as symmetry axis while the possible band edges (where $m_f(E, \rho) = 2$) are in a cone whose axis is the real axis. Therefore we have the following result.

THEOREM 6.4. ([121]) The operator H associated with the differential expression L_n introduced after (6.3) is a finite-band operator whenever n, the order of L_n, is odd.

Finally we turn to the case where L_n is a Picard differential expression. The principal result of this section, Theorem 6.5 below, then shows that algebraic and geometric multiplicities of Floquet multipliers of $L_n y = Ey$ can be different only when E is one of finitely many points.

THEOREM 6.5. ([121]) Suppose the differential expression L_n is Picard. Then there exist n linearly independent solutions of $L_n y = Ey$ which are elliptic of the second kind for all but finitely many values of the spectral parameter E.

SKETCH OF PROOF. The proof is modeled closely after the one of Theorem 4.7. Again, inside a compact set there can be only a finite number of values of E where Floquet multipliers associated with a fundamental period of the coefficients of L_n are degenerate. On the other hand when $|E|$ becomes large we only have to prove that for one of the fundamental periods of the coefficients of L_n all Floquet multipliers of $L_n y = Ey$ are distinct according to Picard's Theorem 4.1.

Assume that the fundamental periods $2\omega_1$ and $2\omega_3$ are such that the angle ϕ between them is less than π/n and assume that z_0 is such that no singularity of $q_0, ..., q_{n-2}$ lies on the line through z_0 and $z_0 + 2\omega_1$ or on the line through z_0 and $z_0 + 2\omega_3$.

Substituting $w(x) = y(2\omega_1 x + z_0)$ and defining $p_k(x) = (2\omega_1)^{n-k} q_k(2\omega_1 x + z_0)$ transforms $L_n y = Ey$ into

$$(6.9) \qquad w^{(n)} + p_{n-2}(x) w^{(n-2)} + ... + p_0(x) y = (2\omega_1)^n Ew.$$

Therefore, Theorem 6.2 implies that all Floquet multipliers associated with the periods $2\omega_1$ ($2\omega_3$) are pairwise distinct provided the spectral parameter $(2\omega_j)^n E$ lies outside the set S_j, $j = 1, 3$, where

$$(6.10) \qquad S_j = \left\{ z : \left| \frac{\mathrm{Im}(z)}{\mathrm{Re}(z)} \right| \leq \frac{\phi}{3} \right\} \cup \{ z : |z| \leq R_j \},$$

with R_j, $j = 1, 3$ being suitable positive constants.

The two sets S_1 and S_3 do not intersect outside a sufficiently large disk D. Hence, for each value of E outside D, Picard's theorem guarantees the existence of n linearly independent solutions of $L_n y = Ey$ which are elliptic functions of the second kind. $\qquad\square$

In particular, when $|E|$ is large and L_n is Picard, we infer that $m_f(E, \rho) = m_g(E, \rho) \leq m_a(E, \rho)$. Moreover, we have shown earlier that necessarily $m_a(E, \rho) = 1$ and $m_f(E, \rho) = 2$ for band edges E with $|E|$ sufficiently large. Thus there are no band edges with sufficiently large absolute values for Picard expressions. One may also show that at most two bands extend to infinity. Hence we have the following final theorem which, in view of Theorem 6.4, has significance only when n is even.

THEOREM 6.6. ([121]) Let L_n be a Picard differential expression and H the associated operator. Then, if $\sigma(H)$ does not contain closed regular analytic arcs, $\sigma(H)$ consists of finitely many analytic arcs which are regular in their interior.

Acknowledgments. F.G. would like to thank all organizers of the AMS meeting at LSU for creating a most stimulating atmosphere and especially for the extraordinary hospitality extended to him.

References

[1] M. Abramowitz and I. A. Stegun, *Handbook of Mathematical Functions*, Dover, New York, 1972.

[2] H. Airault, H. P. McKean, and J. Moser, *Rational and elliptic solutions of the Korteweg-de Vries equation and a related many-body problem*, Commun. Pure Appl. Math. **30** (1977), 95–148.

[3] N. I. Akhiezer, *On the spectral theory of Lamé's equation*, Istor.-Mat. Issled **23** (1978), 77–86, 357. (Russian).

[4] ———, *Elements of the Theory of Elliptic Functions* , Amer. Math. Soc., Providence, RI, 1990.

[5] S. I. Al'ber, *Investigation of equations of Korteweg-de Vries type by the method of recurrence relations*, J. London Math. Soc. **19** (1979), 467–480. (Russian)

[6] F. M. Arscott, *Periodic Differential Equations*, MacMillan, New York, 1964.

[7] M. V. Babich, A. I. Bobenko, and V. B. Matveev, *Reductions of Riemann theta-functions of genus g to theta-functions of lower genus, and symmetries of algebraic curves*, Sov. Math. Dokl. **28** (1983), 304–308.

[8] ———, *Solutions of nonlinear equations integrable in Jacobi theta functions by the method of the inverse problem, and symmetries of algebraic curves* , Math. USSR Izv. **26** (1986), 479–496.

[9] E. D. Belokolos, A. I. Bobenko, V. Z. Enol'skii, A.R. Its, and V. B. Matveev, *Algebro-Geometric Approach to Nonlinear Integrable Equations*, Springer, Berlin, 1994.

[10] E. D. Belokolos, A. I. Bobenko, V. B. Matveev, and V. Z. Enol'skii, *Algebraic-geometric principles of superposition of finite-zone solutions of integrable non-linear equations*, Russian Math. Surv. **41:2** (1986), 1–49.

[11] E. D. Belokolos and V. Z. Enol'skii, *Verdier elliptic solitons and the Weierstrass theory of reduction*, Funct. Anal. Appl. **23** (1989), 46–47.

[12] ———, *Reduction of theta functions and elliptic finite-gap potentials*, Acta Appl. Math. **36** (1994), 87–117.

[13] D. Bennequin, *Hommage à Jean–Louis Verdier: au jardin des systèmes intégrables* , Integrable Systems: The Verdier Memorial Conference (ed. by O. Babelon, P. Cartier, Y. Kosmann-Schwarzbach), Birkhäuser, Boston, 1993, 1–36.

[14] B. Birnir, *Complex Hill's equation and the complex periodic Korteweg-de Vries equations* , Commun. Pure Appl. Math. **39** (1986), 1–49.

[15] ———, *Singularities of the complex Korteweg-de Vries flows*, Commun. Pure Appl. Math. **39** (1986), 283–305.

[16] J. L. Burchnall and T. W. Chaundy, *Commutative ordinary differential operators*, Proc. London Math. Soc. Ser. 2, **21** (1923), 420–440.

[17] ———, *Commutative ordinary differential operators*, Proc. Roy. Soc. London **A 118** (1928), 557–583.

[18] H. Burkhardt, *Elliptische Funktionen*, Verlag von Veit, Leipzig, 2nd ed., 1906.

[19] F. Calogero, *Exactly solvable one-dimensional many-body problems* , Lett. Nuovo Cim. **13** (1975), 411–416.

[20] D. V. Choodnovsky and G. V. Choodnovsky, *Pole expansions of nonlinear partial differential equations*, Nuovo Cim. **40B** (1977), 339–353.

[21] P. L. Christiansen, J. C. Eilbeck, V. Z. Enolskii, and N. A. Kostov, *Quasi-periodic solutions of the coupled nonlinear Schrödinger equations* , Proc. Roy. Soc. London **A 451** (1995), 685–700.

[22] D. V. Chudnovsky, *Meromorphic solutions of nonlinear partial differential equations and many-particle completely integrable systems*, J. Math. Phys. **20** (1979), 2416–2422.

[23] D. V. Chudnovsky and G. V. Chudnovsky, *Appendix I: Travaux de J. Drach (1919)*, Classical and Quantum Models and Arithmetic Problems (ed. by D. V. Chudnovsky and G. V. Chudnovsky), Marcel Dekker, New York, 1984, 445–453.

[24] E. Colombo, G. P. Pirola, and E. Previato, *Density of elliptic solitons*, J. reine angew. Math. **451** (1994), 161–169.

[25] M. L. Da Silva Menezes, *Infinite genus curves with hyperelliptic ends*, Commun. Pure Appl. Math. **42** (1989), 185–212.

[26] L. A. Dickey, *Soliton Equations and Hamiltonian Systems*, World Scientific, Singapore, 1991.

[27] R. Donagi and E. Markman, *Spectral covers, algebraically completely integrable, Hamiltonian systems, and moduli of bundles*, Integrable Systems and Quantum Groups (ed. by R. Donagi, B. Dubrovin, E. Frenkel, and E. Previato), Lecture Notes in Mathematics **1620**, Springer, Berlin, 1996, 1–119.

[28] J. Drach, *Détermination des cas de réduction de l'équation différentielle $d^2y/dx^2 = [\phi(x) + h]y$*, C. R. Acad. Sci. Paris **168** (1919), 47–50.

[29] ———, *Sur l'intégration par quadratures de l'équation $d^2y/dx^2 = [\phi(x) + h]y$* , C. R. Acad. Sci. Paris **168** (1919), 337–340.

[30] B. A. Dubrovin, *Periodic problems for the Korteweg-de Vries equation in the class of finite band potentials*, Funct. Anal. Appl. **9** (1975), 215–223.

[31] _____, *Completely integrable Hamiltonian systems associated with matrix operators and abelian varieties*, Funct. Anal. Appl. **11** (1977), 265–277.

[32] _____, *Theta functions and non-linear equations*, Russ. Math. Surv. **36:2** (1981), 11–92.

[33] _____, *Matrix finite-zones operators*, Revs. Sci. Technology **23** (1983), 20–50.

[34] B. A. Dubrovin and S. P. Novikov, *Periodic and conditionally periodic analogs of the many-soliton solutions of the Korteweg-de Vries equation*, Sov. Phys. JETP **40** (1975), 1058–1063.

[35] J. C. Eilbeck and V. Z. Enol'skii, *Elliptic Baker-Akhiezer functions and an application to an integrable dynamical system*, J. Math. Phys. **35** (1994), 1192–1201.

[36] V. Z. Enol'skii, *On the solutions in elliptic functions of integrable nonlinear equations*, Phys. Lett. **96A** (1983), 327–330.

[37] _____, *On the two-gap Lamé potentials and elliptic solutions of the Kovalevskaja problem connected with them*, Phys. Lett. **100A** (1984), 463–466.

[38] _____, *On solutions in elliptic functions of integrable nonlinear equations associated with two-zone Lamé potentials*, Soc. Math. Dokl. **30** (1984), 394–397.

[39] V. Z. Enol'skii and J. C. Eilbeck, *On the two-gap locus for the elliptic Calogero-Moser model*, J. Phys. **A 28** (1995), 1069–1088.

[40] V. Z. Enol'skii and N. A. Kostov, *On the geometry of elliptic solitons*, Acta Appl. Math. **36** (1994), 57–86.

[41] M. V. Fedoryuk, *Lamé wave functions in the Jacobi form*, J. Diff. Eqs., **23** (1987), 1170–1177.

[42] G. Floquet, *Sur les équations différentielles linéaires à coefficients doublement périodiques*, C. R. Acad. Sci. Paris **98** (1884), 38–39, 82–85.

[43] _____, *Sur les équations différentielles linéaires à coefficients doublement périodiques* , Ann. l'École Normale Sup. **1** (1884), 181–238.

[44] _____, *Addition a un mémorie sur les équations différentielles linéaires*, Ann. l'École Normale Sup. **1** (1884), 405–408.

[45] M. G. Gasymov, *Spectral analysis of a class of second-order non-self-adjoint differential operators*, Funct. Anal. Appl. **14** (1980), 11–15.

[46] M. G. Gasymov, *Spectral analysis of a class of ordinary differential operators with periodic coefficients*, Sov. Math. Dokl. **21** (1980), 718–721.

[47] L. Gatto and S. Greco, *Algebraic curves and differential equations: an introduction*, The Curves Seminar at Queen's, Vol. VIII (ed. by A. V. Geramita), Queen's Papers Pure Appl. Math. **88**, Queen's Univ., Kingston, Ontario, Canada, 1991, B1–B69.

[48] I. M. Gel'fand and L. A. Dikii, *Asymptotic behaviour of the resolvent of Sturm-Liouville equations and the algebra of the Korteweg-de Vries equations*, Russ. Math. Surv. **30:5**, (1975) 77–113.

[49] _____, *Integrable nonlinear equations and the Liouville theorem* , Funct. Anal. Appl. **13** (1979), 6–15.

[50] F. Gesztesy, *On the Modified Korteweg-deVries Equation*, Differential Equations with Applications in Biology, Physics, and Engineering (ed. by J. A. Goldstein, F. Kappel, and W. Schappacher), Marcel Dekker, New York, 1991, 139–183.

[51] F. Gesztesy and R. Ratneseelan, *An alternative approach to algebro-geometric solutions of the AKNS hierarchy*, Rev. Math. Phys., to appear.

[52] F. Gesztesy and W. Sticka, *On a theorem of Picard*, Proc. AMS **126** (1998), 1089–1099.

[53] F. Gesztesy and R. Weikard, *Spectral deformations and soliton equations*, Differential Equations with Applications to Mathematical Physics (ed. by W. F. Ames, E. M. Harrell II, and J. V. Herod), Academic Press, Boston, 1993, 101–139.

[54] _____, *Lamé potentials and the stationary (m)KdV hierarchy* , Math. Nachr. **176** (1995), 73–91.

[55] _____, *Treibich-Verdier potentials and the stationary (m)KdV hierarchy*, Math. Z. **219** (1995), 451–476.

[56] _____, *On Picard potentials*, Diff. Int. Eqs. **8** (1995), 1453–1476.

[57] _____, *Floquet theory revisited*, Differential Equations and Mathematical Physics (ed. by I. Knowles), International Press, Boston, 1995, 67–84.

[58] _____, *A characterization of elliptic finite-gap potentials*, C. R. Acad. Sci. Paris **321** (1995), 837–841.

[59] _____, *Picard potentials and Hill's equation on a torus*, Acta Math. **176** (1996), 73–107.

[60] J. Gray, *Linear Differential Equations and Group Theory from Riemann to Poincaré*, Birkhäuser, Boston, 1986.

[61] S. Greco and E. Previato, *Spectral curves and ruled surfaces: projective models*, The Curves Seminar at Queen's, Vol. VIII (ed. by A. V. Geramita), Queen's Papers Pure Appl. Math. **88**, Queen's Univ., Kingston, Ontario, Canada, 1991, F1–F33.

[62] V. Guillemin and A. Uribe, *Hardy functions and the inverse spectral method*, Commun. PDE **8** (1983), 1455–1474.

[63] G.-H. Halphen, *Traité des Fonctions Elliptiques*, tome 2, Gauthier-Villars, Paris, 1888.

[64] C. Hermite, *Oeuvres*, tome 3, Gauthier-Villars, Paris, 1912.

[65] E. L. Ince, *Further investigations into the periodic Lamé functions*, Proc. Roy. Soc. Edinburgh **60** (1940), 83–99.

[66] E. L. Ince, *Ordinary Differential Equations*, Dover, New York, 1956.

[67] A. R. Its and V. Z. Enol'skii, *Dynamics of the Calogero-Moser system and the reduction of hyperelliptic integrals to elliptic integrals* , Funct. Anal. Appl. **20** (1986), 62–64.

[68] A. R. Its and V. B. Matveev, *Schrödinger operators with finite-gap spectrum and N-soliton solutions of the Korteweg-de Vries equation*, Theoret. Math. Phys. **23** (1975), 343–355.

[69] F. Klein, *Über den Hermite'schen Fall der Lamé'schen Differentialgleichung*, Math. Ann. **40** (1892), 125–129.

[70] N. A. Kostov and V. Z. Enol'skii, *Spectral characteristics of elliptic solitons*, Math. Notes **53** (1993), 287–293.

[71] M. Krause, *Theorie der doppeltperiodischen Funktionen einer veränderlichen Grösse*, Vol. 2, Teubner, Leipzig, 1897.

[72] I. M. Krichever, *Integration of nonlinear equations by the methods of algebraic geometry*, Funct. Anal. Appl. **11** (1977), 12–26.

[73] _____, *Methods of algebraic geometry in the theory of non-linear equations*, Russ. Math. Surv. **32**:6 (1977), 185–213.

[74] _____, *Elliptic solutions of the Kadomtsev-Petviashvili equation and integrable systems of particles*, Funct. Anal. Appl. **14** (1980), 282–290.

[75] _____, *Nonlinear equations and elliptic curves*, Revs. Sci. Technology **23** (1983), 51–90.

[76] _____, *Elliptic solutions of nonlinear integrable equations and related topics*, Acta Appl. Math. **36** (1994), 7–25.

[77] V. B. Matveev and A. O. Smirnov, *Symmetric reductions of the Riemann θ–function and some of their applications to the Schrödinger and Boussinesq equation*, Amer. Math. Soc. Transl. (2) **157** (1993), 227–237.

[78] H. P. McKean, *Boussinesq's equation on the circle*, Commun. Pure Appl. Math. **34** (1981), 599–691.

[79] H. P. McKean and P. van Moerbeke, *The spectrum of Hill's equation*, Invent. Math. **30** (1975), 217–274.

[80] J. Mertsching, *Quasi periodic solutions of the nonlinear Schrödinger equation*, Fortschr. Phys. **85** (1987), 519–536.

[81] G. Mittag-Leffler, *Sur les équations différentielles linéaires à coefficients doublement périodiques*, C. R. Acad. Sci. Paris **90** (1880), 299-300.

[82] O. I. Mokhov, *Commuting differential operators of rank 3, and nonlinear differential equations*, Math. USSR Izv.,· **35** (1990), 629–655.

[83] J. Moser, *Integrable Hamiltonian systems and spectral theory*, Academia Nationale Dei Lincei, Scuola Normale Superiore, Lezione Fermiani, Pisa 1981, preprint, ETH-Zürich, 1982.

[84] _____, *Three integrable Hamiltonian systems connected with isospectral deformations*, Adv. Math. **16** (1975), 197–220.

[85] S. P. Novikov, *The periodic problem for the Korteweg-de Vries equation*, Funct. Anal. Appl. **8** (1974), 236–246.

[86] S. Novikov, S. V. Manakov, L. P. Pitaevskii, and V. E. Zakharov, *Theory of Solitons*, Consultants Bureau, New York, 1984.

[87] M. Ohmiya, *KdV polynomials and Λ–operator*, Osaka J. Math. **32** (1995), 409–430.

[88] M. A. Olshanetsky and A. M. Perelomov, *Classical integrable finite-dimensional systems related to Lie Algebras*, Phys. Rep. **71** (1981), 313–400.

[89] L. A. Pastur and V. A. Tkachenko, *Spectral theory of Schrödinger operators with periodic complex-valued potentials*, Funct. Anal. Appl. **22** (1988), 156–158.

[90] L. A. Pastur and V. A. Tkachenko, *An inverse problem for a class of one-dimensional Schrödinger operators with a complex periodic potential*, Math. USSR Izv. **37** (1991), 611–629.

[91] ———, *Geometry of the spectrum of the one-dimensional Schrödinger equation with a periodic complex-valued potential*, Math. Notes **50** (1991), 1045–1050.

[92] A. M. Perelomov, *Integrable Systems of Classical Mechanics and Lie Algebras*, Vol. 1, Birkhäuser, Basel, 1990.

[93] E. Picard, *Sur une généralisation des fonctions périodiques et sur certaines équations différentielles linéaires*, C. R. Acad. Sci. Paris **89** (1879), 140–144.

[94] ———, *Sur une classe d'équations différentielles linéaires*, C. R. Acad. Sci. Paris **90** (1880), 128–131.

[95] ———, *Sur les équations différentielles linéaires à coefficients doublement périodiques*, J. reine angew. Math. **90** (1881), 281–302.

[96] E. Previato, *The Calogero-Moser-Krichever system and elliptic Boussineq solitons*, Hamiltonian Systems, Transformation Groups and Spectral Transform Methods (ed. by J. Harnard and J. E. Marsden), CRM, Montréal, 1990, 57–67.

[97] ———, *Monodromy of Boussinesq elliptic operators*, Acta Appl. Math. **36** (1994), 49–55.

[98] ———, *Seventy years of spectral curves*, Integrable Systems and Quantum Groups (ed. by R. Donagi, B. Dubrovin, E. Frenkel, and E. Previato), Lecture Notes in Mathematics **1620**, Springer, Berlin, 1996, 419–481.

[99] F. S. Rofe-Beketov, *The spectrum of non-selfadjoint differential operators with periodic coefficients*, Sov. Math. Dokl. **4** (1963), 1563–1566.

[100] R. Schimming, *An explicit expression for the Korteweg–de Vries hierarchy*, Acta Appl. Math. **39** (1995), 489–505.

[101] I. Schur, *Über vertauschbare lineare Differentialausdrücke*, Sitzungsber. Berliner Math. Gesellsch. **4** (1905), 2–8.

[102] G. Segal and G. Wilson, *Loop groups and equations of KdV type*, Publ. Math. IHES **61** (1985), 5–65.

[103] A. O. Smirnov, *Real finite-gap regular solutions of the Kaup-Boussinesq equation*, Theoret. Math. Phys. **66** (1986), 19–31.

[104] ———, *A matrix analogue of Appell's theorem and reductions of multidimensional Riemann theta functions*, Math. USSR Sb. **61** (1988), 379–388.

[105] ———, *Elliptic solutions of the Korteweg-de Vries equation*, Math. Notes **45** (1989), 476–481.

[106] ———, *Real elliptic solutions of the Sine-Gordon equation*, Math. USSR Sb. **70** (1990), 231–240.

[107] ———, *Finite-gap elliptic solutions of the KdV equation*, Acta Appl. Math. **36** (1994), 125–166.

[108] ———, *Solutions of the KdV equation elliptic in t*, Theoret. Math. Phys. **100** (1994), 937–947.

[109] ———, *Elliptic solutions of the nonlinear Schrödinger equation and the modified Korteweg-de Vries equation*, Russ. Acad. Sci. Sb. Math. **82** (1995), 461–470.

[110] I. A. Taimanov, *Elliptic solutions of nonlinear equations*, Theoret. Math. Phys. **84** (1990), 700–706.

[111] ———, *On the two-gap elliptic potentials*, Acta Appl. Math. **36** (1994), 119–124.

[112] V. A. Tkachenko, *Discriminants and generic spectra of non-selfadjoint Hill's operators*, Spectral operator theory and related topics, Adv. Soviet Math. **19**, Amer. Math. Soc., Providence, RI, 1994, 41–71.

[113] A. Treibich, *Compactified Jacobians of Tangential Covers*, Integrable Systems: The Verdier Memorial Conference (ed. by O. Babelon, P. Cartier, Y. Kosmann-Schwarzbach), Birkhäuser, Boston, 1993, 39–60.

[114] ———, *New elliptic potentials*, Acta Appl. Math. **36** (1994), 27–48.

[115] A. Treibich and J.-L. Verdier, *Solitons elliptiques*, The Grothendieck Festschrift, Vol III (ed. by P. Cartier, L. Illusie, N. M. Katz, G. Laumon, Y. Manin, and K. A. Ribet), Birkhäuser, Basel, 1990, 437–480.

[116] ———, *Revêtements tangentiels et sommes de 4 nombres triangulaires*, C. R. Acad. Sci. Paris **311** (1990), 51–54.

[117] _____, *Revêtements exceptionnels et sommes de 4 nombres triangulaires*, Duke Math. J. **68** (1992), 217–236.

[118] A. V. Turbiner, *Lame equation, sl(2) algebra and isospectral deformations*, J. Phys. **A22** (1989), L1–L3.

[119] J.-L. Verdier, *New elliptic solitons*, Algebraic Analysis (ed. by M. Kashiwara and T. Kawai), Academic Press, Boston, 1988, 901–910.

[120] R. S. Ward, *The Nahm equations, finite-gap potentials and Lamé functions*, J. Phys. **A20** (1987), 2679–2683.

[121] R. Weikard, *Picard operators*, Math. Nachr., to appear.

[122] _____, *On second order linear differential equations with inverse square singularities*, Operator Theory: Advances and Applications, Vol. 102: Differential and Integral Operators (ed. by I. Gohberg, R. Mennicken, and C. Tretter), Birkhäuser Verlag, Basel, 1998, 315–324.

[123] _____, *On Hill's equation with a singular complex-valued potential*, Proc. London Math. Soc. **76** (1998), 603-633.

[124] E. T. Whittaker and G. N. Watson, *A Course of Modern Analysis*, Cambridge Univ. Press, Cambridge, 1986.

[125] G. Wilson, *Commuting flows and conservation laws for Lax equations*, Math. Proc. Camb. Phil. Soc. **86** (1979), 131–143.

[126] G. Wilson, *Algebraic curves and soliton equations*, Geometry Today (ed. by E. Arbarello, C. Procesi, and E. Strickland), Birkhäuser, Boston, 1985, 303–329.

DEPARTMENT OF MATHEMATICS, UNIVERSITY OF MISSOURI, COLUMBIA, MISSOURI 65211, USA
E-mail address: `fritz@math.missouri.edu`

DEPARTMENT OF MATHEMATICS, UNIVERSITY OF ALABAMA AT BIRMINGHAM, BIRMINGHAM, ALABAMA 35294-1170, USA
E-mail address: `rudi@math.uab.edu`

Contemporary Mathematics
Volume **221**, 1999

Regularity of the Value Function for Constrained Optimization Problems

Kevin A. Grasse and Jonathan R. Bar-on

ABSTRACT. We consider a parametrized, constrained optimization problem of the form

$$\text{minimize} \quad f(x, \omega) \quad x \in X \quad (\omega \in \Omega \text{ fixed})$$
$$\text{subject to} \quad g(x, \omega) = 0,$$

where X, Y, Ω are sets and $f : X \times \Omega \to \mathbb{R}$, $g : X \times \Omega \to Y$ are mappings (here it is assumed that Y has a distinguished element which we label as 0). Associated to this problem is the "value function" $\mu : \Omega \to [-\infty, \infty]$ defined by

$$\mu(\omega) = \inf\{f(x, \omega) \mid g(x, \omega) = 0\}.$$

Using techniques from set-valued analysis, and assuming appropriate structure on the underlying sets X, Y, Ω, we obtain several results which guarantee various desired regularity properties (e. g., continuity or differentiability) of the value function. These results are illustrated by the presentation of a detailed example in which sufficient conditions are derived for the continuity and differentiability of the value function of a classical optimal control problem.

1. Introduction

This paper is concerned with parametric families of constrained optimization problems and, specifically, with the regularity (e. g., continuity or differentiability) of the optimal value as a function of the parameter. To be more precise, suppose we are given sets X, Y, Ω and a pair of mappings

$$f : X \times \Omega \to \mathbb{R}, \quad g : X \times \Omega \to Y.$$

Assuming that Y has a distinguished element, which we denote by 0, we can define a family of constrained optimization problems parametrized by elements $\omega \in \Omega$:

$$(\text{COP}\omega) \qquad \begin{aligned} &\text{minimize} \quad f_\omega(x), \quad x \in X, \\ &\text{subject to} \quad g_\omega(x) = 0. \end{aligned} \qquad (\omega \in \Omega \text{ fixed})$$

1991 *Mathematics Subject Classification.* Primary 49K40, 49K15; Secondary 93B35.
This work was supported in part by NSF Grant #ECS–9301196.

Here we have used the notation

$$f_\omega(x) = f(x, \omega), \quad g_\omega(x) = g(x, \omega),$$

to indicate that, when the second argument is fixed, a function of two arguments can be viewed as a function of its first argument only. We will always assume that our optimization problems are posed as minimization problems. This results in no loss of generality by virtue of the standard observation that maximizing a function f is equivalent to minimizing $-f$. Associated to (COPω) is the *value function* $\mu : \Omega \to [-\infty, \infty]$ defined by

$$(1) \qquad\qquad \mu(\omega) = \inf\{f_\omega(x) \mid g_\omega(x) = 0\}.$$

We employ the convention that the infimum is assigned the value $-\infty$ when the set over which it is computed is unbounded below, and $+\infty$ when the set over which it is computed is empty. Note that some authors (e. g., [AC] and [AF]) refer to the value function (1) as the *marginal function*.

We present a few examples to illustrate how the problem (COPω) arises in specific situations.

EXAMPLE 1.1 (minimum distance to an embedded submanifold). Let $g : \mathbb{R}^n \to \mathbb{R}^p$ be a smooth submersion ($p < n$) and let $f : \mathbb{R}^n \times \mathbb{R}^n \to \mathbb{R}$ be defined by $f(x, \omega) = |x - \omega|$ (usual Euclidean norm). Then

$$\mu(\omega) = \inf\{f(x, \omega) \mid g(x) = 0\}$$

is the distance function from ω to $g^{-1}(0)$.

EXAMPLE 1.2 (the multivariable phase margin in control theory). Let $L(s)$ be an $m \times m$ transfer function (i. e., an $m \times m$ matrix of rational functions of $s \in \mathbb{C}$ with real coefficients; we assume that $m > 1$ and $L(s)$ is proper with no poles on the imaginary axis). For a certain subset $\Omega \subseteq \mathbb{R} \cup \{\infty\}$ (called the *gain-crossover region*) one can define

$$f : \mathbb{C}^m \times \Omega \to \mathbb{R}, \quad g : \mathbb{C}^m \times \Omega \to \mathbb{R}^2$$

by

$$f(z, \omega) = z^\dagger \left(L(j\omega)^\dagger + L(j\omega)\right) z$$

$$g(z, \omega) = (z^\dagger z - 1, z^\dagger L(j\omega)^\dagger L(j\omega)z - 1)$$

(the † denotes complex-conjugate transpose). Then for $\omega \in \Omega$ the map

$$\mu(\omega) = \min\{f(z, \omega) \mid g(z, \omega) = 0\}$$

is called the *minimum-phase mapping* of the transfer function $L(s)$. One defines, in turn, the *multivariable phase margin* of $L(s)$ via

$$\mathrm{PM}(L) = \inf\left\{\cos^{-1}\left(-\frac{1}{2}\mu(\omega)\right) \mid \omega \in \Omega\right\}.$$

For more details on multivariable phase margin and its associated constrained minimization problem, we refer the reader to [GB] and the references contained therein.

Before proceeding to the next example, we introduce some notation which will be used both in this example and throughout this paper.

NOTATION 1.3. Given Banach spaces $\mathfrak{X}, \mathfrak{Y}$ we let $\mathcal{L}(\mathfrak{X}, \mathfrak{Y})$ denote the space of bounded linear mappings from $\mathfrak{X}$ into $\mathfrak{Y}$ and for integers $k \geq 2$ we let $\mathcal{L}_s^k(\mathfrak{X}, \mathfrak{Y})$ denote the space of bounded, symmetric, k-multilinear mappings of $\mathfrak{X} \times \cdots \times \mathfrak{X}$ (k factors) into $\mathfrak{Y}$. For convenience $\mathcal{L}_s^1(\mathfrak{X}, \mathfrak{Y})$ will be identified with $\mathcal{L}(\mathfrak{X}, \mathfrak{Y})$. Recall that $\mathcal{L}(\mathfrak{X}, \mathfrak{Y})$ and $\mathcal{L}_s^k(\mathfrak{X}, \mathfrak{Y})$ are Banach spaces under naturally defined norms (see [**D**; p. 107]). When $\mathfrak{X} = \mathfrak{Y}$ we will just write $\mathcal{L}(\mathfrak{X})$ instead of $\mathcal{L}(\mathfrak{X}, \mathfrak{X})$. In the finite dimensional case, $\mathcal{L}(\mathbb{R}^m, \mathbb{R}^n)$ is often identified with the space of $m \times n$ real matrices.

EXAMPLE 1.4 (the value function in optimal control theory). Let

$$K : \mathbb{R}^n \to \mathbb{R},\ L : \mathbb{R}^n \times \mathbb{R}^m \to \mathbb{R},\ A : \mathbb{R}^n \to \mathbb{R}^n,\ B : \mathbb{R}^n \to \mathcal{L}(\mathbb{R}^m, \mathbb{R}^n)$$

be sufficiently nice (e. g., C^∞) mappings, and fix a compact compact subinterval $[a, b]$ of $\mathbb{R}$. For $s \in [a, b]$ and $z \in \mathbb{R}^n$ we consider the optimal control problem

$$V(s, z) = \inf \left\{ K(\phi(b)) + \int_s^b L(\phi(\zeta), u(\zeta))\, d\zeta \right\},$$

subject to the ODE constraint

(ACS) $$\dot{\phi}(t) = A(\phi(t)) + B(\phi(t))u(t)$$

with fixed initial condition

(IC) $$\phi(s) = z,$$

where $u(\cdot)$ ranges over some specified space of "admissible controls." For our purposes it is convenient to take $u \in L^2([a, b], \mathbb{R}^m)$ (here we tacitly assume that the right-hand side of the time-dependent ordinary differential equation (ACS) satisfies appropriate growth conditions that guarantee solutions are defined over the entire interval $[a, b]$). Note that (ACS) is a *nonlinear affine control system* on $\mathbb{R}^n$ with controls taking values in $\mathbb{R}^m$. The expression

(CF) $$f(\phi, u, s, z) = K(\phi(b)) + \int_s^b L(\phi(\zeta), u(\zeta))\, d\zeta$$

is a measure of the cost of the control u on the interval $[s, b]$, and thus we call f the *cost functional* for our optimal control problem. To cast this problem in the form of (COPω) we introduce the (Hilbert) space

$$W^{1,2}([a, b], \mathbb{R}^n) = \{\phi : [a, b] \to \mathbb{R}^n \mid \phi \text{ is absolutely continuous and}$$
$$\dot{\phi} \in L^2([a, b], \mathbb{R}^n)\},$$

whose inner product is defined by

$$\langle \phi_1, \phi_2 \rangle_{W^{1,2}} = \langle \phi_1(a), \phi_2(a) \rangle + \int_a^b \langle \dot{\phi}_1(\zeta), \dot{\phi}_2(\zeta) \rangle\, d\zeta,$$

where the unsubscripted inner product is the standard inner product on $\mathbb{R}^n$. Then setting

$$X = W^{1,2}([a, b], \mathbb{R}^n) \times L^2([a, b], \mathbb{R}^m),\ \Omega = [a, b] \times \mathbb{R}^n,$$

and defining $f : X \times \Omega \to \mathbb{R}$ by

$$f((\phi, u), (s, z)) = K(\phi(b)) + \int_s^b L(\phi(\zeta), u(\zeta))\, d\zeta$$

and $g \colon X \times \Omega \to W^{1,2}([a,b], \mathbb{R}^n)$ by

$$g((\phi, u), (s, z))(t) = \phi(t) - z - \int_s^t [A(\phi(\zeta)) + B(\phi(\zeta))u(\zeta)]\,d\zeta,$$

we see that the value function of the above optimal control problem is given by

$$V(\omega) = \inf \{ f_\omega(x) \mid g_\omega(x) = 0 \}$$

where

$$x = (\phi, u) \in X, \quad \omega = (s, z) \in \Omega.$$

Our objective in this paper is to determine conditions under which desired regularity properties of the value function (1) can be inferred from associated regularity properties of the functional $f : X \times \Omega \to \mathbb{R}$ and the constraint mapping $g \colon X \times \Omega \to Y$. Of course, the assignment of regularity properties, such as continuity or differentiability, to f and g presupposes that the underlying spaces X, Y, Ω have the appropriate structure, such as a topological structure or a Banach-space structure. Our presentation will start in a rather general context in which X, Y, Ω are only assumed to be Hausdorff topological spaces. However, we will gradually impose additional structure on the underlying spaces in order to obtain sufficient conditions for the continuity and differentiability of the value function. As is well known, the value function can fail to be continuous even in very simple situations, as the following example shows.

EXAMPLE 1.5. Let (x_1, x_2) be the usual coordinates in $\mathbb{R}^2$, let ω be a scalar parameter, and consider the constrained optimization problem

$$\text{minimize} \quad x_1$$
$$\text{subject to} \quad x_1^2 + x_2^2 = 1,$$
$$\frac{x_1^2}{1 + \omega^2} + x_2^2 = 1.$$

It is easy to see that

$$\mu(\omega) = \min\{x_1 \mid x_1^2 + x_2^2 = x_1^2/(1 + \omega^2) + x_2^2 = 1\}$$

is given by

$$\mu(\omega) = \begin{cases} 0, & \omega \neq 0, \\ -1, & \omega = 0, \end{cases}$$

so μ is discontinuous at $\omega = 0$.

The above example suggests that one must expect sufficient conditions for the differentiability of the value function to be quite strong, especially if the problem is posed in an infinite dimensional setting as in Example 1.4. The challenge is to find sufficient conditions for continuity or differentiability of the value function that are satisfied in a reasonably large class of interesting examples.

For related work on this general type of problem, we refer the reader to [BC1], [BC2], [BC3], [Le], and [LM].

The remainder of this paper is organized as follows. In Sect. 2 we collect some known facts that are essential for the derivation of our results. Section 3 presents a sufficient condition for the continuity of the value function, while Section 4 contains a sufficient condition for the differentiability of the value function. In Sect. 5 a detailed application of these results is presented dealing with the value function of the optimal control problem introduced in Ex. 1.4.

2. Preliminaries and background results

The study of the regularity properties of the value function of the parametrized family of constrained optimization problems COPω is facilitated through the use of techniques in set-valued analysis. We cite [AC] and [AF] as representative of general references on this topic. Here we only give brief overview of those definitions and results that will be essential for our development.

NOTATION 2.1. For a nonempty set X we let $\mathcal{P}(X)$ denote the set of all subsets of X and we let $\mathcal{P}_0(X)$ denote the set of all nonempty subsets of X.

DEFINITION 2.2. Let X, Ω be nonempty sets. A *set-valued* mapping from Ω to X is a mapping $G: \Omega \to \mathcal{P}(X)$.

The notion of a set-valued mapping allows us to concisely redefine the value function (1) of the problem COPω as follows. Associated to the constraint mapping $g: X \times \Omega \to Y$ is the set-valued mapping $G: \Omega \to \mathcal{P}(X)$ defined by

$$G(\omega) = \{x \in X \mid g_\omega(x) = 0\} = (g_\omega)^{-1}(0)$$

(recall that 0 is some distinguished element of Y). With this notation the value function μ defined in (1) can be redefined as

(2) $$\mu(\omega) = \inf\{f_\omega(x) \mid x \in G(\omega)\}.$$

This formulation offers the advantage that continuity properties of μ can be related to continuity properties of the set-valued mapping G. For completeness, we re-mind the reader of the definitions of the basic continuity properties of a set-valued mapping.

DEFINITION 2.3. Let X, Ω be Hausdorff topological spaces, let $G: \Omega \to \mathcal{P}_0(X)$ be a set-valued mapping, and fix $\omega_0 \in \Omega$.
1. G is *uppersemicontinuous* (usc) at ω_0 if for every open subset V of X for which $G(\omega_0) \subseteq V$ there exists an open neighborhood U of ω_0 in Ω such that $G(\omega) \subseteq V$ for every $\omega \in U$. G is usc on a subset S of Ω if it is usc at every point of S.
2. G is *lowersemicontinuous* (lsc) at ω_0 if for every open subset V of X for which $G(\omega_0) \cap V \neq \emptyset$ there exists an open neighborhood U of ω_0 in Ω such that $G(\omega) \cap V \neq \emptyset$ for every $\omega \in U$. G is lsc on a subset S of Ω if it is lsc at every point of S.

To formulate a sufficient condition for uppersemicontinuity of a set-valued mapping we need another definition.

DEFINITION 2.4. Let X, Ω be Hausdorff topological spaces and let $G: \Omega \to \mathcal{P}_0(X)$ be a set-valued mapping. The *graph* of G is defined by

$$\text{graph}(G) = \{(x, \omega) \in X \times \Omega \mid x \in G(\omega)\} \subseteq X \times \Omega.$$

The following proposition and corollary give useful sufficient conditions for a set-valued mapping to be usc.

PROPOSITION 2.5. *Let X, Ω be Hausdorff topological spaces and let $F, G : \Omega \to \mathcal{P}_0(X)$ be set-valued mappings such that $F(y) \cap G(y) \neq \emptyset$ for every $\omega \in \Omega$. Define a set-valued mapping $F \cap G : \Omega \to \mathcal{P}_0(X)$ by $(F \cap G)(\omega) = F(\omega) \cap G(\omega)$ and suppose that for some $\omega_0 \in \Omega$ we have:*

1. *F is usc at ω_0;*
2. *$F(\omega_0)$ is compact;*
3. *graph(G) is a closed subset of $X \times \Omega$.*

Then $F \cap G$ is usc at ω_0.

PROOF. See [**AC**; Sect. 1.1, Thm. 1]. □

COROLLARY 2.6. *Let X, Y, Ω be Hausdorff topological spaces with X compact, let $g : X \times \Omega \to Y$ be a mapping such that for some $\bar{y} \in Y$ $g^{-1}(\bar{y})$ is closed in $X \times \Omega$ and the following condition holds:*

$$\forall \omega \in \Omega \quad \exists x \in X \text{ such that } g(x, \omega) = \bar{y}.$$

Then the correspondence $\omega \mapsto G(\omega) = \{x \in X \mid g(x, \omega) = \bar{y}\}$ defines set-valued mapping $G : \Omega \to \mathcal{P}_0(X)$; moreover, G is usc and $G(\omega)$ is a nonempty compact subset of X for every $\omega \in \Omega$.

PROOF. It is obvious from the definitions that graph(G) $= g^{-1}(\bar{y})$, so graph(G) is closed in $X \times \Omega$ by assumption. The set-valued mapping $F : \Omega \to \mathcal{P}_0(X)$ defined by $F(\omega) = X$ for every $\omega \in \Omega$ is evidently usc with compact values, since X is assumed to be compact. Moreover, $F \cap G = G$, so G is usc by the Proposition. □

We can now state a relationship between the semicontinuity properties of the set-valued mapping G and the value function μ in (2). Once again, the reader is referred to [**AC**; Sect. 1.2, Thm. 4 and Thm. 5] for the details.

PROPOSITION 2.7. *Let X, Ω be Hausdorff topological spaces, let $f : X \times \Omega \to \mathbb{R}$ be lsc (i. e., for every $a \in \mathbb{R}$ the pre-image $f^{-1}(a, \infty)$ is an open subset of $X \times \Omega$), let $G : \Omega \to \mathcal{P}_0(X)$ be a set-valued mapping, and define $\mu : \Omega \to [-\infty, \infty)$ by*

$$\mu(\omega) = \inf\{f_\omega(x) \mid x \in G(\omega)\}.$$

Suppose that for some $\omega_0 \in \Omega$ the mapping G is usc at ω_0 and the set $G(\omega_0)$ is compact. Then μ is lsc at ω_0.

REMARK 2.8. As would be expected, there is a result dual to Prop. 2.7 which states that μ is usc whenever f is usc and G is lsc (see [**AC**]). We omit the formal statement of this fact since in our results the uppersemicontinuity of μ will be obtained by direct arguments. However, it will be convenient to state explicitly a property of the mapping $g : X \times \Omega \to Y$ that guarantees the lower semicontinuity of the associated set-valued mapping $\omega \mapsto G(\omega) = \{x \in X \mid g(x, \omega) = \bar{y}\}$, for it is this property to which we will appeal in our arguments.

DEFINITION 2.9. Let X, Y, Ω be Hausdorff topological spaces and let $g : X \times \Omega \to Y$ be a mapping. A point $y \in Y$ is called an *implicit-mapping value* of g if for every $(x, \omega) \in g^{-1}(y)$ there exist an open neighborhood V of ω in Ω and a continuous mapping $\alpha : V \to X$ such that $\alpha(\omega) = x$ and $g(\alpha(\nu), \nu) = y$ for every $\nu \in V$.

The existence of an implicit-mapping value appears as one of the hypotheses in our main theorem on the continuity of the value function (see Thm. 3.2 in the next section). In many cases, the existence of implicit-mapping values is established via the implicit function theorem. For reference, we present a statement of the implicit function theorem (Thm. 2.12) that is slightly more general than what we have been able to locate in existing references (e. g., [**D**; p. 270]). This is done to enhance the potential applicability of our continuity theorem (Thm. 3.2) in those cases where the implicit function theorem is used to verify the implicit-mapping value hypothesis. Unfortunately, the proof of Thm. 2.12, as stated here, would require a significant digression from the topic of this paper and thus is omitted. However, the reader will note that, in the special case where $\mathfrak{X}, \mathfrak{Y}$, and Ω are Hilbert spaces and the mapping g is C^1 on W, Thm. 2.12 is an immediate corollary of the standard implicit function theorem. We also note that there are useful situations in which the hypotheses of the standard implicit function theorem are not satisfied, but the existence of implicit mapping values can still be established by direct arguments. One such situation is presented in Thm. 5.3.

NOTATION 2.10. Given Banach spaces $\mathfrak{X}, \mathfrak{Y}$, an open subset $W \subseteq \mathfrak{X}$, and a C^k mapping $F: W \to \mathfrak{Y}$ ($k \in \mathbb{N}$), we let $D^i F: W \to \mathcal{L}_s^i(\mathfrak{X}, \mathfrak{Y})$, $1 \leq i \leq k$, denote the ith Frechet derivative of F. For $x_0, x \in \mathfrak{X}$ and $i \geq 2$ the notation $D^i F(x_0)(x)^i$ will stand for the i-multilinear mapping $D^i F(x_0)$ evaluated on the i-tuple $(x, \ldots, x) \in \mathfrak{X}^i$. If Ω is a set and $F: \mathfrak{X} \times \Omega \to \mathfrak{Y}$ is a mapping with the property that for some fixed $\omega \in \Omega$ the mapping $x \mapsto F(x, \omega)$ is C^k, then we let $(D_1)^i F(x, \omega)$ denote the ith *partial* Frechet derivative of this mapping at the point (x, ω) with respect to its first variable x.

DEFINITION 2.11. Let $\mathfrak{X}, \mathfrak{Y}$ be Banach spaces, let Ω be a Hausdorff topological space, and let $W \subseteq \mathfrak{X} \times \Omega$ be an open set. A mapping $g: W \to \mathfrak{Y}$ is called C^k *in its first variable* ($k \in \mathbb{N}$) if the following conditions hold:
1. g is continuous;
2. for every $\omega \in \Omega$ the mapping $x \mapsto g(x, \omega)$ is C^k Frechet differentiable at every point of its domain;
3. for $1 \leq i \leq k$ the partial derivatives $(D_1)^i g: W \to \mathcal{L}_s^i(\mathfrak{X}, \mathfrak{Y})$ are continuous.

THEOREM 2.12. *Let $\mathfrak{X}, \mathfrak{Y}$ be Banach spaces, let Ω be a Hausdorff topological space, let $W \subseteq \mathfrak{X} \times \Omega$ be open, and let $g: W \to \mathfrak{Y}$ be C^1 in its first variable. Suppose that for some $(x_0, \omega_0) \in W$ the Frechet derivative $D_1 g(x_0, \omega_0): \mathfrak{X} \to \mathfrak{Y}$ is surjective and set $y_0 = g(x_0, \omega_0)$. Then there exist an open neighborhood $U_0 \times V_0$ of (y_0, ω_0) in $\mathfrak{Y} \times \Omega$ and a continuous mapping $\beta: U_0 \times V_0 \to \mathfrak{X}$ such that $\beta(y_0, \omega_0) = x_0$ and $g(\beta(y, \omega), \omega) = y$ for every $(y, \omega) \in U_0 \times V_0$.*

COROLLARY 2.13. *Let $\mathfrak{X}, \mathfrak{Y}$ be Banach spaces, let Ω be a Hausdorff topological space, and let $g: \mathfrak{X} \times \Omega \to \mathfrak{Y}$ be C^1 in its first variable. Suppose that for every $(x, \omega) \in g^{-1}(0)$ the Frechet derivative $D_1 g(x, \omega): \mathfrak{X} \to \mathfrak{Y}$ is surjective. Then $0 \in \mathfrak{Y}$ is an implicit mapping value of g.*

PROOF. For each $(x, \omega) \in g^{-1}(0)$ the Theorem yields an open neighborhood $U \times V$ of $(0, \omega)$ in $\mathfrak{Y} \times \Omega$ and a continuous mapping $\beta: U \times V \to \mathfrak{X}$ satisfying $\beta(0, \omega) = x$ and $g(\beta(y, \nu), \nu) = y$ for every $(y, \nu) \in U \times V$. The mapping $\alpha: V \to \mathfrak{X}$ defined by $\alpha(\nu) = \beta(0, \nu)$ is obviously continuous and satisfies $\alpha(\omega) = x$ and $g(\alpha(\nu), \nu) = 0$ for every $\nu \in V$, so the proof is complete. $\qquad\square$

The final proposition of this section will be used in the proof of Thm. 4.4.

PROPOSITION 2.14. *Let* $3, \Omega$ *be Banach spaces and let* $H : 3 \times \Omega \to 3$ *be a* C^1 *mapping such that for some* $(z_0, \omega_0) \in 3 \times \Omega$ *the partial Frechet derivative* $D_1 H(x_0, \omega_0) : 3 \to 3$ *is a linear homeomorphism of* 3. *Then there exist open neighborhoods* U_0 *of* z_0 *and* V_0 *of* ω_0 *such that for every* $\omega \in V_0$ *the mapping* $z \mapsto H(z, \omega)$ *is a* C^1 *diffeomorphism of* U_0 *onto its image* $H(U_0, \omega)$.

PROOF. Let $\tilde{H} : 3 \times \Omega \to 3 \times \Omega$ be the mapping defined by $\tilde{H}(z, \omega) = (H(z, \omega), \omega)$. Obviously, $\tilde{H}$ is C^1 and the assumption that $D_1 H(z_0, \omega_0)$ is a linear homeomorphism of 3 implies that $D\tilde{H}(z_0, \omega_0)$ is a linear homeomorphism of $3 \times \Omega$. By the inverse function theorem, there exists an open neighborhood $U_0 \times V_0$ of (z_0, ω_0) in $3 \times \Omega$ such that $\tilde{H}$ is a diffeomorhism of $U_0 \times V_0$ onto its image $\tilde{H}(U_0 \times V_0)$. It is clear that U_0 and V_0 satisfy the conclusion of the Proposition. $\square$

3. A sufficient condition for the continuity of the value function

NOTATION/REMARKS 3.1. If $\mathfrak{X}$ is a Banach space, then we let $\mathfrak{X}^*$ denote the dual space of $\mathfrak{X}$ and we let $\mathfrak{X}_{\mathrm{wk}}$ denote the underlying vector space of $\mathfrak{X}$ equipped with the weakest topology for which every element of $\mathfrak{X}^*$ is continuous (i. e., the *weak topology on* $\mathfrak{X}$). Recall that $\mathfrak{X}$ is *reflexive* precisely when the natural embedding of $\mathfrak{X}$ into $\mathfrak{X}^{**}$ is surjective. A well known result from functional analysis states that a Banach space $\mathfrak{X}$ is reflexive if and only if its closed unit ball is compact in the weak topology. Examples of reflexive Banach spaces include Hilbert spaces and the Lebesgue spaces L^p for $1 < p < \infty$. By homogeneity it follows immediately that every closed ball of finite radius in a reflexive Banach space is weakly compact. For $x \in \mathfrak{X}$ and $r \in \mathbb{R}$ positive we let $B(x, r)$ (resp., $\overline{B}(x, r)$) denote the open (resp., closed) ball in $\mathfrak{X}$ centered at x of radius r.

THEOREM 3.2. *Let* $\mathfrak{X}$ *be a reflexive Banach space, let* $\mathfrak{Y}$ *be an arbitrary Banach space, and let* Ω *be a Hausdorff topological space. Let*

$$f : \mathfrak{X} \times \Omega \to \mathbb{R}, \quad g : \mathfrak{X} \times \Omega \to \mathfrak{Y}$$

be mappings and set $\Omega_0 = \{\omega \in \Omega \mid (g_\omega)^{-1}(0) \neq \emptyset\}$. *Suppose that:*
  1. $f : \mathfrak{X} \times \Omega \to \mathbb{R}$ *is usc;*
  2. *for every* $r > 0$ *the restriction* $f|_{\overline{B}(0,r) \times \Omega} : \overline{B}(0, r) \times \Omega \to \mathbb{R}$ *is lsc with respect to the topology that* $\overline{B}(0, r) \times \Omega$ *inherits from* $\mathfrak{X}_{\mathrm{wk}} \times \Omega$;
  3. *for every* $r > 0$ *the set* $g^{-1}(0) \cap (\overline{B}(0, r) \times \Omega)$ *is closed in* $\mathfrak{X}_{\mathrm{wk}} \times \Omega$;
  4. $0 \in \mathfrak{Y}$ *is an implicit-mapping value of* g;
  5. *for every* $\omega \in \Omega_0$ *at least one of the following conditions holds:*
    (C1) *for every real number* $\gamma > 0$ *there exist an open neighborhood* N *of* ω *and a real number* $r > 0$ *such that* $\|x\| \geq r, \nu \in N$, *and* $g(x, \nu) = 0$ *imply* $f(x, \nu) \geq \gamma$;
    (C2) *there exist a real number* $r > 0$ *and an open neighborhood* N *of* ω *such that* $\nu \in N$ *implies* $(g_\nu)^{-1}(0) \subseteq \overline{B}(0, r)$.
Then Ω_0 *is an open subset of* Ω, *for every* $\omega \in \Omega_0$ *the infimum of the set* $\{f_\omega(x) \mid x \in (g_\omega)^{-1}(0)\}$ *is finite and is achieved at some point of* $\mathfrak{X}$, *and the value function*

$$\mu : \Omega_0 \to \mathbb{R}, \quad \mu(\omega) = \min\{f_\omega(x) \mid x \in (g_\omega)^{-1}(0)\}$$

is continuous.

PROOF. We proceed under the assumption that $g^{-1}(0) \neq \emptyset$ since otherwise the theorem holds vacuously. We first show that Ω_0 is an open subset of Ω and the value function

$$\mu : \Omega_0 \to \mathbb{R}, \quad \mu(\omega) = \inf\{f_\omega(x) \mid x \in (g_\omega)^{-1}(0)\}$$

is usc on Ω_0; later in the proof we will show that the infimum is achieved, so that we can replace "inf" by "min" in the definition of μ. Fix $\omega_0 \in \Omega_0$ and let γ be an arbitrary real number such that $\gamma > \mu(\omega_0)$. By definition of $\mu(\omega_0)$ there exists $x_0 \in (g_{\omega_0})^{-1}(0)$ (so that $g(x_0, \omega_0) = 0$) satisfying

$$\mu(\omega_0) \leq f(x_0, \omega_0) < \gamma.$$

Since f is usc on the domain $\mathfrak{X} \times \Omega$ there exist open neighborhoods W_0 of x_0 in $\mathfrak{X}$ and N_0 of ω_0 in Ω such that

(3) $$(x, \omega) \in W_0 \times N_0 \Rightarrow f(x, \omega) < \gamma.$$

Since $g(x_0, \omega_0) = 0$ and $0 \in \mathfrak{Y}$ is an implicit-mapping value of g, there exist an open neighborhood V_0 of ω_0 in Ω and a continuous mapping $\alpha : V_0 \to \mathfrak{X}$ such that $\alpha(\omega_0) = x_0$ and $g(\alpha(\omega), \omega) = 0$ for every $\omega \in V_0$. In particular, for $\omega \in V_0$ we have $\alpha(\omega) \in (g_\omega)^{-1}(0)$, so we conclude that $V_0 \subseteq \Omega_0$. Because $\omega_0 \in \Omega_0$ was arbitrary, this proves that Ω_0 is an open subset of Ω. Moreover, since $\alpha : V_0 \to \mathfrak{X}$ is continuous and satisfies $\alpha(\omega_0) = x_0 \in W_0$, we can shrink the neighborhood V_0 of ω_0 if necessary so as to obtain $\alpha(\omega) \in W_0$ for every $\omega \in V_0$. From this and (3) we obtain

(4)
$$\omega \in V_0 \cap N_0 \Rightarrow (\alpha(\omega), \omega) \in W_0 \times N_0$$
$$\Rightarrow \mu(\omega) \leq f(\alpha(\omega), \omega) < \gamma,$$

whence μ is usc at ω_0.

We next show that, under either of the assumptions (C1) or (C2), the infimum of the set $\{f_{\omega_0}(x) \mid x \in (g_{\omega_0})^{-1}(0)\}$ is finite, the infimum is achieved, and the value function μ is lsc at ω_0. Here ω_0 is the fixed, but arbitrary, element of Ω_0 chosen in the previous paragraph and we continue to use the notation developed in that paragraph. As an intermediate step, we show that there exists $r > 0$ and an open neighborhood N_1 of ω_0 contained in Ω_0 (recall that Ω_0 is open) such that $\omega \in N_1$ implies

(5) $$\inf\{f_\omega(x) \mid x \in (g_\omega)^{-1}(0)\} = \inf\{f_\omega(x) \mid x \in (g_\omega)^{-1}(0) \cap \overline{B}(0, r)\}.$$

If (C2) holds, this is easy to establish. Indeed for the $r > 0$ and open neighborhood N of ω_0 given by (C2), we can set $N_1 = N \cap \Omega_0$ and observe that for every $\omega \in N_1$ we have $(g_\omega)^{-1}(0) \subseteq \overline{B}(0, r)$ so that $(g_\omega)^{-1}(0) = (g_\omega)^{-1}(0) \cap \overline{B}(0, r)$ and equality in (5) is obvious. On the other hand, if (C1) holds, then there exists $r > 0$ and an open neighborhood N of ω_0 such that

(6) $$\|x\| \geq r, \omega \in N \text{ and } g(x, \omega) = 0 \Rightarrow f(x, \omega) \geq \gamma,$$

where $\gamma > \mu(\omega_0)$ is as chosen in the previous paragraph. There is no loss of generality in assuming that $r > \|x_0\|$, where $x_0 \in (g_{\omega_0})^{-1}(0)$ is as in the previous paragraph. Let N_1 be the open neighborhood of ω_0 given by $N_1 = N \cap V_0 \cap N_0$, where V_0 and N_0 are as determined above. Then $N_1 \subseteq \Omega_0$, because $V_0 \subseteq \Omega_0$, and

$$\omega \in N_1 \Rightarrow \alpha(\omega) \in (g_\omega)^{-1}(0) \text{ and } f(\alpha(\omega), \omega) < \gamma$$

(we have used (4)). Furthermore, since $\alpha(\omega_0) = x_0 \in B(0, r)$ and α is continuous, we can shrink N_1 to a smaller open neighborhood of ω_0, if necessary, so as to obtain

$$\omega \in N_1 \Rightarrow \alpha(\omega) \in B(0, r).$$

For $\omega \in N_1$ it is obvious that the left-hand side of (5) does not exceed the right-hand side. If it is the case that for some $\omega \in N_1$ the left-hand side of (5) is strictly less than the right-hand side, then there exists $\bar{x} \in (g_\omega)^{-1}(0)$ such that

$$(7) \qquad f(\bar{x}, \omega) < \inf\{f_\omega(x) \mid x \in (g_\omega)^{-1}(0) \cap \overline{B}(0, r)\} \leq f(\alpha(\omega), \omega) < \gamma.$$

However, the first inequality in (7) forces $\|\bar{x}\| > r$, so (6) yields $f(\bar{x}, \omega) \geq \gamma$, which is obviously incompatible with the last inequality in (7). Thus we must have equality in (5) whenever $\omega \in N_1$.

Since $\mathfrak{X}$ is reflexive, the closed ball $\overline{B}(0, r)$ is a compact subset of $\mathfrak{X}_{\mathrm{wk}}$. It is easy to see that Assumption 3 implies that for every $\omega \in \Omega$ the set $\overline{B}(0, r) \cap (g_\omega)^{-1}(0)$ is a closed subset of $\mathfrak{X}_{\mathrm{wk}}$. We define the mapping

$$\tilde{g} : \overline{B}(0, r) \times N_1 \to \mathfrak{Y}, \quad \tilde{g}(x, \omega) = g(x, \omega),$$

(here $N_1 \subseteq \Omega_0$ is the open neighborhood of ω_0 determined in the previous paragraph) and proceed to show that it satisfies the hypotheses of Cor. 2.6 when its domain $\overline{B}(0, r) \times N_1$ is given the relative topology induced from $\mathfrak{X}_{\mathrm{wk}} \times \Omega$. Note first that

$$(8) \qquad (\tilde{g})^{-1}(0) = \left[g^{-1}(0) \cap \left(\overline{B}(0, r) \times \Omega\right)\right] \cap \left[\overline{B}(0, r) \times N_1\right],$$

so $(\tilde{g})^{-1}(0)$ is closed in $\overline{B}(0, r) \times N_1$ (in the relative topology inherited from $\mathfrak{X}_{\mathrm{wk}} \times \Omega$) since the first bracketed term in (8) is closed in $\mathfrak{X}_{\mathrm{wk}} \times \Omega$ by Assumption 3. Next note that for every $\omega \in N_1$ we have $\tilde{g}(\alpha(\omega), \omega) = g(\alpha(\omega), \omega) = 0$ since $(\alpha(\omega), \omega) \in \overline{B}(0, r) \times N_1$. Thus from Cor. 2.6 we infer that the set-valued mapping

$$G : N_1 \to \mathcal{P}_0(\overline{B}(0, r)), \quad G(\omega) = (\tilde{g}_\omega)^{-1}(0),$$

is usc with compact values when $\overline{B}(0, r)$ is given the relative weak topology inherited from $\mathfrak{X}_{\mathrm{wk}}$. Observe that the set-valued mapping G can be defined equivalently by $G(\omega) = (g_\omega)^{-1}(0) \cap \overline{B}(0, r)$. Thus for $\omega \in N_1$ the equality in (5) yields

$$(9) \qquad \mu(\omega) = \inf\{f_\omega(x) \mid x \in G(\omega)\},$$

and since $f : \overline{B}(0, r) \times N_1 \to \mathbb{R}$ lsc with respect to the topology that $\overline{B}(0, r) \times N_1$ inherits from $\mathfrak{X}_{\mathrm{wk}} \times \Omega$, we infer from Prop. 2.7 that μ is lsc on N_1. In particular μ is lsc at ω_0. Moreover, the infimum in (9) is achieved because $f_{\omega_0} : \overline{B}(0, r) \to \mathbb{R}$ is weakly lsc and $G(\omega_0)$ is a weakly compact subset of $\overline{B}(0, r)$. Since ω_0 was an arbitrary element of Ω_0, the proof is complete. □

REMARK 3.3. Assumption 3.2.4 holds if $g : \mathfrak{X} \times \Omega \to \mathfrak{Y}$ is C^1 in its first variable and if for every $(x, \omega) \in g^{-1}(0)$ the Frechet derivative $D_1 g(x, \omega) : \mathfrak{X} \to \mathfrak{Y}$ is surjective. This is an immediate consequence of Cor. 2.13.

There is an interesting special case of Thm. 3.2 involving an unconstrained parametrized optimization problem, which we state as a corollary.

COROLLARY 3.4. *Let $\mathfrak{X}$ be a reflexive Banach space, let Ω be a Hausdorff topological space, and let $f:\mathfrak{X} \times \Omega \to \mathbb{R}$ be a mapping such that:*

1. *$f:\mathfrak{X} \times \Omega \to \mathbb{R}$ is usc;*
2. *for every $r > 0$ the restriction $f|_{\overline{B}(0,r)\times\Omega}:\overline{B}(0,r)\times\Omega \to \mathbb{R}$ is lsc with respect to the topology that $\overline{B}(0,r) \times \Omega$ inherits from $\mathfrak{X}_{\mathrm{wk}} \times \Omega$;*
3. *for every $\omega \in \Omega$ and for every real number $\gamma > 0$ there exist an open neighborhood N of ω and a real number $r > 0$ such that $\|x\| \geq r$ and $\nu \in N$ imply $f(x,\nu) \geq \gamma$.*

Then for every $\omega \in \Omega$ the infimum of the set $\{f_\omega(x) \mid x \in \mathfrak{X}\}$ is finite and is achieved at some point of $\mathfrak{X}$ and the value function

$$\mu:\Omega \to \mathbb{R}, \quad \mu(\omega) = \min\{f_\omega(x) \mid x \in \mathfrak{X}\}$$

is continuous.

PROOF. This follows immediately from Thm. 3.2 by taking $\mathfrak{Y}$ to be the (zero-dimensional) space $\mathfrak{Y} = \{0\}$ and letting $g:\mathfrak{X} \times \Omega \to \mathfrak{Y}$ be the zero map. $\square$

REMARK 3.5. It is well known that when $\mathfrak{X}$ is infinite dimensional $\mathfrak{X}_{\mathrm{wk}}$ is not first-countable as a topological space with the weak topology. This may cause some difficulty in verifying Assumption 3.2.3 since ostensibly one cannot appeal to sequential arguments to verify this assumption; i. e., in general one cannot infer closedness of $g^{-1}(0) \cap \overline{B}(0,r)$ by establishing the implication that if $\{(x_n,\omega_n) \mid n \in \mathbb{N}\}$ is a sequence in $\overline{B}(0,r) \times \Omega$ satisfying $g(x_n,\omega_n) = 0 \ \forall n \in \mathbb{N}$ and $(x_n,\omega_n) \to (\bar{x},\bar{\omega})$, then $g(\bar{x},\bar{\omega}) = 0$. Rather one would expect to have to use nets instead of sequences. However, there is one useful situation where sequences do suffice.

COROLLARY 3.6. *The conclusion of Theorem 3.2 continues to hold if we replace Assumptions 3.2.2 and 3.2.3, respectively, by*

3.2.2′ *for every $r > 0$ the restriction $f|_{\overline{B}(0,r)\times\Omega}:\overline{B}(0,r)\times\Omega \to \mathbb{R}$ is sequentially lsc with respect to the topology that $\overline{B}(0,r) \times \Omega$ inherits from $\mathfrak{X}_{\mathrm{wk}} \times \Omega$;*

3.2.3′ *for every $r > 0$ the set $g^{-1}(0) \cap \left(\overline{B}(0,r) \times \Omega\right)$ is sequentially closed in $\mathfrak{X}_{\mathrm{wk}} \times \Omega$,*

and if we assume that the reflexive Banach space $\mathfrak{X}$ is separable and the topological space Ω is first countable.

PROOF. We will show that condition 3.2.2′ (resp., 3.2.3′) implies condition 3.2.2 (resp., 3.2.3) under the additional assumptions that $\mathfrak{X}$ is separable and Ω is first countable. Since $\overline{B}(0,r)\times\Omega$ is closed in $\mathfrak{X}_{\mathrm{wk}}\times\Omega$ (due to the fact that $\overline{B}(0,r) \subseteq \mathfrak{X}_{\mathrm{wk}}$ is compact), to obtain 3.2.3 from 3.2.3′ it suffices to show that $g^{-1}(0)\cap\left(\overline{B}(0,r)\times\Omega\right)$ is closed in $\overline{B}(0,r) \times \Omega$ whenever it is sequentially closed. However, because $\overline{B}(0,r)$ is a weakly compact subset of $\mathfrak{X}_{\mathrm{wk}}$ (due to the reflexivity of $\mathfrak{X}$) and $\mathfrak{X}$ is separable, we can infer from a well known fact in functional analysis (see [**DS**; p. 434]) that the weak topology on $\overline{B}(0,r)$ is metrizable. In particular, $\overline{B}(0,r)$ with the induced weak topology is first countable. Consequently, the product space $\overline{B}(0,r) \times \Omega$ is first countable, and we conclude that a subset of $\overline{B}(0,r) \times \Omega$ is closed if and only if it is sequentially closed. Likewise, the first countability of $\overline{B}(0,r) \times \Omega$ (in the topology induced from $\mathfrak{X}_{\mathrm{wk}} \times \Omega$) implies that a mapping defined on $\overline{B}(0,r) \times \Omega$ is lsc if and only if it is sequentially lsc. This proves the Corollary. $\square$

4. A sufficient condition for the differentiability of the value function

The statement of our sufficient condition for the differentiability of the value function will require some preliminary discussion to properly motivate its hypotheses. Consider, for the moment, the (unparametrized) constrained optimization problem

$$\text{(COP)} \qquad \begin{aligned} &\text{minimize} \quad f(x) \\ &\text{subject to} \quad g(x) = 0 \end{aligned}$$

where $\mathfrak{X}, \mathfrak{Y}$ are Banach spaces, and the mappings $f : \mathfrak{X} \to \mathbb{R}$ and $g : \mathfrak{X} \to \mathfrak{Y}$ are at least C^1. Then the well known Lagrange multiplier theorem states that if $x_0 \in g^{-1}(0)$ is a *regular point* of g (by which we mean that the map $Dg(x_0) : \mathfrak{X} \to \mathfrak{Y}$ is surjective) which satisfies

$$(10) \qquad f(x_0) = \min\{f(x) \mid x \in g^{-1}(0)\},$$

then there exists $\lambda \in \mathfrak{Y}^*$ such that the following first-order condition holds:

$$\text{(FO)} \qquad Df(x_0) + \lambda \circ Dg(x_0) = 0.$$

In addition, if f and g are C^2, then it is also necessary that the following second-order condition holds:

$$\text{(SO)} \qquad \begin{aligned} &v \in \mathfrak{X} \text{ and } Dg(x_0)v = 0 \\ &\Rightarrow \big(D^2 f(x_0) + \lambda \circ D^2 g(x_0)\big)(v, v) \geq 0. \end{aligned}$$

DEFINITION 4.1. We call a point $x_0 \in g^{-1}(0)$ satisfying (10) a *nondegenerate minimum point* of (COP) if the maps f and g in (COP) are C^2, x_0 is a regular point of g, so that (FO) holds for some $\lambda \in \mathfrak{Y}^*$, and if the following strengthened version of (SO) holds:

$$\text{(SSO)} \qquad \begin{aligned} &\exists \, \sigma > 0 \text{ such that } v \in \mathfrak{X} \text{ and } Dg(x_0)v = 0 \\ &\Rightarrow \big(D^2 f(x_0) + \lambda \circ D^2 g(x_0)\big)(v, v) \geq \sigma \|v\|^2. \end{aligned}$$

REMARK 4.2. In an infinite-dimensional setting, the conditions (FO) and (SO) are somewhat easier to apply in Hilbert spaces than in arbitrary Banach spaces because, when $\mathfrak{Y}$ (the target space of the constraint mapping) is a Hilbert space, the "Lagrange multiplier" λ can be viewed to be an element of $\mathfrak{Y}$ instead of its dual $\mathfrak{Y}^*$ via the natural identification of $\mathfrak{Y}$ with $\mathfrak{Y}^*$ afforded by the Riesz representation theorem. This identification eliminates the need to obtain a precise characterization of $\mathfrak{Y}^*$. In order to facilitate the proof of our sufficient condition for the differentiability of the value function, it will be convenient to introduce some notation that formalizes the identification of a Hilbert space with its dual space. We do this in the next paragraph.

NOTATION/DEFINITIONS 4.3. Let $\mathfrak{X}$ be a (real) Hilbert space with inner product $\langle \cdot, \cdot \rangle$. For $x \in \mathfrak{X}$ we let $x^\tau \in \mathfrak{X}^*$ denote the continuous linear functional defined by $x^\tau(z) = \langle z, x \rangle$ ($z \in \mathfrak{X}$). Similarly (and with some abuse of the notation) for $\lambda \in \mathfrak{X}^*$ we let $\lambda^\tau \in \mathfrak{X}$ denote the unique element satisfying $\lambda(z) = \langle z, \lambda^\tau \rangle$ for every $z \in \mathfrak{X}$; the Riesz representation theorem assures the existence and uniqueness of λ^τ. Furthermore, both of the correspondences $x \mapsto x^\tau$ of $\mathfrak{X} \to \mathfrak{X}^*$ and $\lambda \mapsto \lambda^\tau$ of $\mathfrak{X}^* \to \mathfrak{X}$ are (real) linear isometries. We recall that if $\mathfrak{X}, \mathfrak{Y}$ are Hilbert spaces, then every continuous linear mapping $T \in \mathcal{L}(\mathfrak{X}, \mathfrak{Y})$ determines a continuous linear mapping $T^{\tau} \in \mathcal{L}(\mathfrak{Y}, \mathfrak{X})$ characterized by the relation

$$\langle Tx, y \rangle = \langle x, T^{\tau}y \rangle \quad \forall\, x \in \mathfrak{X},\ \forall\, y \in \mathfrak{Y};$$

we refer to T^{τ} as the *Hilbert-space adjoint* of T. A continuous linear operator $T \in \mathcal{L}(\mathfrak{X})$ is called *symmetric* (or *self-adjoint*) precisely when $T = T^{\tau}$. If $f : \mathfrak{X} \to \mathbb{R}$ is Frechet differentiable at $x \in \mathfrak{X}$, then the Frechet derivative $Df(x)$ is, by definition, an element of $\mathfrak{X}^*$. We define the *gradient* of f at x by $\nabla f(x) = (Df(x))^{\tau} \in \mathfrak{X}$. If $f : \mathfrak{X} \to \mathbb{R}$ is twice differentiable at $x \in \mathfrak{X}$, then the second derivative $D^2 f(x) : \mathfrak{X} \times \mathfrak{X} \to \mathbb{R}$ is a continuous, symmetric, bilinear mapping. Standard Hilbert space results (see, e. g., [**La**; pp. 158–159]) show that there exists a unique $\nabla^2 f(x) \in \mathcal{L}(X)$, called the *Hessian* of f at x, such that

$$D^2 f(x)(z, w) = \langle z, \nabla^2 f(x)w \rangle \quad \forall\, z, w \in \mathfrak{X}.$$

Note that $\nabla^2 f(x)$ is symmetric as an operator because $D^2 f(x)$ is symmetric as a bilinear mapping. Furthermore, it is easy to see that $D(\nabla f)(x) = \nabla^2 f(x)$.

THEOREM 4.4. *Let $\mathfrak{X}, \mathfrak{Y}$ be Hilbert spaces, let Ω be an open subset of a Banach space, and let*

$$f : \mathfrak{X} \times \Omega \to \mathbb{R}, \quad and \quad g : \mathfrak{X} \times \Omega \to \mathfrak{Y},$$

be C^2 mappings. Assume that $g^{-1}(0) \neq \emptyset$, set $\Omega_0 = \{\omega \in \Omega \mid (g_\omega)^{-1}(0) \neq \emptyset\}$, and assume that for every $\omega \in \Omega_0$

$$\inf\{f_\omega(x) \mid x \in (g_\omega)^{-1}(0)\}$$

is finite and is achieved at a point of $(g_\omega)^{-1}(0)$ and the value function $\mu : \Omega_0 \to \mathbb{R}$ given by

$$\mu(\omega) = \min\{f_\omega(x) \mid x \in \mathfrak{X} \text{ and } g_\omega(x) = 0\}$$

is continuous. Further assume that for every $(x, \omega) \in g^{-1}(0)$ the partial Frechet derivative $D_1 g(x, \omega) : \mathfrak{X} \to \mathfrak{Y}$ is surjective (which implies Ω_0 is open in Ω) and assume that for some $\omega_0 \in \Omega_0$ the following conditions hold:

1. *there exists a unique $x_0 \in (g_{\omega_0})^{-1}(0)$ such that $\mu(x_0) = f(x_0, \omega_0)$;*
2. *x_0 is a nondegenerate minimum point for the constrained minimization problem at ω_0;*
3. *if $\{\omega_k \mid k \in \mathbb{N}\} \subseteq \Omega$ is a sequence such that $\omega_k \to \omega_0$ and if $(x_k) \subset \mathfrak{X}$ is such that $g(x_k, \omega_k) = 0$ and $\mu(\omega_k) = f(x_k, \omega_k)$, then (x_k) has a strongly convergent subsequence.*

Then there exists a $\delta > 0$ such that μ is C^1 for $\omega \in B(\omega_0, \delta)$.

PROOF. Let $F : \mathfrak{X} \times \mathfrak{Y} \times \Omega \to \mathfrak{X} \times \mathfrak{Y}$ be the mapping

(11) $$F(x, y, \omega) = (\nabla_1 f(x, \omega) + D_1 g(x, \omega)^{\tau} y, g(x, \omega));$$

here $\nabla_1 f(x, \omega) = (D_1 f(x, \omega))^{\tau}$ and we recall that the subscript 1 is used to indicate the partial Frechet derivative computed with respect to the first variable while the second variable is held fixed. By assumption, we have

$$f(x_0, \omega_0) = \mu(\omega_0) = \min\{f_{\omega_0}(x) \mid x \in (g_{\omega_0})^{-1}(0)\},$$

and, since x_0 is a regular point of g_{ω_0}, the Lagrange multiplier theorem yields a $\lambda_0 \in \mathfrak{Y}^*$ satisfying

(12) $$D_1 f(x_0, \omega_0) + \lambda_0 \circ D_1 g(x_0, \omega_0) = 0 \in \mathfrak{X}^*.$$

Through the identification of a Hilbert space with its dual, as discussed above, we can rewrite (12) in the equivalent form

$$(13) \qquad \nabla_1 f(x_0, \omega_0) + D_1 g(x_0, \omega_0)^\tau (\lambda_0)^\tau = 0 \in \mathfrak{X}.$$

In terms of the mapping F defined in (11), equation (13) and the constraint relation $g(x_0, \omega_0) = 0$ yield

$$F(x_0, (\lambda_0)^\tau, \omega_0) = (0, 0).$$

Since f, g are assumed to be C^2, it is clear that F is C^1. For fixed $\omega \in \Omega$ we let

$$\overline{D}F(x, y, \omega) \colon \mathfrak{X} \times \mathfrak{Y} \to \mathfrak{X} \times \mathfrak{Y}$$

denote the Frechet derivative of the mapping $(x, y) \mapsto F(x, y, \omega)$ (ω fixed). A quick check of the definitions shows that for $(x, \omega) \in \mathfrak{X} \times \Omega$ we have

$$\nabla_1 f(x, \omega) + D_1 g(x, \omega)^\tau (\lambda)^\tau = \nabla_1 (f + \lambda \circ g)(x, \omega).$$

From this and the relation

$$D(\nabla_1 (f + \lambda \circ g))(x, \omega) = (\nabla_1)^2 (f + \lambda \circ g)(x, \omega)$$

one verifies that the Frechet derivative $\overline{D}F(x, y, \omega) \in \mathcal{L}(\mathfrak{X} \times \mathfrak{Y})$ is given by the formula

$$\overline{D}F(x, y, \omega)(h, k) = \big((\nabla_1)^2 (f + \lambda \circ g)(x, \omega)h + D_1 g(x, \omega)^\tau k, D_1 g(x, \omega)h \big).$$

Since x_0 is by assumption a nondegenerate minimum point of the constrained minimization problem at ω_0, we have that $D_1 g(x_0, \omega_0) \colon \mathfrak{X} \to \mathfrak{Y}$ is surjective and for some $\sigma > 0$

$$v \in \mathfrak{X} \text{ and } D_1 g(x_0, \omega_0)v = 0$$
$$\Rightarrow \big((D_1)^2 f(x_0, \omega_0) + \lambda \circ (D_1)^2 g(x_0, \omega_0) \big)(v, v) \geq \sigma \|v\|^2$$
$$\Rightarrow \langle v, (\nabla_1)^2 (f + \lambda_0 \circ g)(x_0, \omega_0)v \rangle \geq \sigma \|v\|^2.$$

It follows from Thm. A.1 in the Appendix that $\overline{D}F(x_0, (\lambda_0)^\tau, \omega_0)$ is a linear homeomorphism of $\mathfrak{X} \times \mathfrak{Y}$ with itself. Consequently, by Prop. 2.14 there exists $\delta > 0$ such that for every $\omega \in B(\omega_0, \delta)$ the mapping $(x, y) \mapsto F(x, y, \omega)$ is injective on the set $B(x_0, \delta) \times B((\lambda_0)^\tau, \delta)$. We may assume $\delta > 0$ is small enough to ensure that $B(\omega_0, \delta) \subseteq \Omega_0$.

Because $F(x_0, (\lambda_0)^\tau, \omega_0) = (0, 0)$ and $\overline{D}F(x_0, (\lambda_0)^\tau, \omega_0)$ is a linear homeomorphism of $\mathfrak{X} \times \mathfrak{Y}$ with itself, the implicit function theorem yields a $\zeta > 0$ and C^1 mappings

$$\alpha \colon B(\omega_0, \zeta) \to \mathfrak{X}, \quad \beta \colon B(\omega_0, \zeta) \to \mathfrak{Y}$$

such that $\alpha(\omega_0) = x_0$, $\beta(\omega_0) = (\lambda_0)^\tau$, and

$$(14) \qquad F(\alpha(\omega), \beta(\omega), \omega) = (0, 0) \quad \forall \omega \in B(\omega_0, \zeta).$$

There is no loss of generality in assuming that $\zeta > 0$ is sufficiently small that $\zeta \leq \delta$ and

$$\alpha(B(\omega_0, \zeta)) \subseteq \overline{B}(x_0, \delta/2), \quad \beta(B(\omega_0, \zeta)) \subseteq \overline{B}((\lambda_0)^\tau, \delta/2).$$

CLAIM. There exists $\eta \in (0, \zeta)$ such that the value function μ satisfies

$$\omega \in B(\omega_0, \eta) \Rightarrow \mu(\omega) = f(\alpha(\omega), \omega).$$

Observe that the claim immediately implies the desired conclusion of the Theorem since both α and f are known to be C^1.

PROOF OF CLAIM. The proof is by contradiction. If the assertion of the claim is false, then there exists a sequence $\{\omega_k | k \in \mathbb{N}\} \subseteq B(\omega_0, \zeta)$ such that $\omega_k \to \omega_0$ and

(15) $$\mu(\omega_k) \neq f(\alpha(\omega_k), \omega_k).$$

By (14) for every $k \in \mathbb{N}$ we have $F(\alpha(\omega_k), \beta(\omega_k), \omega_k) = (0, 0)$, so in particular $g(\alpha(\omega_k), \omega_k) = 0$. The definition of μ yields for every $k \in \mathbb{N}$

$$\mu(\omega_k) = \min\{f(x, \omega_k) \mid g(x, \omega_k) = 0\} \leq f(\alpha(\omega_k), \omega_k),$$

so from this and (15) we obtain

(16) $$\mu(\omega_k) < f(\alpha(\omega_k), \omega_k) \quad \forall k \in \mathbb{N}.$$

For every $k \in \mathbb{N}$ choose $x_k \in \mathfrak{X}$ so that

$$\mu(\omega_k) = f(x_k, \omega_k) \quad \text{and} \quad g(x_k, \omega_k) = 0$$

(recall that the minimum is assumed to be achieved). Since $D_1 g(x_k, \omega_k) : \mathfrak{X} \to \mathfrak{Y}$ is surjective by assumption, the Lagrange multiplier theorem yields $\lambda_k \in \mathfrak{Y}^*$ such that

$$D_1 f(x_k, \omega_k) + \lambda_k \circ D_1 g(x_k, \omega_k) = 0, \quad g(x_k, \omega_k) = 0,$$

from which we obtain

$$F(x_k, (\lambda_k)^\tau, \omega_k) = (0, 0) \quad \forall k \in \mathbb{N}.$$

Observe that we must have

(17) $$k \in \mathbb{N} \Rightarrow (x_k, (\lambda_k)^\tau) \notin \overline{B}(x_0, \delta/2) \times \overline{B}((\lambda_0)^\tau, \delta/2),$$

since if (17) is violated for some $k \in \mathbb{N}$ the equality

$$F(x_k, (\lambda_k)^\tau, \omega_k) = 0 = F(\alpha(\omega_k), \beta(\omega_k), \omega_k)$$

would force

$$x_k = \alpha(\omega_k), \quad (\lambda_k)^\tau = \beta(\omega_k),$$

because $(x, y) \mapsto F(x, y, \omega_k)$ is injective on the set $\overline{B}(x_0, \delta/2) \times \overline{B}((\lambda_0)^\tau, \delta/2)$. But the choice of x_k yields

$$\mu(\omega_k) = f(x_k, \omega_k) = f(\alpha(\omega_k), \omega_k),$$

which contradicts (16). Thus (17) holds for all $k \in \mathbb{N}$.

By Assumption 3 $\{x_k | k \in \mathbb{N}\}$ has a strongly convergent subsequence $\{x_{k_q} | q \in \mathbb{N}\}$, so that $x_{k_q} \to \bar{x}$ as $q \to \infty$. For every $q \in \mathbb{N}$ the relation

(18) $$D_1 f(x_{k_q}, \omega_{k_q}) + \lambda_{k_q} \circ D_1 g(x_{k_q}, \omega_{k_q}) = 0$$

implies that

$$\lambda_{k_q} = -D_1 f(x_{k_q}, \omega_{k_q}) \circ D_1 g(x_{k_q}, \omega_{k_q})^\tau \circ \left(D_1 g(x_{k_q}, \omega_{k_q}) \circ D_1 g(x_{k_q}, \omega_{k_q})^\tau\right)^{-1}$$

(recall that the surjectivity of $D_1g(x,\omega):\mathfrak{X} \to \mathfrak{Y}$ implies that $D_1g(x,\omega) \circ D_1g(x,\omega)^\tau$ is a linear homeomorphism of $\mathfrak{Y}$). Since $x_{k_q} \to \bar{x}$ and $\omega_{k_q} \to \omega_0$, the continuity of the partial derivatives D_1f and D_1g allows us to infer from the previous displayed equation that λ_{k_q} converges in $\mathfrak{Y}^*$; i. e., there exists $\bar{\lambda} \in \mathfrak{Y}^*$ such that $\lambda_{k_q} \to \bar{\lambda}$. Thus we have

$$(x_{k_q}, (\lambda_{k_q})^\tau) \to (\bar{x}, (\bar{\lambda})^\tau),$$

and (17) implies

(19) $$(\bar{x}, (\bar{\lambda})^\tau) \notin B(x_0, \delta/2) \times B((\lambda_0)^\tau, \delta/2).$$

However, if we let $q \to \infty$ in the equations

$$\mu(\omega_{k_q}) = f(x_{k_q}, \omega_{k_q}), \quad g(x_{k_q}, \omega_{k_q}) = 0,$$

we obtain

$$\mu(\omega_0) = f(\bar{x}, \omega_0) \quad \text{and} \quad g(\bar{x}, \omega_0) = 0.$$

Due to Assumption 1, there is a unique $x \in (g_{\omega_0})^{-1}(0)$ for which $\mu(\omega_0) = f(x, \omega_0)$, so we deduce that $x_0 = \bar{x}$. Next let $q \to \infty$ in (18) to obtain

(20) $$D_1f(x_0, \omega_0) + \bar{\lambda} \circ D_1g(x_0, \omega_0) = 0.$$

Since $D_1g(x_0, \omega_0)$ is surjective, a comparison of (20) with (12) yields that $\bar{\lambda} = \lambda_0$. Thus $(x_0, (\lambda_0)^\tau) = (\bar{x}, (\bar{\lambda})^\tau)$, which evidently contradicts (19). Thus the claim is proved, and so is the Theorem.   □

REMARK 4.5. In practice the assumptions in Thm. 4.4 that the infimum of the set $\{f_\omega(x)|x \in (g_\omega)^{-1}(0)\}$ is finite and achieved for every $\omega \in \Omega_0$ and the value function μ is continuous can be verified by Thm. 3.2.

REMARK 4.6. If, in addition to the other hypotheses of Thm. 4.4, the mappings f, g are assumed to be C^k for $3 \leq k < \infty$ (resp., $k = \infty$), then the mapping F defined (11) and its associated implicit mapping α defined in (14) are C^{k-1} (resp., C^∞), so the same proof shows that the value function μ is C^{k-1} (resp., C^∞).

REMARK 4.7. If assumption 4.4.1 is violated, one can give simple finite dimensional examples where the value function fails μ to be differentiable at ω_0.

In finite-dimensional Hilbert spaces (e. g., $\mathbb{R}^n$) the weak and strong topologies coincide and every closed, bounded set is compact. This allows us to state simpler forms of Thm. 4.4. One such version, in which Assumption 4.4.3 is dispensed with, is presented as the next corollary.

COROLLARY 4.8. Let

$$f:\mathbb{R}^n \times \mathbb{R}^m \to \mathbb{R}, \quad g:\mathbb{R}^n \times \mathbb{R}^m \to \mathbb{R}^p$$

be C^2 mappings. Suppose that for every $(x, \omega) \in g^{-1}(0)$ the partial Frechet derivative $D_1g(x, \omega):\mathbb{R}^n \to \mathbb{R}^m$ is surjective and for every $\omega \in \mathbb{R}^m$ the preimage $g_\omega^{-1}(0)$ is a nonempty compact subset of $\mathbb{R}^n$. Let

$$\mu(\omega) = \min\{f_\omega(x) \mid x \in \mathbb{R}^n \text{ and } g_\omega(x) = 0\}$$

denote the value function. Suppose that for some $\omega_0 \in \mathbb{R}^m$ it is the case that:
1. there exists a unique $x_0 \in (g_{\omega_0})^{-1}(0)$ such that $\mu(\omega_0) = f(x_0, \omega_0)$;
2. x_0 is a nondegenerate minimum point for the constrained minimization problem at ω_0.

Then there exists a $\delta > 0$ such that the value function μ is C^1 for $\omega \in B(\omega_0, \delta)$.

5. An application: the value function in optimal control theory

In the remainder of this paper, our main results, Thms. 3.2 and 4.4, are applied to obtain conditions that guarantee regularity of the value function in an optimal control problem. These results are not new and have been obtained by other authors using other methods (see, e. g., [CF] and [CL] for differentiability results, as well as [FR] and [Y] for related regularity results such as Lipschitz continuity). Our purpose here is to show that these results can also be obtained as a consequence of the general theorems derived in this paper.

Example 1.4 gave a brief description of the value function in optimal control theory and how it can be viewed as a parametrized constrained optimization problem. For simplicity, we imposed minimal hypotheses on the control system and the cost functional in that example. However, it is well known (see, e. g., [B; pp. 40–46]) that even the existence of optimal controls requires rather restrictive assumptions on the control system and cost functional. Furthermore, these assumptions must be made even more restrictive to ensure that the value function exhibits specified regularity properties. Thus our immediate task is to impose hypotheses on the control system and cost functional that will enable us to verify the hypotheses of Thms. 3.2 and 4.4. Along the way, we will review certain well known facts in varying degrees of detail. We make no claim that our hypotheses are the weakest possible hypotheses required to yield the results obtained. In some cases we have chosen simplicity over generality.

5.1 Continuous dependence of control-system trajectories on the control. We begin by reviewing a useful qualitative property of the solutions of the control system

$$(\text{ACS}) \qquad \dot{\phi}(t) = A(\phi(t)) + B(\phi(t))u(t).$$

The desired property requires the imposition of a rather strong assumption on the mappings $A: \mathbb{R}^n \to \mathbb{R}^n$, $B: \mathbb{R}^n \to \mathcal{L}(\mathbb{R}^m, \mathbb{R}^n)$ that appear on the right-hand side of (ACS). Specifically, instead of assuming that the mappings A, B are continuously differentiable, we will assume that they are globally Lipschitz; that is, we will assume

$$(\text{GL}) \qquad \begin{aligned} &\exists K > 0 \text{ such that } z, z' \in \mathbb{R}^n \\ &\Rightarrow |A(z) - A(z')| \le K|z - z'|, \quad |B(z) - B(z')| \le K|z - z'| \end{aligned}$$

(here $|\cdot|$ denotes any convenient norm on a finite-dimensional vector space). One useful consequence of (GL)—well known from the theory of ODEs—is that for every compact interval $[a, b]$, every $u \in L^2([a, b], \mathbb{R}^m)$, and every $(s, z) \in [a, b] \times \mathbb{R}^n$ there is a unique solution of (ACS) satisfying $\phi(s) = z$ that is defined on the entire interval $[a, b]$ (so solutions cannot escape to infinity in finite time). Actually this is true for every $u \in L^1([a, b], \mathbb{R}^m)$, but we state it for L^2 controls because we will need the Hilbert-space structure for later considerations. Thus for each

$$(s, z, u) \in [a, b] \times \mathbb{R}^n \times L^2([a, b], \mathbb{R}^m)$$

there exists a unique mapping $t \mapsto \xi(t, s, z, u)$ of $[a, b]$ into $\mathbb{R}^n$ with $\xi(\cdot, s, z, u) \in W^{1,2}([a, b], \mathbb{R}^n)$ such that $\xi(\cdot, s, z, u)$ is the solution of (ACS) on the interval $[a, b]$

satisfying $\xi(s, s, z, u) = z$. The mapping

$$\xi : [a, b] \times [a, b] \times \mathbb{R}^n \times L^2([a, b], \mathbb{R}^m) \to \mathbb{R}^n$$

is sometimes called the *global flow* of (ACS). Associated to the global flow ξ is the *trajectory mapping*

$$\Phi : [a, b] \times \mathbb{R}^n \times L^2([a, b], \mathbb{R}^m) \to W^{1,2}([a, b], \mathbb{R}^n)$$

given by

(TM) $$\Phi(s, z, u)(t) = \xi(t, s, z, u).$$

As one would expect, the trajectory mapping is continuous on its domain $[a, b] \times \mathbb{R}^n \times L^2([a, b], \mathbb{R}^m)$ when L^2 is given its norm topology. Somewhat surprising, however, is the fact that this mapping remains (sequentially) continuous when $L^2([a, b], \mathbb{R}^m)$ is given the weak topology and we view trajectories as elements of the Banach space $C([a, b], \mathbb{R}^n)$ of continuous mappings from $[a, b]$ into $\mathbb{R}^n$ with the supremum norm. The next theorem states this fact formally. Observe that we do not assume any differentiability of the mappings A and B.

THEOREM 5.1. *If the mappings A and B in the control system (ACS) satisfy the global Lipshitz condition (GL), then the trajectory mapping*

$$\Phi : [a, b] \times \mathbb{R}^n \times L^2([a, b], \mathbb{R}^m) \to C([a, b], \mathbb{R}^n)$$

defined by (TM) is sequentially continuous when $L^2([a, b], \mathbb{R}^m)$ is equipped with its weak topology. That is, if $\{(s_k, z_k, u_k) | k \in \mathbb{N}\}$ is a sequence in $[a, b] \times \mathbb{R}^n \times L^2([a, b], \mathbb{R}^m)$ such that $s_k \to s \in [a, b]$, $z_k \to z \in \mathbb{R}^n$ (in the standard topologies), and $u_k \to u \in L^2([a, b], \mathbb{R}^m)$ in the weak topology, then $\Phi(s_k, z_k, u_k) \to \Phi(s, z, u)$ uniformly on $[a, b]$.

PROOF. See [**LS**].   □

5.2 Continuity of the value function of the optimal control problem. In this section we use Thm 3.2 to show that, under appropriate assumptions, the value function of the optimal control problem of Example 1.4 is continuous. Thus, we fix $b > 0$ and for each $(s, z) \in [0, b] \times \mathbb{R}^n$ consider the optimal control problem of minimizing the cost functional

(CF) $$f(\phi, u, s, z) = K(\phi(b)) + \int_s^b L(\phi(\zeta), u(\zeta)) \, d\zeta$$

subject to the control-system constraint (ACS), where the mappings $A : \mathbb{R}^n \to \mathbb{R}^n$ and $B : \mathbb{R}^n \to \mathcal{L}(\mathbb{R}^m, \mathbb{R}^n)$ satisfy the global Lipschitz condition (GL). To be assured of the existence of an optimal control we must impose some conditions on the mappings

$$K : \mathbb{R}^n \to \mathbb{R}, \quad L : \mathbb{R}^n \times \mathbb{R}^m \to \mathbb{R},$$

that appear in the cost functional. Specifically, we assume that both K and L are continuous,

(21) $$K(z) \geq 0 \ \forall z \in \mathbb{R}^n,$$

and that L has the form

(22) $$L(z, w) = L_0(z) + \langle w, L_1(z) \rangle + \frac{1}{2} \langle w, L_2(z) w \rangle,$$

where $\langle \cdot, \cdot \rangle$ denotes the standard inner product on $\mathbb{R}^m$ and the mappings

(23) $$L_0 : \mathbb{R}^n \to \mathbb{R}, \ L_1 : \mathbb{R}^n \to \mathbb{R}^m, \ L_2 : \mathbb{R}^n \to \mathcal{L}(\mathbb{R}^m, \mathbb{R}^m),$$

satisfy the following conditions:

(24) $$L_0(z) \geq 0 \ \forall z \in \mathbb{R}^n;$$

(25) $$\exists \, \eta_1 > 0 \text{ such that } |L_1(z)| \leq \eta_1 \ \forall z \in \mathbb{R}^n;$$

and

(26) $$\forall z \in \mathbb{R}^n \text{ the matrix } L_2(z) \text{ is symmetric and } \exists \, \eta_2 > 0 \text{ such that}$$
$$(z, w) \in \mathbb{R}^n \times \mathbb{R}^m \Rightarrow \langle w, L_2(z)w \rangle \geq \eta_2 |w|^2.$$

PROPOSITION 5.2. *Under the assumptions that the mappings*

$$K : \mathbb{R}^n \to \mathbb{R}, \quad L : \mathbb{R}^n \times \mathbb{R}^m \to \mathbb{R},$$

are continuous and satisfy conditions (21)–(26), for every choice of compact interval $[a, b]$ the mapping

$$\Lambda : C([a, b], \mathbb{R}^n) \times L^2([a, b], \mathbb{R}^m) \times [a, b] \to \mathbb{R}$$

defined by

$$\Lambda(\phi, u, s) = \int_s^b L(\phi(\zeta), u(\zeta)) \, d\zeta$$

has the following properties:
1. *Λ is continuous when $L^2([a, b], \mathbb{R}^m)$ is equipped with the strong topology;*
2. *Λ is sequentially weakly lowersemicontinuous when $L^2([a, b], \mathbb{R}^m)$ is equipped with the weak topology.*

PROOF. The routine proof is omitted. We note in passing that the weak lower semicontinuity of Λ is due, in essence, to the convexity of $L(z, w)$ in the variable w. $\qquad\square$

We reformulate the optimal control problem introduced at the beginning of this subsection in the following manner. Choose a real number $a < 0$. Although we are interested in the optimal control problem on the interval $[0, b]$, with $b > 0$ fixed previously, we extend the interval of consideration to $[a, b]$ for technical reasons which will become apparent shortly. Suppose that the mappings A and B in (ACS) satisfy the global Lipschitz condition (GL) and let Φ denote the trajectory mapping of (ACS) as defined in (TM). We then define

$$\tilde{f} : L^2([a, b], \mathbb{R}^m) \times (a, b] \times \mathbb{R}^n \to \mathbb{R}$$

by

$$\tilde{f}(u, s, z) = K(\Phi(s, z, u)(b)) + \int_s^b L(\Phi(s, z, u)(\zeta), u(\zeta)) \, d\zeta.$$

In this fashion for each $(s, z) \in [0, b] \times \mathbb{R}^n$ the problem of minimizing the cost functional (CF) subject to the control-system constraint (ACS) is equivalent to minimizing the value of $\tilde{f}(u, s, z)$ subject to $u \in L^2([a, b], \mathbb{R}^m)$. Observe that the reformulated problem is (at least for the moment) an *unconstrained* minimization problem due to the incorporation of the control system constraint into the cost

functional via the use of the trajectory mapping (TM). To show both existence of
the optimal control and the continuity of the value function, we would like to appeal
to Cor. 3.4 since we seem to have disposed of the constraint. However, after some
thought one realizes that there is little hope of satisfying the coercivity condition
3.4.3. The basic problem here is that the value of $\tilde{f}(u, s, z)$ depends only on the
values of the control u on the interval $[s, b]$; thus the L^2-norm of u on the interval
$[a, b]$ could be quite large, but have no effect on the value of $\tilde{f}(u, s, z)$ (say, if u
were large on the subinterval $[a, s]$, but zero on the subinterval $(s, b]$). A similar
problem occurs with the point $s = b$, since $\tilde{f}(u, b, z) = K(z)$ is independent of u.
We remedy this problem by introducing a modified constraint mapping which will
allow us to apply Thm. 3.2, and in particular verify assumption 3.2.5.(C1).

THEOREM 5.3. *Let $[a, b]$ be a fixed compact interval with $a < 0$ and $b > 0$.
Assume that the control system (ACS) satisfies the global Lipschitz condition (GL)
and let*

$$K: \mathbb{R}^n \to \mathbb{R}, \quad L: \mathbb{R}^n \times \mathbb{R}^m \to \mathbb{R},$$

be continuous mappings that satisfy conditions (21)–(26). Define the cost functional

$$f: W^{1,2}([a, b], \mathbb{R}^n) \times L^2([a, b], \mathbb{R}^m) \times [a, b] \times \mathbb{R}^n \to \mathbb{R}$$

by

$$f(\phi, u, s, z) = K(\phi(b)) + \int_s^b L(\phi(\zeta), u(\zeta)) \, d\zeta$$

and define

$$g: W^{1,2}([a, b], \mathbb{R}^n) \times L^2([a, b], \mathbb{R}^m) \times [a, b] \times \mathbb{R}^n \to W^{1,2}([a, b], \mathbb{R}^n)$$

by

$$g(\phi, u, s, z)(t) = \phi(t) - z - \int_s^t [A(\phi(\zeta)) + B(\phi(\zeta))u(\zeta)] \, d\zeta,$$

so that (ϕ, u) satisfies (ACS) with initial condition $\phi(s) = z$ if and only if

$$g(\phi, u, s, z)(t) = 0 \; \forall t \in [a, b].$$

For every $(s, z) \in [0, b] \times \mathbb{R}^n$ consider the optimal control problem

$$\text{minimize} \quad f(\phi, u, s, z)$$
$$\text{subject to} \quad g(\phi, u, s, z) = 0.$$

*Then for every $(s, z) \in [0, b] \times \mathbb{R}^n$ there exists an optimal control-trajectory pair
(ϕ^*, u^*); i. e., $g(\phi^*, u^*, s, z) = 0$ and*

$$f(\phi^*, u^*, s, z) = \min\{f(\phi, u, s, z) \mid g(\phi, u, s, z) = 0\}.$$

Furthermore, the value function $V: [0, b] \times \mathbb{R}^n \to \mathbb{R}$ defined by

(27) $$V(s, z) = \min\{f(\phi, u, s, z) \mid g(\phi, u, s, z) = 0\}$$

is continuous.

PROOF. Define mappings

$$\tilde{f}, \tilde{g}: L^2([a,b], \mathbb{R}^m) \times [0,b] \times \mathbb{R}^n \to \mathbb{R}$$

by

(28)
$$\tilde{f}(u, s, z) = K(\Phi(s, z, u)(b)) + \int_s^b L(\Phi(s, z, u)(\varsigma), u(\varsigma)) \, d\varsigma$$
$$= f(\Phi(s, z, u), u, s, z),$$

and

(29)
$$\tilde{g}(u, s, z) = \int_a^s |u(\varsigma)|^2 \, d\varsigma.$$

Rather than attempt to apply Thm. 3.2 directly to the cost functional f and the constraint map g, we will apply it to $\tilde{f}, \tilde{g}$ for reasons which will become apparent during the proof. Observe that the definition of $\tilde{f}$ incorporates into the cost functional f the constraint $g = 0$ by means of the trajectory mapping (TM). Thus it is clear that the value function can be defined equivalently by

(30) $$V(s, z) = \min\{\tilde{f}(u, s, z) \mid u \in L^2([a,b], \mathbb{R}^m) \text{ and } \tilde{g}(u, s, z) = 0\}.$$

Observe that the constraint $\tilde{g} = 0$ does not in fact have any effect on the value of $\tilde{f}$; we introduce it to ensure that $\tilde{f}$ satisfies the requisite coercivity condition.

We proceed to show that the constrained optimization problem defined by (30) satisfies the hypotheses of Thm. 3.2 when applied to the mappings

$$\tilde{f}: \mathfrak{X} \times \Omega \to \mathbb{R}, \quad \tilde{g}: \mathfrak{X} \times \Omega \to \mathbb{R},$$

with $\mathfrak{X} = L^2([a,b], \mathbb{R}^m), \Omega = [0,b) \times \mathbb{R}^n$ (for technical reasons we will handle continuity at points (b, z) in a separate argument), and $\mathfrak{Y} = \mathbb{R}$. First note that for every $(s, z) \in \Omega$ we have $\tilde{g}(0, s, z) = 0$ so that the zero element of $L^2([a,b], \mathbb{R}^m)$ is in $(\tilde{g}_{(s,z)})^{-1}(0)$, whence

$$\Omega_0 = \{(s, z) \in \Omega \mid (\tilde{g}_{(s,z)})^{-1}(0) \neq \emptyset\} = \Omega.$$

From Prop. 5.2.1 and the remarks preceding the statement of Thm. 5.1 we easily deduce that $\tilde{f}$ is continuous on its domain when $\mathfrak{X} = L^2([a,b], \mathbb{R}^m)$ is given the strong topology. Furthermore, Thm. 5.1 and Prop. 5.2.2 imply that $\tilde{f}$ is sequentially weakly lsc when $L^2([a,b], \mathbb{R}^m)$ is given the weak topology. Since $L^2([a,b], \mathbb{R}^m)$ is separable, Cor. 3.6 yields the lowersemicontinuity of the restriction of $\tilde{f}$ to $\overline{B}(0, r) \times \Omega$ when $\overline{B}(0, r)$ is equipped with the weak topology (note that Ω as defined here is obviously first countable).

We now turn our attention to the properties of the mapping $\tilde{g}$. It is clear that $u \in \tilde{g}^{-1}(0)$ if and only if $u(t) = 0$ almost everywhere on the interval $[a, s]$. Thus $\tilde{g}_{(s,z)}^{-1}(0)$ is a strongly closed vector subspace of $L^2([a,b], \mathbb{R}^m)$, and consequently $\tilde{g}_{(s,z)}^{-1}(0)$ is also weakly closed in $L^2([a,b], \mathbb{R}^m)$. From this and the remarks in the previous paragraph it follows by a routine argument that $\tilde{g}^{-1}(0) \cap (\overline{B}(0, r) \times \Omega)$ is closed in $L^2([a,b], \mathbb{R}^m)_{\text{wk}} \times \Omega$.

Next we claim that $0 \in \mathbb{R}$ is an implicit mapping value of $\tilde{g}$. To this end let $(\bar{u}, \bar{s}, \bar{z}) \in \tilde{g}^{-1}(0)$ and define $\alpha : [0, b) \times \mathbb{R}^n \to L^2([a, b], \mathbb{R}^m)$ as follows: for $s \leq \bar{s}$ let $\alpha(s, z) = \bar{u}$ and for $s > \bar{s}$ let

$$\alpha(s, z)(t) = \begin{cases} \bar{u}(t), & t \leq \bar{s} \text{ or } t \geq s, \\ 0, & \bar{s} < t < s. \end{cases}$$

One verifies that α is continuous, $\alpha(\bar{s}, \bar{z}) = \bar{u}$ and $\tilde{g}(\alpha(s, z), s, z) = 0$ for every $(s, z) \in [0, b) \times \mathbb{R}^n$, whence 0 is an implicit mapping value as desired.

We now verify that $\tilde{f}$ satisfies the coercivity condition (C1) in the statement of Thm. 3.2. For $u \in L^2([a, b], \mathbb{R}^m)$ and $(s, z) \in [0, b) \times \mathbb{R}^n$ we note using (21), (24), (25), (26) and the Cauchy-Schwarz inequality that

$$
\begin{aligned}
(31) \quad \tilde{f}(u, s, z) &\geq \int_s^b \left[\langle u(\zeta), L_1(\Phi(u, s, z)(\zeta)) \rangle \right. \\
&\qquad\qquad \left. + \langle u(\zeta), L_2(\Phi(u, s, z)(\zeta)) u(\zeta) \rangle \right] d\zeta \\
&\geq \int_s^b \eta_2 |u(\zeta)|^2 \, d\zeta - \int_s^b \eta_1 |u(\zeta)| \, d\zeta \\
&\geq \int_s^b \eta_2 |u(\zeta)|^2 \, d\zeta - \left\{ \int_s^b (\eta_1)^2 \, d\zeta \right\}^{1/2} \left\{ \int_s^b |u(\zeta)|^2 \, d\zeta \right\}^{1/2} \\
&\geq \eta_2 \int_s^b |u(\zeta)|^2 \, d\zeta - \eta_1 (b-a)^{1/2} \left\{ \int_s^b |u(\zeta)|^2 \, d\zeta \right\}^{1/2} \\
&\geq \eta_2 \|u\|_s^2 - \eta_3 \|u\|_s,
\end{aligned}
$$

where $\|u\|_s^2 = \int_s^b |u(\zeta)|^2 \, d\zeta$ is the square of the L^2 norm of u restricted to the interval $[s, b]$ (so that $\|\cdot\|_a$ is the L^2 norm on $L^2([a, b], \mathbb{R}^m)$) and $\eta_3 = \eta_1(b-a)^{1/2}$. Observe that the positive constants η_2, η_3 do not depend on $(s, z) \in [0, b) \times \mathbb{R}^n$. Let $\gamma > 0$ be given and choose a real number $r > 0$ with the property that $\zeta \geq r \Rightarrow \eta_1 \zeta^2 - \eta_3 \zeta \geq \gamma$. Then for $(s, z) \in [0, b) \times \mathbb{R}^n$ and $u \in L^2([a, b], \mathbb{R}^m)$ with $\|u\|_a \geq r$ and $\tilde{g}(u, s, z) = 0$ we see that

$$\int_a^b |u(\zeta)|^2 \, d\zeta \geq r^2 \text{ and } \int_a^s |u(\zeta)|^2 \, d\zeta = 0$$

$$\Rightarrow \int_s^b |u(\zeta)|^2 \, d\zeta = \|u\|_s^2 \geq r^2$$

$$\Rightarrow \tilde{f}(u, s, z) \geq \eta_2 \|u\|_s^2 - \eta_3 \|u\|_s \geq \gamma.$$

Thus condition (C1) of Thm. 3.2 is satisfied, so we infer that the value function of the constrained minimization problem determined by the functions $\tilde{f}, \tilde{g}$ is continuous on the domain $[0, b) \times \mathbb{R}^n$.

To finish the proof it remains to show continuity of V at points of the form (b, z) with $z \in \mathbb{R}^n$. The definition of V yields $V(b, z) = K(z)$, so the restriction of V to $\{b\} \times \mathbb{R}^n$ is continuous on $\{b\} \times \mathbb{R}^n$ by the assumed continuity of K. Consequently, to show that V is continuous at a fixed (b, z) it suffices to show that if $\{(s_k, z_k) | k \in \mathbb{N}\}$ is any sequence in $[0, b) \times \mathbb{R}^n$ with $(s_k, z_k) \to (b, z)$, then

$V(s_k, z_k) \to V(b, z)$. Letting θ denote the zero function in $L^2([a, b], \mathbb{R}^m)$, we see from the definition of V that for every $k \in \mathbb{N}$

$$(32) \qquad V(s_k, z_k) \leq K(\Phi(s_k, z_k, \theta)(b)) + \int_{s_k}^{b} L_0(\Phi(s_k, z_k, \theta)(\zeta)) \, d\zeta.$$

It is clear that the mapping

$$(s, z) \mapsto K(\Phi(s, z, \theta)(b)) + \int_s^b L_0(\Phi(s, z, \theta)(\zeta)) \, d\zeta$$

is continuous on $[0, b] \times \mathbb{R}^n$, so from (32) we infer that

$$(33) \qquad \limsup_{k \to \infty} V(s_k, z_k) \leq K(z) = V(b, z).$$

On the other hand, for every $k \in \mathbb{N}$ choose a control $u_k^* \in L^2([a, b], \mathbb{R}^m)$ such that $\tilde{g}(u_k^*, s_k, z_k) = 0$ (which forces $u_k^*(t) = 0$ for almost every $t \in [a, s_k]$) and $V(s_k, z_k) = \tilde{f}(u_k^*, s_k, z_k)$. We claim that the sequence $\{u_k^*\}$ is bounded in $L^2([a, b], \mathbb{R}^m)$. To see this note that (33) implies that there exists $B > 0$ such that $V(s_k, z_k) \leq B$ for every $k \in \mathbb{N}$. Hence for each $k \in \mathbb{N}$ (31) yields

$$B \geq V(s_k, z_k) = \tilde{f}(u_k^*, s_k, z_k) \geq \eta_2 ||u_k^*||_{s_k}^2 - \eta_3 ||u_k^*||_{s_k}$$
$$= \eta_2 ||u_k^*||_a^2 - \eta_3 ||u_k^*||_a,$$

whence it must be the case that the sequence of L^2 norms of the functions u_k^* on the interval $[a, b]$ has a uniform upper bound. Thus $\{u_k^*\}$ is bounded in $L^2([a, b], \mathbb{R}^m)$, and we can extract a subsequence $\{u_{k_q}^*\} \subseteq L^2([a, b], \mathbb{R}^m)$ such that $u_{k_q}^* \to v$ weakly in $L^2([a, b], \mathbb{R}^m)$. Since $\tilde{f}$ is sequentially weakly lsc, we have

$$(34) \qquad \liminf_{q \to \infty} V(s_{k_q}, z_{k_q}) = \liminf_{q \to \infty} \tilde{f}(u_{k_q}^*, s_{k_q}, z_{k_q}) \geq \tilde{f}(v, b, z) = K(z) = V(z, b).$$

The inequalities (33) and (34) yield

$$\lim_{q \to \infty} V(s_{k_q}, z_{k_q}) = V(b, z).$$

Now, if it were not the case that $V(s_k, z_k) \to V(b, z)$, then there would exist $\epsilon > 0$ and a subsequence $\{(s_{k_q}, z_{k_q}) | q \in \mathbb{N}\}$ such that

$$(35) \qquad |V(s_{k_q}, z_{k_q}) - V(b, z)| \geq \epsilon \quad \forall q \in \mathbb{N}.$$

However, the preceding argument, applied to the subsequence $\{(s_{k_q}, z_{k_q}) | q \in \mathbb{N}\}$, would yield another subsequence $\{(s_{k_{q_p}}, z_{k_{q_p}}) | p \in \mathbb{N}\}$ such that

$$\lim_{p \to \infty} V(s_{k_{q_p}}, z_{k_{q_p}}) = V(b, z),$$

which would then contradict (35). Hence, $V(s_k, z_k) \to V(b, z)$ whenever $(s_k, z_k) \to (b, z)$, so V is continuous at points of the form (b, z). This fact and the previously established continuity of V on the set $[0, b) \times \mathbb{R}^n$ yield the continuity of V on the set $[0, b] \times \mathbb{R}^n$.

Since the value function of the constrained minimization problem (30) coincides with the value function of the optimal control problem (27), the proof of Thm. 5.3 is complete. $\qquad \square$

5.3 Control systems viewed as a constraint in a space of functions. Let

$$A:\mathbb{R}^n \to \mathbb{R}^n, \quad B:\mathbb{R}^n \to \mathcal{L}(\mathbb{R}^m, \mathbb{R}^n)$$

be C^k mappings, where $1 \le k \le \infty$, and fix a compact interval $[a, b]$. In Example 1.4 we made the standard observation that a pair $(\phi, u) \in W^{1,2}([a, b], \mathbb{R}^n) \times L^2([a, b], \mathbb{R}^m)$ solves the ordinary differential equation (ACS) with initial condition $\phi(s) = z$ if and only if for every $t \in [a, b]$ we have

$$\phi(t) - z - \int_s^t [A(\phi(\zeta)) + B(\phi(\zeta))u(\zeta)]\, d\zeta = 0.$$

Following the development in Example 1.4, we let

$$\mathfrak{X} = W^{1,2}([a, b], \mathbb{R}^n) \times L^2([a, b], \mathbb{R}^m), \quad \mathfrak{Y} = W^{1,2}([a, b], \mathbb{R}^n), \quad \Omega = [a, b] \times \mathbb{R}^n,$$

and we let $g:\mathfrak{X} \times \Omega \to \mathfrak{Y}$ be defined by

$$(36) \qquad g((\phi, u), (s, z))(t) = \phi(t) - z - \int_s^t [A(\phi(\zeta)) + B(\phi(\zeta))u(\zeta)]\, d\zeta.$$

Observe that g does indeed map into $\mathfrak{Y} = W^{1,2}([a, b], \mathbb{R}^n)$ because u is an L^2 function and the control appears linearly. The fact that (ϕ, u) is a trajectory-control pair of the affine control system (ACS) (with initial condition $\phi(s) = z$) is then embodied by the constraint $g((\phi, u), (s, z)) = 0$. For the purpose of studying the differentiability properties of g, we introduce the related mapping $H_s:\mathfrak{X} \to \mathfrak{Y}$ defined by

$$(37) \qquad H_s(\phi, u)(t) = \int_s^t [A(\phi(\zeta)) + B(\phi(\zeta))u(\zeta)]\, d\zeta, \ t \in [a, b].$$

One can use routine arguments (which we omit; see, e. g., [**BM**] and [**G**] for related, though admittedly not identical, results) to prove the following theorem.

THEOREM 5.4. *The assumption that the mappings $A:\mathbb{R}^n \to \mathbb{R}^n$ and $B:\mathbb{R}^n \to \mathcal{L}(\mathbb{R}^m, \mathbb{R}^n)$ are C^k, $1 \le k \le \infty$, implies that for every $s \in [a, b]$ the mapping H_s defined in (37) is C^k Frechet differentiable. Moreover, for $1 \le i \le k$ (i finite) the ith Frechet derivative of H_s satisfies*

$$D^i H_s(\phi, u)(\psi, v)^i(t) = \int_s^t \Big[D^i A(\phi(\zeta))(\psi(\zeta))^i + D^i B(\phi(\zeta))(\psi(\zeta))^i u(\zeta)$$
$$iD^{i-1} B(\phi(\zeta))(\psi(\zeta))^{i-1} v(\zeta) \Big]\, d\zeta$$

for $t \in [a, b]$.

The following corollary is an immediate consequence of this theorem.

COROLLARY 5.5. *If the mappings*

$$A:\mathbb{R}^n \to \mathbb{R}^n, \quad B:\mathbb{R}^n \to \mathcal{L}(\mathbb{R}^m.\mathbb{R}^n)$$

are C^k, then the mapping $g : \mathfrak{X} \to \mathfrak{Y}$ defined in (36) is C^k in its first variable $x = (\phi, u) \in \mathfrak{X}$. Moreover, for $((\phi, u), (s, z)) \in \mathfrak{X} \times \Omega$ the partial Frechet derivative

$$D_1 g((\phi, u), (s, z)):\mathfrak{X} \to \mathfrak{Y}$$

with respect to the first variable $x = (\phi, u)$ is given by

$$D_1 g((\phi, u), (s, z))(\psi, v)(t) = \psi(t) - \int_s^t \left[DA(\phi(\zeta))\psi(\zeta) + DB(\phi(\zeta))(\psi(\zeta))u(\zeta) \right.$$
$$\left. + B(\phi(\zeta))v(\zeta) \right] d\zeta$$

As it happens, the partial derivative $D_1 g((\phi, u), (s, z))$ is always surjective. This fact is a direct consequence of a standard property of Fredholm operators. The next proposition and its corollaries present the essential details.

PROPOSITION 5.6. *Let $E : [a, b] \to \mathcal{L}(\mathbb{R}^n)$ be an $n \times n$-matrix valued mapping whose entries are in $L^2([a, b], \mathbb{R})$, fix $s \in [a, b]$, and define*

$$T : W^{1,2}([a, b], \mathbb{R}^n) \to W^{1,2}([a, b], \mathbb{R}^n)$$

by

$$T(\psi)(t) = \int_s^t E(\zeta)\psi(\zeta)\, d\zeta.$$

Then T is a compact linear operator on $W^{1,2}([a, b], \mathbb{R}^n)$.

PROOF. This result is certainly well known, but we include the proof as it is very short. Let $\{\psi_k | k \in \mathbb{N}\}$ be a bounded sequence in $W^{1,2}([a, b], \mathbb{R}^n)$. The definition of the $W^{1,2}$ norm implies immediately that the sequence $\{\psi_k(a)\}$ is bounded in $\mathbb{R}^n$ and the sequence $\{\dot{\psi}_k\}$ is bounded in $L^2([a, b], \mathbb{R}^m)$. If we choose $\beta > 0$ such that $\|\dot{\psi}_k\|_{L^2} \leq \beta$ for every $k \in \mathbb{N}$, then for every $k \in \mathbb{N}$ and every pair $t, t' \in [a, b]$ the Cauchy-Schwarz inequality yields

$$|\psi_k(t') - \psi_k(t)| = \left| \int_t^{t'} \dot{\psi}_k(\zeta)\, d\zeta \right|$$
$$\leq \left| \int_t^{t'} 1^2\, d\zeta \right|^{1/2} \left| \int_t^{t'} |\dot{\psi}_k(\zeta)|^2\, d\zeta \right|^{1/2}$$
$$\leq \beta |t' - t|^{1/2},$$

whence the family of continuous functions $\{\psi_k\}$ is both uniformly bounded in the supremum norm and equicontinuous on $[a, b]$. By Ascoli's theorem there exists a subsequence $\{\psi_{k_q}\}$ that converges uniformly on $[a, b]$ to a continuous function ψ. It is then obvious that the sequence $\{E\psi_{k_q}\} \subseteq L^2([a, b], \mathbb{R}^m)$ is Cauchy in the L^2 norm. From the equality

$$\|T(\psi_{k_p}) - T(\psi_{k_q})\|^2_{W^{1,2}} = \left| \int_s^a E(\zeta)\psi_{k_p}(\zeta)\, d\zeta - \int_s^a E(\zeta)\psi_{k_q}(\zeta)\, d\zeta \right|^2$$
$$+ \|E\psi_{k_p} - E\psi_{k_q}\|^2_{L^2}$$

we infer that the sequence $\{T(\psi_{k_q})\}$ is Cauchy in the $W^{1,2}$ norm, and thus converges by the completeness of $W^{1,2}([a, b], \mathbb{R}^n)$. This completes the proof. $\square$

COROLLARY 5.7. *For $T : W^{1,2}([a, b], \mathbb{R}^n) \to W^{1,2}([a, b], \mathbb{R}^n)$ as defined in the Proposition the mapping $S : W^{1,2}([a, b], \mathbb{R}^n) \to W^{1,2}([a, b], \mathbb{R}^n)$ defined by $S(\psi) = \psi + T(\psi)$ is a linear homeomorphism of $W^{1,2}([a, b], \mathbb{R}^n)$.*

PROOF. Since S is defined to be a compact perturbation of the identity (i. e., a Fredholm operator), by standard properties of Fredholm operators (see [**D**; p. 322]) it is enough to show S is injective. But $S(\psi_1) = S(\psi_2)$ implies that $\psi_1 - \psi_2$ satisfies

$$\dot\psi_1(t) - \dot\psi_2(t) = -E(t)(\psi_1(t) - \psi_2(t)), \quad \psi_1(s) - \psi_2(s) = 0,$$

so we conclude that $\psi_1 = \psi_2$ by the uniqueness of solutions of initial value problems involving linear ODEs. $\square$

COROLLARY 5.8. *Let*

$$\mathfrak{X} = W^{1,2}([a,b], \mathbb{R}^n) \times L^2([a,b], \mathbb{R}^m), \quad \mathfrak{Y} = W^{1,2}([a,b], \mathbb{R}^n), \quad \Omega = [a,b] \times \mathbb{R}^n,$$

and let $g : \mathfrak{X} \times \Omega \to \mathfrak{Y}$ be as defined in (36). Then under the assumption that the mappings $A : \mathbb{R}^n \to \mathbb{R}^n$ and $B : \mathbb{R}^n \to \mathcal{L}(\mathbb{R}^m, \mathbb{R}^n)$ are C^1 it is the case that for every $((\phi, u), (s, z)) = (x, \omega) \in \mathfrak{X} \times \Omega$ the partial Frechet derivative $D_1 g(x, \omega) : \mathfrak{X} \to \mathfrak{Y}$ is surjective.

PROOF. Fix $((\phi, u), (s, z)) = (x, \omega) \in \mathfrak{X} \times \Omega$ and let $S : \mathfrak{Y} \to \mathfrak{Y}$ be defined by

$$S(\psi)(t) = \psi(t) - \int_s^t \left[DA(\phi(\zeta))\psi(\zeta) + DB(\phi(\zeta))(\psi(\zeta))u(\zeta) \right] d\zeta.$$

By Cor. 5.7 S is a linear homeomorphism of $\mathfrak{Y} = W^{1,2}([a,b], \mathbb{R}^n)$ and, in particular, is surjective. Thus for every $\psi' \in \mathfrak{Y}$ there exists $\psi \in \mathfrak{Y}$ such that $S(\psi) = \psi'$. From the formula for $D_1 g((\phi, u), (s, z))$ given in Cor. 5.5 we infer that

$$D_1 g((\phi, u), (s, z))(\psi, 0) = S(\psi) = \psi',$$

whence $D_1 g((\phi, u), (s, z))$ is surjective as required. $\square$

5.4 Some consequences of the standard necessary conditions for optimality. In Sect. 5.2 we proved the continuity of the value function of the optimal control problem in which the cost functional (CF) is minimized subject to the control system constraint (ACS) and the initial condition (IC); in particular we proved that an optimal control exists for each fixed initial condition $(s, z) \in [0, b] \times \mathbb{R}^n$. These results were obtained under the assumptions that $[a, b]$ is a fixed compact interval with $a < 0$ and $b > 0$, the mappings A and B in the control system (ACS) satisfy the global Lipschitz condition (GL), and the mappings

$$K : \mathbb{R}^n \to \mathbb{R}, \quad L : \mathbb{R}^n \times \mathbb{R}^m \to \mathbb{R},$$

in the cost functional (CF) are continuous and satisfy conditions (21)–(26). Now we impose the additional assumption that the mappings A, B, K, and L are C^1. This allows us to apply the Pontryagin maximum principle to obtain some useful information about the optimal controls for this problem. To this end we form the Hamiltonian

(38) $$H(z, p, w) = L(z, w) + \langle p, A(z) + B(z)w \rangle,$$

where $z, p \in \mathbb{R}^n$ and $w \in \mathbb{R}^m$. Owing to the assumed form of L in (22), we can write H as

(39) $$H(z, p, w) = L_0(z) + \langle w, L_1(z) \rangle + \frac{1}{2}\langle w, L_2(z)w \rangle + \langle p, A(z) \rangle + \langle p, B(z)w \rangle.$$

Since H is quadratic in w and $L_2(z)$ is positive definite, hence invertible, for all $z \in \mathbb{R}^n$ by (26), for each fixed $(z, p) \in \mathbb{R}^n \times \mathbb{R}^n$ the mapping $w \mapsto H(z, p, w)$ has a unique global minimum at the point $w \in \mathbb{R}^m$ which satisfies the equation $D_3 H(z, p, w) = 0$ (D_3 denotes the partial derivative with respect to the third variable w). An elementary computation shows that

$$(40) \qquad D_3 H(z, p, w) = 0 \iff w = -\big(L_2(z)\big)^{-1}\big[L_1(z) + B(z)^\tau p\big] = h(z, p),$$

where the superscript τ denotes transpose and the last equality defines the function h. Observe that h is C^1 due to the assumed continuous differentiability of the mappings comprising the control system and the cost functional.

For a fixed initial condition (s, z) in $[0, b] \times \mathbb{R}^n$ let $(\phi^*, u^*) \in W^{1,2}([a, b], \mathbb{R}^n) \times L^2([a, b], \mathbb{R}^m)$ be an optimal trajectory-control pair for the system (ACS) that minimizes the cost functional (CF) (our assumptions guarantee that such an optimal pair exists by the results of Sect. 5.2). By the Pontryagin Maximum Principle there exists an absolutely continuous mapping $p^* : [s, b] \to \mathbb{R}^n$ such that ϕ^*, u^*, and p^* satisfy the following ODE system on the interval $[s, b]$:

$$(41) \qquad \begin{aligned} \dot{\phi}^*(t) &= A(\phi^*(t)) + B(\phi^*(t))u^*(t), \\ \dot{p}^*(t) &= -\big[DA(\phi^*(t)) + DB(\phi^*(t))u^*(t)\big]^\tau p^*(t) - \big(D_1 L(\phi^*(t), u^*(t))\big)^\tau, \end{aligned}$$

where ϕ^* and p^* satisfy the boundary conditions

$$(42) \qquad \phi^*(s) = z, \quad p^*(b) = \nabla K(\phi^*(b)).$$

Furthermore, we must have

$$(43) \qquad u^*(t) = h(\phi^*(t), p^*(t)) \text{ a. e. for } t \in [s, b],$$

where h is the C^1 mapping defined in (40). One immediate consequence of (43) is that the optimal control u^*, a priori known only to be an L^2 function on $[s, b]$, must in fact be absolutely continuous on $[s, b]$. A less immediate consequence of these necessary conditions is given in the next theorem, which we will play a crucial role in the verification of Assumption 4.4.3 when we use Thm. 4.4 to deduce differentiability of the value function of the optimal control problem.

THEOREM 5.9. *Assume that the mappings A and B in the control system (ACS) are C^1 and satisfy the global Lipschitz condition (GL), and assume that the mappings*

$$K : \mathbb{R}^n \to \mathbb{R}, \quad L : \mathbb{R}^n \times \mathbb{R}^m \to \mathbb{R},$$

in the cost functional (CF) are C^1 and satisfy conditions (21)–(26). Fix the initial time s at 0, let $\{z_k\} \subseteq \mathbb{R}^n$ be a sequence such that $z_k \to z \in \mathbb{R}^n$, and for each $k \in \mathbb{N}$ let

$$(\phi_k, u_k) \in W^{1,2}([0, b], \mathbb{R}^n) \times L^2([0, b], \mathbb{R}^m)$$

be an optimal trajectory-control pair that minimizes (CF) subject to the control-system constraint (ACS) with initial condition $\phi(0) = z_k$. Then there exists a subsequence

$$\{\phi_{k_q}, u_{k_q}\} \subseteq W^{1,2}([0, b], \mathbb{R}^n) \times L^2([0, b], \mathbb{R}^m)$$

that is strongly convergent to an element (ϕ, u).

PROOF. We note first that the existence of the optimal trajectory control pair (ϕ_k, u_k) for every initial condition $(0, z_k)$ follows from the discussion in Sect. 5.2. By the remarks preceding the statement of this theorem, for each $k \in \mathbb{N}$ there exists an absolutely continuous mapping $p_k : [0, b] \to \mathbb{R}^n$ such that a. e. on $[0, b]$ we have

$$\dot{\phi}_k(t) = A(\phi_k(t)) + B(\phi_k(t))u_k(t),$$
$$\dot{p}_k(t) = -\left[DA(\phi_k(t)) + DB(\phi_k(t))u_k(t)\right]^\tau p_k(t) - \left(D_1 L(\phi_k(t), u_k(t))\right)^\tau,$$
$$u_k(t) = h(\phi_k(t), u_k(t)),$$

where ϕ_k and p_k satisfy the boundary conditions

$$\phi_k(0) = z_k, \quad p_k(b) = \nabla K(\phi_k(b)).$$

By the definition of the value function V for this optimal control problem and by the inequality (31) in the proof of Thm. 5.3, we see that

(44) $$V(0, z_k) = \tilde{f}(u_k, 0, z_k) \geq \eta_2 \|u_k\|_0^2 - \eta_3 \|u_k\|_0.$$

Here $\tilde{f}$ is as defined in (28), η_2, η_3 are appropriate positive real numbers, and $\|\cdot\|_0$ denotes the L^2 norm on the interval $[0, b]$. Since the assumptions of Thm. 5.3 are in place, the continuity of the value function is assured, so the convergence of the sequence $\{z_k\}$ implies the convergence, and in particular, boundedness, of the sequence $\{V(0, z_k)\}$. It follows from (44) that the sequence $\{u_k\}$ is bounded in $L^2([0, b], \mathbb{R}^m)$, so there exists a weakly convergent subsequence $\{u_{k_q}\}$. Thm. 5.1 asserts the continuity of the trajectory mapping Φ with respect to the (sequential) weak topology on L^2, and our definitions imply that $\Phi(0, z_k, u_k) = \phi_k$, so we infer that the subsequence $\{\phi_{k_q}\}$ converges uniformly to an element of $C([0, b], \mathbb{R}^n)$.

Since the subsequence $\{u_{k_q}\}$ is bounded in $L^2([0, b], \mathbb{R}^m)$ and the sequence $\{\phi_{k_q}\}$ is uniformly bounded in the supremum norm on $[0, b]$, an examination of the differential equations satisfied by the mappings p_k and the Gronwall inequality show that the sequence $\dot{p}_k$ is uniformly bounded in the space $L^1([0, b], \mathbb{R}^n)$. Thus the sequence of absolutely continuous mappings $\{p_{k_q}\}$ is uniformly bounded in the supremum norm, and hence, from an examination of the ODEs satisfied by the functions p_{k_q}, is also uniformly bounded in the total variation norm. We can invoke the Helly selection principle to extract yet another subsequence, which we continue to denote by $\{p_{k_q}\}$, that converges pointwise on $[0, b]$. These facts and the formula $u_{k_q}(t) = h(\phi_{k_q}(t), p_{k_q}(t))$ (for $t \in [0, b]$) show that the sequence $\{u_{k_q}\}$ converges pointwise on $[0, b]$ and is uniformly bounded in the supremum norm. Hence the dominated convergence theorem implies that $\{u_{k_q}\}$ converges in the L^2 norm, from which we infer that the sequence $\{\phi_{k_q}\}$ actually converges in the $W^{1,2}$ norm, so the proof of the theorem is complete. $\square$

REMARK 5.10. It was not essential to fix the initial time at $s = 0$ in Thm. 5.9, but we have only stated the version of the theorem that is essential for our differentiability result. For more comments on allowing the initial time to vary, see Rem. 5.12.

5.5 Differentiability of the value function of the optimal control problem. We conclude our discussion of the optimal control problem by using Thm. 4.4 to deduce differentiability of the value function under appropriate assumptions on the cost functional and control system. Throughout this section we assume that the mappings

$$A : \mathbb{R}^n \to \mathbb{R}^n, \quad B : \mathbb{R}^n \to \mathcal{L}(\mathbb{R}^m, \mathbb{R}^n)$$

in (ACS) satisfy (GL) and are at least be C^2. We also assume that the mappings

$$K : \mathbb{R}^n \to \mathbb{R}, \quad L : \mathbb{R}^n \times \mathbb{R}^m \to \mathbb{R},$$

are at least C^2 and satisfy (21)–(26). Our attention will be focused on the optimal control problem of minimizing

$$K(\phi(b)) + \int_0^b L(\phi(\zeta), u(\zeta)) \, d\zeta$$

with

$$(\phi, u) \in W^{1,2}([0, b], \mathbb{R}^n) \times L^2([0, b], \mathbb{R}^m)$$

subject to the constraint

$$\dot{\phi}(t) = A(\phi(t)) + B(\phi(t))u(t),$$

where the initial condition $\phi(0) = z$ and terminal time $b > 0$ are fixed in advance (see also Rem. 5.12). We define mappings

$$f : W^{1,2}([0, b], \mathbb{R}^n) \times L^2([0, b], \mathbb{R}^m) \times \mathbb{R}^n \to \mathbb{R}$$

by

$$f(\phi, u, z) = K(\phi(b)) + \int_0^b L(\phi(\zeta), u(\zeta)) \, d\zeta$$

and

$$g : W^{1,2}([0, b], \mathbb{R}^n) \times L^2([0, b], \mathbb{R}^m) \times \mathbb{R}^n \to W^{1,2}([0, b], \mathbb{R}^n)$$

by

$$g(\phi, u, z)(t) = \phi(t) - z - \int_0^t [A(\phi(\zeta)) + B(\phi(\zeta))u(\zeta)] \, d\zeta.$$

So as to be consistent with the notation of Thm. 4.4 we let

$$\mathfrak{X} = W^{1,2}([0, b], \mathbb{R}^n) \times L^2([0, b], \mathbb{R}^m), \quad \mathfrak{Y} = W^{1,2}([0, b], \mathbb{R}^n), \quad \Omega = \mathbb{R}^n,$$

so that f, g, as defined, are mappings

$$f : \mathfrak{X} \times \Omega \to \mathbb{R}, \quad g : \mathfrak{X} \times \Omega \to \mathfrak{Y}.$$

Thm. 5.4 implies immediately that g is C^2 on the domain $\mathfrak{X} \times \Omega$ (note that the additional variable $z \in \Omega = \mathbb{R}^n$ poses no problem here). Arguments analogous to those used in the proof of Thm. 5.4 can be used to establish that f is C^2 on the domain $\mathfrak{X} \times \Omega$. Thus both f and g are C^2, and we also infer from Cor. 5.8 that the partial derivative

$$D_1 g(x, \omega) : \mathfrak{X} \to \mathfrak{Y}$$

is surjective for every $(x, \omega) \in \mathfrak{X} \times \Omega$. Furthermore, since the assumptions of Thm. 5.3 are in force, we may assert that the value function $V : \mathbb{R}^n \to \mathbb{R}$ defined by

(45) $$V(z) = \inf\{f(\phi, u, z) \mid g(\phi, u, z) = 0\}$$

is continuous (we again remind the reader that the initial time is fixed at $s = 0$ here), and for each $z \in \mathbb{R}^n$ there exists an optimal control-trajectory pair

$$(\phi_0, u_0) \in W^{1,2}([0, b], \mathbb{R}^n) \times L^2([0, b], \mathbb{R}^m)$$

for which the infimum in (45) is achieved; that is, $V(z) = f(\phi_0, u_0, z)$.

We comment on the form of the basic necessary conditions for constrained optimization problems when applied to the optimal control problem. Since the functional f and the constraint g are C^2, since

$$f(\phi_0, u_0, z) = \min\{f(\phi, u, z) \mid g(\phi, u, z) = 0\},$$

and since $D_1 g(x_0, z)$ is surjective (where $x = (\phi_0, u_0)$), the Lagrange multiplier theorem yields a $\lambda_0 \in W^{1,2}([0, b], \mathbb{R}^n)$ such that for all $(\phi, u) \in \mathfrak{X}$

$$D_1 f((\phi_0, u_0), z)(\phi, u) + \langle \lambda_0, D_1 g((\phi_0, u_0), z)(\phi, u) \rangle = 0$$

(the inner product here is the $W^{1,2}$ inner product defined in Ex. 1.4). Routine computations show that the Lagrange multiplier λ_0 is given by

$$\lambda_0(t) = -p_0(0) - \int_0^t p_0(\zeta)\, d\zeta,$$

where $p_0(t)$ is the solution of the so-called adjoint equation

$$(46) \qquad \dot{p}_0(t) = -\big[DA(\phi_0(t)) + DB(\phi_0(t))u_0(t)\big]^\tau p_0(t) - \big(D_1 L(\phi_0(t), u_0(t))\big)^\tau,$$

that satisfies the terminal condition $p_0(b) = \nabla K(\phi_0(b))$. In particular, for $\psi \in W^{1,2}([0, b], \mathbb{R}^n)$ the definition of the $W^{1,2}$ inner product yields

$$\langle \lambda_0, \psi \rangle = -\langle p_0(0), \psi(0) \rangle - \int_0^b \langle p_0(\zeta), \dot{\psi}(\zeta) \rangle\, d\zeta.$$

To apply Thm. 4.4 we must be able to identify when (ϕ_0, u_0) is a nondegenerate minimum point of the constrained optimization problem in the sense of Def. 4.1. Routine computations show that, for the problem under consideration here, the second derivative of $f + (\lambda_0)^\tau \circ g$ with respect to $x = (\phi, u) \in \mathfrak{X}$ at (ϕ_0, u_0, z) when evaluated on $(\psi, v)^2 \in \mathfrak{X} \times \mathfrak{X}$ is given by the formula

$$\big((D_1)^2 f((\phi_0, u_0), z) + (\lambda_0)^\tau \circ (D_1)^2 g((\phi_0, u_0), z)\big)(\psi, v)^2$$

$$= \langle \psi(b), \nabla^2 K(\phi_0(b))\psi(b) \rangle + \int_0^b \Big[D^2 L(\phi_0(\zeta), u_0(\zeta))(\psi(\zeta), v(\zeta))^2$$

$$+ \langle -p_0(\zeta), -D^2 F(\phi_0(\zeta), u_0(\zeta))(\psi(\zeta), v(\zeta))^2 \rangle \Big]\, d\zeta,$$

where we have used $F: \mathbb{R}^n \times \mathbb{R}^m \to \mathbb{R}^n$ as shorthand for the mapping $F(z, w) = A(z) + B(z)w$ and we recall the mapping $L: \mathbb{R}^n \times \mathbb{R}^m \to \mathbb{R}$ has the form given in (22). In terms of the Hamiltonian H defined in (38) we can write the preceding formula a bit more concisely as

$$\big((D_1)^2 f((\phi_0, u_0), z) + (\lambda_0)^\tau \circ (D_1)^2 g((\phi_0, u_0), z)\big)(\psi, v)^2$$

$$(47)$$
$$= \langle \psi(b), \nabla^2 K(\phi_0(b))\psi(b) \rangle + \int_0^b \overline{D}^2 H(\phi_0(\zeta), p_0(\zeta), u_0(\zeta))(\psi(\zeta), v(\zeta))^2\, d\zeta,$$

where $\overline{D}H$ denotes the partial derivative of H with respect to the pair of variables (z, w) with p fixed.

THEOREM 5.11. *Assume that the mappings*

$$A:\mathbb{R}^n \to \mathbb{R}^n, \quad B:\mathbb{R}^n \to \mathcal{L}(\mathbb{R}^m, \mathbb{R}^n)$$

in (ACS) satisfy (GL) and are at least C^2. Also assume that the mappings

$$K:\mathbb{R}^n \to \mathbb{R}, \quad L:\mathbb{R}^n \times \mathbb{R}^m \to \mathbb{R},$$

are at least C^2, satisfy (21)–(26), and further assume that the mapping K is convex. Define mappings

$$f:W^{1,2}([0,b],\mathbb{R}^n) \times L^2([0,b],\mathbb{R}^m) \times \mathbb{R}^n \to \mathbb{R}$$

by

$$f(\phi, u, z) = K(\phi(b)) + \int_0^b L(\phi(\zeta), u(\zeta))\, d\zeta$$

and

$$g:W^{1,2}([0,b],\mathbb{R}^n) \times L^2([0,b],\mathbb{R}^m) \times \mathbb{R}^n \to W^{1,2}([0,b],\mathbb{R}^n)$$

by

$$g(\phi, u, z)(t) = \phi(t) - z - \int_0^t [A(\phi(\zeta)) + B(\phi(\zeta))u(\zeta)]\, d\zeta.$$

Then for each $z \in \mathbb{R}^n$ the infimum

$$\inf\{f(\phi, u, z) \mid g(\phi, u, z) = 0\}$$

is achieved at some $(\phi, u) \in W^{1,2}([0,b],\mathbb{R}^n) \times L^2([0,b],\mathbb{R}^m)$ and the value function $V:\mathbb{R}^n \to \mathbb{R}$ defined by

$$V(z) = \min\{f(\phi, u, z) \mid g(\phi, u, z) = 0\}$$

is continuous. Futhermore, given $z_0 \in \mathbb{R}^n$, if there exists a unique optimal trajectory-control pair (ϕ_0, u_0) such that $V(z_0) = f(\phi_0, u_0, z_0)$ and if for a. e. $t \in [0,b]$ the Hessian of H with respect to (z, w) satisfies

(48)
$$\nabla^2_{(z,w)} H(\phi_0(t), p_0(t), u_0(t)) \geq 0$$

(i. e., is positive semidefinite), where p_0 is the solution of the adjoint equation (46) satisfying the terminal condition $p_0(b) = \nabla K(\phi_0(b))$, then there exists $\delta > 0$ such that the value function V is C^1 for $z \in B(z_0, \delta)$.

PROOF. As we noted above, to be consistent with the notation of Thm. 4.4 we let

$$\mathfrak{X} = W^{1,2}([0,b],\mathbb{R}^n) \times L^2([0,b],\mathbb{R}^m), \quad \mathfrak{Y} = W^{1,2}([0,b],\mathbb{R}^n), \quad \Omega = \mathbb{R}^n.$$

Then f, g, as defined, are mappings

$$f:\mathfrak{X} \times \Omega \to \mathbb{R}, \quad g:\mathfrak{X} \times \Omega \to \mathfrak{Y},$$

and we have already noted that f and g are C^2, for every $(x, z) \in \mathfrak{X} \times \Omega$ (where $x = (\phi, u)$) the partial derivative $D_1 g(x, z)$ is surjective, and the value function V is continuous. It remains to show that Assumptions 1–3 of Thm. 4.4 are satisfied.

Assumption 4.4.1 is, of course, satisfied by our assumption here that for the given $z_0 \in \mathbb{R}^n$ there is only one optimal trajectory-control pair (ϕ_0, u_0). To show that $x_0 = (\phi_0, u_0) \in \mathfrak{X}$ is a nondegenerate minimum point for the constrained mimimization problem

$$\text{minimize } f(x, z) \quad \text{subject to } g(x, z) = 0,$$

we let $p_0 \in \mathfrak{Y}$ be the Lagrange multiplier for the problem at the constrained minimum point $x_0 = (\phi_0, u_0)$. As noted previously, $p_0 \in W^{1,2}([0, b], \mathbb{R}^n)$ is the solution of the adjoint equation (46) satisfying the terminal condition $p_0(b) = \nabla K(\phi_0(b))$. In light of equation (47), to show that (ϕ_0, u_0) is nondegenerate (i. e., satisfies condition (SSO)), we must verify that there exists $\sigma > 0$ such that

(49)

$$(\psi, v) \in \mathfrak{X} \text{ and } D_1 g((\phi_0, u_0), z_0)(\psi, v) = 0$$
$$\Rightarrow \langle \psi(b), \nabla^2 K(\phi_0(b)) \psi(b) \rangle$$
$$+ \int_0^b \langle (\psi(\zeta), v(\zeta)), (\nabla_{(z,w)}^2) H(\phi_0(\zeta), p_0(\zeta), u_0(\zeta))(\psi(\zeta), v(\zeta)) \rangle \, d\zeta,$$
$$\geq \sigma ||(\psi, v)||^2.$$

From the formula for the partial derivative $D_1 g$ given in Cor. 5.5 it is evident that

$$D_1 g((\phi_0, u_0), z_0)(\psi, v) = 0$$

if and only if (ψ, v) satisfies $\psi(0) = 0$ and

(LVCS) $\dot{\psi}(t) = DA(\phi_0(t))\psi(t) + DB(\phi_0(t))(\psi(t))u_0(t) + B(\phi_0(t))v(t)$

for almost every $t \in [0, b]$. The equation (LVCS) is the well known *linear variational control system* of (ACS) along the control-trajectory pair (ϕ_0, u_0). We also note that from the formula for the Hamiltonian H in (40) we have

$$(\nabla_w)^2 H(z, p, w) = L_2(z) > 0 \quad \forall z \in \mathbb{R}^n,$$

so we can apply Theorem A.2 in the appendix to infer the existence of a $\sigma > 0$ satisfying

(50) $\int_0^b \langle (\psi(\zeta), v(\zeta)), \nabla_{(z,w)}^2 H(\phi_0(\zeta), p_0(\zeta), u_0(\zeta))(\psi(\zeta), v(\zeta)) \rangle \, d\zeta \geq \sigma ||(\psi, v)||^2,$

whenever (ψ, v) satisfies the equation (LVCS) with initial condition $\psi(0) = 0$. Since $\nabla^2 K(\phi(b))$ is positive semidefinite by the assumed convexity of K, we deduce (49) from (50).

Finally, we note that the verification of Assumption 4.4.3 is a direct consequence of Thm. 5.9. Hence the proof is complete. $\quad\square$

REMARK 5.12. In the proof of Thm. 5.11, we kept the initial time fixed at $s = 0$ because the cost functional (CF) and the constraint (36) are not continuously differentiable functions of s when the controls are only assumed to be L^2 functions. Thus, had we attempted to allow the initial time to vary in the interval $[0, b]$ (as we did in the proof of the continuity theorem, Thm. 5.3), we would not have been able to verify the differentiability assumptions of Thm. 4.4 with respect to the parameter space $\Omega = (a, b) \times \mathbb{R}^n$. Nevertheless, under the hypotheses of Thm. 5.11, we can infer that the time-dependent value function (27) is C^1 on the domain $[0, b) \times \mathbb{R}^n$ by appealing to the fixed initial time case (Thm. 5.11) and a standard change of variable (see, e. g., [B; p. 27]). Specifically, for every $s \in [0, b)$ and every initial condition $\phi(s) = z$, we can transform the problem of minimizing (CF) subject to the control system constraint (ACS) on the interval $[s, b]$ to an equivalent problem on the fixed time interval $[0, b]$ by the change of time scale

$$t = \sigma(t') = \frac{b-s}{b} t' + s.$$

This results in the transformed optimal control problem of minimizing

(TCF) $\qquad K(\psi(t')) + \int_0^b \rho(t') \left[A(\psi(t')) + B(\psi(t'))v(t') \right] dt'$

subject to the control-system constraint

(TACS) $\qquad \begin{aligned} & \frac{d\psi}{dt'}(t') = \rho(t') \left[A(\psi(t')) + B(\psi(t'))v(t') \right], \\ & \frac{d\rho}{dt'}(t') = 0, \hspace{3.5cm} t \in [0, b] \\ & \frac{d\sigma}{dt'}(t') = \rho(t'), \end{aligned}$

with fixed initial condition

$$\psi(0) = z, \quad \rho(0) = (b-s)/b, \quad \sigma(0) = s.$$

The augmented state variables ρ and σ in the transformed system are scalars, and the controls and trajectories of the transformed system are related to the controls and trajectories of the original system by

$$v(t') = u(\sigma(t')), \quad \psi(t') = \phi(\sigma(t')).$$

It must be acknowledged that the transformed system does not quite meet the hypotheses of Thm. 5.11 as stated (e. g., the transformed ODE (TACS) does not satisfy the global Lipschitz condition even when A and B do), but through minor modifications a differentiability proof along the lines of that given in Thm. 5.11 can be given for the transformed system. We leave the details as an exercise. A shortcoming of this approach is that it does not appear to yield information about the differentiability of the time-dependent value function (27) at points of the form (b, z).

References

[AC]   J. P. Aubin and A. Cellina, *Differential inclusions*, Springer-Verlag, New York, NY, 1984.

[AF]   J. P. Aubin and H. Frankowska, *Set-valued analysis*, Birkäuser, Boston, MA, 1990.

[B]   L. D. Berkovitz, *Optimal control theory*, Springer-Verlag, New York, NY, 1974.

[BC1]   J. F. Bonnans and R. Cominetti, *Perturbed optimization in Banach spaces I: a general theory based on weak directional constraint qualification*, SIAM J. Control Optim. **34** (1996), 1151–1171.

[BC2]   J. F. Bonnans and R. Cominetti, *Perturbed optimization in Banach spaces II: a theory based on a strong directional constraint qualification*, SIAM J. Control Optim. **34** (1996), 1172–1189.

[BC3]   J. F. Bonnans and R. Cominetti, *Perturbed optimization in Banach spaces III: semi-infinite optimization*, SIAM J. Control Optim. **34** (1996), 1555–1568.

[BM]   R. M. Bianchini and A. Margheri, *First-order differentiability of the flow of a system with L^p controls*, J. Optim. Theory Appl. **89** (1996), 293–310.

[CF]   P. Cannarsa and H. Frankowska, *Some characterizations of optimal trajectories in control theory*, SIAM J. Control Optim. **29** (1991), 1322–1347.

[CL]   F. H. Clarke and P. D. Loewen, *The value function in optimal control: sensitivity, controllability, and time optimality*, SIAM J. Control Optim. **24** (1986), 243–263.

[D]   J. Dieudonné, *Foundations of modern analysis*, Academic Press, New York, NY, 1969.

[DS]   N. Dunford and J. T. Schwartz, *Linear operators, Part I*, Interscience, New York, NY, 1957.

[FR]   W. H. Fleming and R. W. Rishel, *Deterministic and stochastic optimal control*, Springer-Verlag, New York, NY, 1975.

[G]   K. A. Grasse, *Controllability and accessibility in nonlinear control systems*, PhD Dissertation, University of Illinois at Urbana-Champaign, Urbana, IL, 1979.

[GB]   K. A. Grasse and J. R. Bar-on, *Regularity properties of the phase for multivariable systems*, SIAM J. Control Optim. **35** (1997), 1366–1386.

[La]   S. Lang, *Analysis II*, Addison Wesley, Reading, MA, 1969.

[Le]   E. S. Levitin, *Perturbation theory in mathematical programming and its applications*, John Wiley & Sons, New York, NY, 1994.

[LM]   F. Lempio and H. Maurer, *Differential stability in infinite-dimensional nonlinear programming*, Appl. Math. Optim. **6** (1980), 139–152.

[SL]   H. J. Sussmann and W. Liu, *A characterization of continuous dependence of trajectories with respect to the input for control-affine systems*, Preprint.

[Y]   R. Yue, *Lipschitz continuity of the value function in a class of optimal control problems governed by ordinary differential equations*, in *Control theory, stochastic analysis, and applications (Hangzhou, 1991)*, World Sci. Publishing, River Edge, NJ, 1991, pp. 125–136.

Appendix

This appendix contains two theorems whose proofs would have unduly disrupted the presentation in the main body of the paper. The first theorem presents a purely operator-theoretic result for which we have not been able to find a reference. This result is of essential importance in the proof of Thm. 4.4.

THEOREM A.1.   *Let $\mathfrak{X}, \mathfrak{Y}$ be (real) Hilbert spaces, let $M \in \mathcal{L}(\mathfrak{X})$ be symmetric, and let $N \in \mathcal{L}(\mathfrak{X}, \mathfrak{Y})$ be surjective. Suppose in addition that there exists $\sigma > 0$ satisfying*

$$\text{(A1)} \qquad x \in \mathfrak{X} \text{ and } Nx = 0 \Rightarrow \langle x, Mx \rangle \geq \sigma ||x||^2.$$

Then the mapping $T : \mathfrak{X} \times \mathfrak{Y} \to \mathfrak{X} \times \mathfrak{Y}$ defined by

$$T(x, y) = (Mx + N^\tau y, Nx)$$

is symmetric and is a linear homeomorphism of $\mathfrak{X} \times \mathfrak{Y}$ with itself.

PROOF.   The linearity and continuity of T are immediate consequences of the corresponding properties of M and N. It is an easy exercise to verify that T is self adjoint (with respect to the inner product $\langle (x, y), (x', y') \rangle = \langle x, x' \rangle + \langle y, y' \rangle$ on $\mathfrak{X} \times \mathfrak{Y}$). We also make note of the well known fact that the surjectivity of $N : \mathfrak{X} \to \mathfrak{Y}$ implies that the linear operator $NN^\tau : \mathfrak{Y} \to \mathfrak{Y}$ is a homeomorphism of $\mathfrak{Y}$ (here $N^\tau : \mathfrak{Y} \to \mathfrak{X}$ is the Hilbert-space adjoint of N). By the open mapping theorem, to show that T is a homeomorphism, it suffices to show that T is a bijection. To this end we first prove that T is injective.

Let $(x, y) \in \mathfrak{X} \times \mathfrak{Y}$ be such that $T(x, y) = (0, 0)$. From the definition of T we obtain

$$\text{(A2)} \qquad Mx + N^\tau y = 0, \quad Nx = 0.$$

The second equation in (A2) yields

$$0 = \langle Nx, y \rangle = \langle x, N^\tau y \rangle,$$

so, taking the inner product of both sides of the first equation in (A2) with x, we obtain $\langle x, Mx \rangle = 0$. Since $Nx = 0$, the assumed property (A1) of M yields

$$0 = \langle x, Mx \rangle \geq \sigma ||x||^2,$$

whence $x = 0$. Hence the first equation in (A2) becomes $N^\tau y = 0$, so that $(NN^\tau)y = 0$. The invertibility of NN^τ implies that $y = 0$, and we conclude that T is injective (if $\mathfrak{X}$ and $\mathfrak{Y}$ were both finite dimensional, then this would be sufficient to deduce the conclusion of the Theorem).

To show that T is surjective, we will first prove that T has closed range. Thus let $(\xi, \eta) \in \overline{\text{ran}(T)}$ and choose sequences $\{x_k | k \in \mathbb{N}\} \subseteq \mathfrak{X}$ and $\{y_k | k \in \mathbb{N}\} \subseteq \mathfrak{Y}$ such that $T(x_k, y_k) \to (\xi, \eta)$. Set

(A3) $$\xi_k = Mx_k + N^\tau y_k, \quad \eta_k = Nx_k,$$

so that $\xi_k \to \xi$ and $\eta_k \to \eta$. For every $k \in \mathbb{N}$ we let

$$z_k = N^\tau (NN^\tau)^{-1} \eta_k, \quad w_k = x_k - z_k.$$

Since the sequence $\{\eta_k\}$ converges to η and the linear mapping $N^\tau (NN^\tau)^{-1}$ is continuous, we deduce that the sequence $\{z_k\}$ converges to an element, call it z, of $\mathfrak{X}$. That is,

$$z = \lim_{k \to \infty} z_k = \lim_{k \to \infty} N^\tau (NN^\tau)^{-1} \eta_k = N^\tau (NN^\tau)^{-1} \eta.$$

Observe that for every $k \in \mathbb{N}$ we have $Nz_k = \eta_k = Nx_k$ and $w_k = x_k - z_k$, so we obtain

(A4) $$\{w_k \mid k \in \mathbb{N}\} \subseteq \ker N.$$

For $k \in \mathbb{N}$ let

(A5) $$\zeta_k = Mw_k + N^\tau y_k.$$

Since

$$\zeta_k = Mw_k + N^\tau y_k = Mx_k - Mz_k + N^\tau y_k = \xi_k - Mz_k,$$

and both of the sequences $\{\xi_k\}$, $\{z_k\}$ are convergent in $\mathfrak{X}$, we infer that $\{\zeta_k\}$ is also convergent in $\mathfrak{X}$. From (A4) and (A5) it follows that for every pair of integers $k, \ell \in \mathbb{N}$ we have

(A6)(a) $$M(w_k - w_\ell) + N^\tau (y_k - y_\ell) = \zeta_k - \zeta_\ell,$$
(A6)(b) $$N(w_k - w_\ell) = 0.$$

The second equation yields

$$0 = \langle N(w_k - w_\ell), y_k - y_\ell \rangle = \langle w_k - w_\ell, N^\tau (y_k - y_\ell) \rangle,$$

so if we take the inner product of both sides of (A6)(a) with $w_k - w_\ell$ we obtain

$$\langle w_k - w_\ell, M(w_k - w_\ell) \rangle = \langle w_k - w_\ell, \zeta_k - \zeta_\ell \rangle.$$

This equality, (A6)(b), and the assumed property (A1) of M yield

$$\sigma ||w_k - w_\ell||^2 \leq \langle w_k - w_\ell, M(w_k - w_\ell) \rangle = \langle w_k - w_\ell, \zeta_k - \zeta_\ell \rangle$$
$$\leq ||w_k - w_\ell|| \, ||\zeta_k - \zeta_\ell||,$$

where the last inequality follows from the Cauchy-Schwarz inequality. Hence for every $k, \ell \in \mathbb{N}$ we have

$$\|w_k - w_\ell\| \leq \frac{1}{\sigma}\|\zeta_k - \zeta_\ell\|,$$

so the convergence of $\{\zeta_k\}$ implies that the sequence $\{w_k\}$ is Cauchy and itself convergent. Consequently, the sequence $\{x_k\}$ is convergent since by definition $x_k = w_k + z_k$ and the sequence $\{z_k\}$ is known to converge. From (A3) we obtain for every $k \in \mathbb{N}$

$$y_k = (NN^\tau)^{-1}N(\xi_k - Mx_k),$$

so the convergence of the sequences $\{x_k\}$ and $\{\xi_k\}$ and the continuity of the linear mappings in this expression allows us to infer that the sequence $\{y_k\}$ also converges. If we let $x = \lim_{k\to\infty} x_k$ and $y = \lim_{k\to\infty} y_k$, then the relation $T(x_k, y_k) \to (\xi, \eta)$ yields $T(x, y) = (\xi, \eta)$, so $(\xi, \eta) \in \operatorname{ran}(T)$. Since (ξ, η) was an arbitrary element of $\operatorname{ran}(T)$, we infer that $\operatorname{ran}(T)$ is closed.

The surjectivity now follows immediately from the standard Hilbert space formula

$$\overline{\operatorname{ran}(T^\tau)} = [\ker(T)]^\perp,$$

the symmetry of T (so $T = T^\tau$), the closedness of $\operatorname{ran}(T)$ (so that $\overline{\operatorname{ran}(T^\tau)} = \overline{\operatorname{ran}(T)} = \operatorname{ran}(T)$), and the already proved injectivity of T (so that $[\ker(T)]^\perp = \{(0,0)\}^\perp = \mathfrak{X} \times \mathfrak{Y}$). This completes the proof. $\square$

The second theorem in this appendix is crucial in the proof of the Thm. 5.11.

THEOREM A.2. *Let $[a, b]$ be a compact interval and let $\mathcal{H}: [a, b] \to \mathcal{L}(\mathbb{R}^{n+m})$ be a symmetric, positive semidefinite $(n + m) \times (n + m)$ matrix valued function whose entries are Lebesgue measurable functions of $t \in [a, b]$. Further suppose that for each $t \in [a, b]$ $\mathcal{H}(t)$ has the block structure*

$$\mathcal{H}(t) = \begin{bmatrix} Q(t) & P(t)^\tau \\ P(t) & R(t) \end{bmatrix},$$

where:

1. *$Q(t)$ is a symmetric $n \times n$ matrix with entries in $L^1([a, b], \mathbb{R})$;*
2. *$P(t)$ is an $m \times n$ matrix with entries in $L^2([a, b], \mathbb{R})$;*
3. *$R(t)$ is a symmetric $m \times m$ matrix with entries in $L^\infty([a, b], \mathbb{R})$ and there exists $\beta > 0$ such that $\langle w, R(t)w \rangle \geq \beta \|w\|^2$ for every $w \in \mathbb{R}^m$ and $t \in [a, b]$.*

Define a quadratic form $\mathcal{S}$ on the Hilbert space $W^{1,2}([a, b], \mathbb{R}^n) \times L^2([a, b], \mathbb{R}^m)$ by

$$\mathcal{S}(\psi, v) = \int_a^b [\psi(t)^\tau, v(t)^\tau] \begin{bmatrix} Q(t) & P(t)^\tau \\ P(t) & R(t) \end{bmatrix} \begin{bmatrix} \psi(t) \\ v(t) \end{bmatrix} dt.$$

Further let $\mathcal{A}: [a, b] \to \mathcal{L}(\mathbb{R}^n)$ be a measurable $n \times n$ matrix valued function with entries in $L^2([a, b], \mathbb{R})$, let $\mathcal{B}: [a, b] \to \mathcal{L}(\mathbb{R}^m, \mathbb{R}^n)$ be a measurable $n \times m$ matrix valued function with entries in $L^\infty([a, b], \mathbb{R})$, and consider the initial value problem

(IVP) $\dot{\psi}(t) = \mathcal{A}(t)\psi(t) + \mathcal{B}(t)v(t), \quad \psi(a) = 0.$

Then:

(a) *If $(\psi, v) \in W^{1,2}([a, b], \mathbb{R}^n) \times L^2([a, b], \mathbb{R}^m)$ solves (IVP) and satisfies $\mathcal{S}(\psi, v) = 0$, then $(\psi, v) = (0, 0)$;*

(b) *there exists $\sigma > 0$ such that if $(\psi, v) \in W^{1,2}([a, b], \mathbb{R}^n) \times L^2([a, b], \mathbb{R}^m)$ solves (IVP) then $\mathcal{S}(\psi, v) \geq \sigma\|(\psi, v)\|^2$.*

PROOF. To prove assertion (a) we begin by selecting an $m \times m$ positive-definite, symmetric matrix function $D(t)$ that satisfies $D(t)^2 = R(t)$ for every $t \in [a, b]$ and whose entries are in $L^\infty([a, b], \mathbb{R})$. By assumption, the matrix function $\mathcal{H}(t)$ is positive semidefinite on $[a, b]$, so for every $t \in [a, b]$ and for every $(z, w) \in \mathbb{R}^n \times \mathbb{R}^m$ we have

(A7)
$$
\begin{aligned}
0 \leq [z^\tau, w^\tau] & \begin{bmatrix} Q(t) & P(t)^\tau \\ P(t) & R(t) \end{bmatrix} \begin{bmatrix} z \\ w \end{bmatrix} \\
&= z^\tau Q(t) z + 2 z^\tau P(t)^\tau w + w^\tau R(t) w \\
&= z^\tau Q(t) z - z^\tau P(t)^\tau R(t)^{-1} P(t) z \\
&\quad + z^\tau P(t)^\tau [D(t)^{-1}]^\tau D(t)^{-1} P(t) z + 2 z^\tau P(t)^\tau w + w^\tau D(t)^\tau D(t) w \\
&= z^\tau \left(Q(t) - P(t)^\tau R(t)^{-1} P(t) \right) z + |D(t) w + D(t)^{-1} P(t) z|^2.
\end{aligned}
$$

In particular, for $z \in \mathbb{R}^n$ arbitrary, we can take $w = -R(t)^{-1} P(t) z$ to obtain

$$
D(t) w + D(t)^{-1} P(t) z = 0,
$$

which upon substitution in (A7) yields

$$
0 \leq z^\tau \left(Q(t) - P(t)^\tau R(t)^{-1} P(t) \right) z.
$$

We infer that the matrix $Q(t) - P(t)^\tau R(t)^{-1} P(t)$ is positive semidefinite on $[a, b]$. Consequently, if for an arbitrary pair $(z, w) \in \mathbb{R}^n \times \mathbb{R}^m$ we have

(A8)
$$
[z^\tau, w^\tau] \begin{bmatrix} Q(t) & P(t)^\tau \\ P(t) & R(t) \end{bmatrix} \begin{bmatrix} z \\ w \end{bmatrix} = 0,
$$

then (A8) implies that both

$$
z^\tau \left(Q(t) - P(t)^\tau R(t)^{-1} P(t) \right) z = 0
$$

and

$$
D(t) w + D(t)^{-1} P(t) z = 0.
$$

In particular, (A8) implies that

(A9)
$$
w = -R(t)^{-1} P(t) z.
$$

Now let $(\psi, v) \in W^{1,2}([a, b], \mathbb{R}^n) \times L^2([a, b], \mathbb{R}^m)$ be a pair which satisfies (IVP) and for which $\mathcal{S}(\psi, v) = 0$. Since the matrix $\mathcal{H}(t)$ is positive semidefinite on $[a, b]$, the fact that $\mathcal{S}(\psi, v) = 0$ yields

$$
[\psi(t)^\tau, v(t)^\tau] \begin{bmatrix} Q(t) & P(t)^\tau \\ P(t) & R(t) \end{bmatrix} \begin{bmatrix} \psi(t) \\ v(t) \end{bmatrix} = 0 \text{ a. e. on } [a, b],
$$

whence we deduce from (A9) that

(A10)
$$
v(t) = -R(t)^{-1} P(t) \psi(t) \quad \text{for } t \in [a, b]
$$

(after modifying v on a set of measure zero if necessary).

Let $X : [a, b] \times [a, b] \to \mathcal{L}(\mathbb{R}^n)$ denote the fundamental matrix solution of the linear homogeneous ODE $\dot{z} = \mathcal{A}(t) z$; that is, for each $s \in [a, b]$ we have $X(s, s) = I_n$ (the $n \times n$ identity matrix) and for fixed $s \in [a, b]$

$$
\frac{\partial}{\partial t} X(t, s) = \mathcal{A}(t) X(t, s) \text{ a. e. on } [a, b].
$$

The assumption that (ψ, v) solves (IVP) allows us to write

$$\psi(t) = \int_a^t X(t, \zeta)\mathcal{B}(\zeta)v(\zeta)\, d\zeta, \quad t \in [a, b],$$

and the substitution of this expression in (A10) yields

(A11) $\qquad v(t) = -R(t)^{-1}P(t) \int_a^t X(t, \zeta)\mathcal{B}(\zeta)v(\zeta)\, d\zeta \quad t \in [a, b].$

For $t \in [a, b]$ we define

$$\Gamma(t) = X(a, t)\mathcal{B}(t), \quad W(t) = R(t)^{-1}P(t)X(t, a).$$

Multiplying both sides of (A11) by $\Gamma(t)$ and re-arranging terms, we obtain

(A12) $\qquad \Gamma(t)v(t) + \Gamma(t)W(t)\int_a^t \Gamma(\zeta)v(\zeta)\, d\zeta = 0.$

For $t \in [a, b]$ let $\Psi(t)$ denote the absolutely continuous, $n \times n$ matrix valued solution of the initial-value problem

$$\dot{\Psi}(t) = \Psi(t)\Gamma(t)W(t), \quad \Psi(a) = I_n.$$

Note that $\Psi(t)$ is invertible for every $t \in [a, b]$. We multiply both sides of (A12) by $\Psi(t)$ to obtain

$$\Psi(t)\Gamma(t)v(t) + \Psi(t)\Gamma(t)W(t)\int_a^t \Gamma(\zeta)v(\zeta), d\zeta = 0$$

for every $t \in [a, b]$. The right-hand side of the above expression is seen to be the derivative (almost everywhere) of the matrix function $\Psi(t)\int_a^t \Gamma(\zeta)v(\zeta)\, d\zeta$, so we obtain

$$\frac{d}{dt}\Psi(t)\int_a^t \Gamma(\zeta)v(\zeta)\, d\zeta = 0 \quad \forall t \in [a, b]$$

$$\Rightarrow \Psi(t)\int_a^t \Gamma(\zeta)v(\zeta)\, d\zeta \text{ is constant on } [a, b]$$

$$\Rightarrow \Psi(t)\int_a^t \Gamma(\zeta)v(\zeta)\, d\zeta \equiv \Psi(a)\int_a^a \Gamma(\zeta)v(\zeta)\, d\zeta = 0$$

$$\Rightarrow \int_a^t \Gamma(\zeta)v(\zeta)\, d\zeta \equiv 0$$

$$\Rightarrow \Gamma(t)v(t) = X(a, t)\mathcal{B}(t)v(t) = 0 \quad \text{a. e. on } [a, b]$$

The insertion of this last expression into (A11) yields $v(t) = 0$ a. e. on $[a, b]$, whence $v = 0$ (as an element of L^2) and obviously $\psi = 0$ as well. This completes the proof of assertion (a).

To prove (b), by the homogeneity of the quadratic form $\mathcal{S}$ it is sufficient to prove that there exists $\sigma > 0$ such that for $(\psi, v) \in W^{1,2}([a, b], \mathbb{R}^n) \times L^2([a, b], \mathbb{R}^m)$ we have

(A13) $\qquad \|(\psi, v)\|^2 = \|\psi\|_{W^{1,2}}^2 + \|v\|_{L^2}^2 = 1 \text{ and } (\psi, v) \text{ satisfies (IVP)}$

$\qquad\qquad \Rightarrow \mathcal{S}(\psi, v) \geq \sigma.$

The existence of $\sigma > 0$ satisfying (A13) will be established by an indirect argument. Indeed, if no such $\sigma > 0$ exists, then for every $k \in \mathbb{N}$ there exists a solution (ψ_k, v_k) of (IVP) which satisfies

(A14) $$||\psi_k||^2_{W^{1,2}} + ||v_k||^2_{L^2} = 1$$

and

(A15) $$\mathcal{S}(\psi_k, v_k) < 1/k.$$

We observe that $\psi_k(a) = 0$, since (ψ_k, v_k) solves (IVP), so from the definition of the $W^{1,2}$ norm it is clear that

$$||\psi||_{W^{1,2}} = ||\dot{\psi}||_{L^2}.$$

Since $\{(\psi_k, v_k)|k \in \mathbb{N}\}$ is a bounded sequence in the Hilbert space $W^{1,2}([a,b],\mathbb{R}^n) \times L^2([a,b],\mathbb{R}^m)$, we can extract a weakly convergent subsequence $\{(\psi_{k_q}, v_{k_q})|q \in \mathbb{N}\}$, say

$$(\psi_{k_q}, v_{k_q}) \xrightarrow{wk} (\bar{\psi}, \bar{v}) \in W^{1,2}([a,b],\mathbb{R}^n) \times L^2([a,b],\mathbb{R}^m).$$

Note that (A15) implies

(A16) $$\mathcal{S}(\psi_{k_q}, v_{k_q}) \to 0 \text{ as } q \to \infty.$$

It must be the case that $\bar{\psi}(a) = 0$, because $\{\psi \in W^{1,2}([a,b],\mathbb{R}^n) \mid \psi(a) = 0\}$ is a weakly closed set in $W^{1,2}([a,b],\mathbb{R}^n)$ (since it is a strongly closed vector subspace) containing the sequence $\{\psi_{k_q}\}$ and $\bar{\psi}$ is the weak limit of this sequence. It follows from the definition of the $W^{1,2}$ inner product that

$$\psi_{k_q} \xrightarrow{wk} \bar{\psi} \text{ in } W^{1,2}([a,b],\mathbb{R}^n) \Rightarrow \dot{\psi}_{k_q} \xrightarrow{wk} \dot{\bar{\psi}} \text{ in } L^2([a,b],\mathbb{R}^m).$$

Since $v_{k_q} \xrightarrow{wk} \bar{v}$ in L^2, the definition of weak convergence in L^2 and the variation of parameters formula expressing ψ_{k_q} in terms of v_{k_q} readily imply that for every $t \in [a,b]$ we have

(A17)
$$\psi_{k_q}(t) = \int_a^t X(t,\zeta)\mathcal{B}(\zeta)v_{k_q}(\zeta)\,d\zeta$$
$$\to \int_a^t X(t,\zeta)\mathcal{B}(\zeta)\bar{v}(\zeta)\,d\zeta \text{ as } q \to \infty.$$

But by virtue of the weak convergence $\dot{\psi}_{k_q} \xrightarrow{wk} \dot{\bar{\psi}}$ in L^2 we also have for every $t \in [a,b]$

$$\psi_{k_q}(t) - \bar{\psi}(t) = \int_a^t \left(\dot{\psi}_{k_q}(\zeta) - \dot{\bar{\psi}}(\zeta)\right)d\zeta$$
$$= \int_a^b \chi_{[a,t]}(\zeta)\left(\dot{\psi}_{k_q}(\zeta) - \dot{\bar{\psi}}(\zeta)\right)d\zeta$$
$$\to 0 \quad \text{as } q \to \infty.$$

From this and (A17) we deduce that

(A18) $$\bar{\psi}(t) = \int_a^t X(t,\zeta)\mathcal{B}(\zeta)\bar{v}(\zeta)\,d\zeta \quad \forall\, t \in [a,b].$$

Since the sequence $\{\psi_{k_q}\}$ is bounded in $W^{1,2}([a,b],\mathbb{R}^n)$ the argument in the proof of Prop. 5.6 shows that we can extract a uniformly convergent subsequence;

to avoid the notational complexities of dealing with subsequences of subsequences, we persist in denoting the uniformly convergent subsequence of $\{\psi_{k_q}\}$ by $\{\psi_{k_q}\}$. It follows that $\{\psi_{k_q}\}$ converges uniformly to $\bar{\psi}$ on the interval $[a, b]$.

From the definition of the quadratic form $\mathcal{S}$ and the computations in (A7) we see that for every $q \in \mathbb{N}$ we have

$$\mathcal{S}(\psi_{k_q}, v_{k_q}) = \int_a^b \Big\{ (\psi_{k_q}(t)^\tau \big(Q(t) - P(t)^\tau R(t)^{-1} P(t)\big) \psi_{k_q}(t) \\ + |D(t) v_{k_q}(t) + D(t)^{-1} P(t) \psi_{k_q}(t)|^2 \Big\} \, dt.$$

Since $\mathcal{S}(\psi_{k_q}, v_{k_q}) \to 0$ and since the matrix

$$Q(t) - P(t)^\tau R(t)^{-1} P(t)$$

is positive semidefinite for almost every $t \in [a, b]$, we infer that

$$\lim_{q \to \infty} \int_a^b |D(t) v_{k_q}(t) + D(t)^{-1} P(t) \psi_{k_q}(t)|^2 \, dt = 0;$$

equivalently,

$$Dv_{k_q} + D^{-1} P\psi_{k_q} \to 0 \text{ in the } L^2 \text{ norm as } q \to \infty.$$

Thus we can extract yet another subsequence, still denoted by $\{(\psi_{k_q}, v_{k_q})\}$, with the properties that

(A19) $Dv_{k_q} + D^{-1} P\psi_{k_q} \to 0$ pointwise a. e. on $[a, b]$ as $q \to \infty$

and there exists a function $\alpha \in L^2([a, b], \mathbb{R})$ such that

$$|Dv_{k_q} + D^{-1} P\psi_{k_q}| \le \alpha \text{ pointwise a. e. on } [a, b].$$

By the uniform convergence of ψ_{k_q} to $\bar{\psi}$ on $[a, b]$, the invertibility of $D(t)$, and (A19), we infer that

(A20) $v_{k_q} \to -R^{-1} P\bar{\psi}$ pointwise a. e. on $[a, b]$.

Moreover, almost everywhere on the interval $[a, b]$ we have the estimate

$$|v_{k_q} + R^{-1} P\psi_{k_q}| \le |D^{-1}|\alpha.$$

Since the entries of $D(t)^{-1}$ are in $L^\infty([a, b], \mathbb{R})$, we see that $|D^{-1}|\alpha$ is an L^2 function on $[a, b]$. This yields the inequality

$$|v_{k_q}| \le |v_{k_q} + R^{-1} P\psi_{k_q}| + |-R^{-1} P\psi_{k_q}| \\ \le |D^{-1}|\alpha + |R^{-1} P\psi_{k_q}|$$

pointwise a. e. on $[a, b]$, so the uniform boundedness of the sequence $\{\psi_{k_q}\}$ allows us to apply the dominated convergence theorem to deduce from (A20) that

$$\Rightarrow v_{k_q} \to -R^{-1} P\bar{\psi} \text{ strongly in } L^2$$
$$\Rightarrow v_{k_q} \to -R^{-1} P\bar{\psi} \text{ weakly in } L^2$$
$$\Rightarrow \bar{v} = -R^{-1} P\bar{\psi},$$

where the last equality follows from the facts that $v_{k_q} \xrightarrow{wk} \bar{v}$ and weak limits are unique. Note also that $(\bar{\psi}, \bar{v})$ satisfies (IVP) as a consequence of (A18).

Thus far, we have succeeded in extracting a subsequence $\{(\psi_{k_q}, v_{k_q})\}$ of the original sequence $\{(\psi_k, v_k)\}$ with the property that $\psi_{k_q} \to \bar{\psi}$ uniformly on $[a, b]$ and $v_{k_q} \to \bar{v}$ strongly in L^2. Indeed, it is not difficult to see that we actually have $\psi_{k_q} \to \bar{\psi}$ in the $W^{1,2}$ norm, because the fact that the pairs (ψ_{k_q}, v_{k_q}) satisfy (IVP) and their known convergence properties yield

$$\dot{\psi}_{k_q} = \mathcal{A}\psi_{k_q} + \mathcal{B}v_{k_q} \to \mathcal{A}\bar{\psi} + \mathcal{B}\bar{v} = \dot{\bar{\psi}}$$

in the L^2 norm. Since $\bar{\psi}(a) = \psi_{k_q}(a) = 0$, we infer that

$$||\psi_{k_q} - \bar{\psi}||_{W^{1,2}} = ||\dot{\psi}_{k_q} - \dot{\bar{\psi}}||_{L^2} \to 0,$$

so that $\psi_{k_q} \to \bar{\psi}$ in the $W^{1,2}$ norm.

From the definition of the quadratic form $\mathcal{S}$ it follows easily that

$$0 = \lim_{q \to \infty} \mathcal{S}(\psi_{k_q}, v_{k_q}) = \mathcal{S}(\bar{\psi}, \bar{v}),$$

Since $(\bar{\psi}, \bar{v})$ satisifies (IVP) the result of part (a) implies that $\bar{\psi} = 0$ and $\bar{v} = 0$. However, the relations

$$\psi_{k_q} \to \bar{\psi} \quad \text{in } W^{1,2}([a, b], \mathbb{R}^n),$$

$$v_{k_q} \to \bar{v} \quad \text{in } L^2([a, b], \mathbb{R}^m),$$

and (from (A14))

$$||\psi_{k_q}||_{W^{1,2}}^2 + ||v_{k_q}||_{L^2}^2 = 1,$$

yield, when $q \to \infty$,

$$||\bar{\psi}||_{W^{1,2}}^2 + ||\bar{v}||_{L^2}^2 = 1.$$

This obviously contradicts the just established fact that $\bar{\psi} = 0$ and $\bar{v} = 0$, and shows that the assumption that (b) is false is untenable. This completes the proof. $\square$

DEPARTMENT OF MATHEMATICS, UNIVERSITY OF OKLAHOMA, NORMAN, OKLAHOMA, 73019
E-mail address: `kgrasse@ou.edu`

HUGHES MISSILE SYSTEMS COMPANY, 805, M/S K6, P. O. BOX 11337, TUCSON, ARIZONA 85734–1337
E-mail address: `jrbar-on@CCGATE.HAC.COM`

Contemporary Mathematics
Volume **221**, 1999

ON PERTURBATIONS OF DOMINATED SEMIGROUPS

VITALI LISKEVICH AND AMIR MANAVI

Let A be a linear operator in a Banach space X. We consider the abstract Cauchy problem

$$\frac{du}{dt} = Au, \quad u(0) = f, \quad f \in \mathcal{D}(A). \tag{1}$$

Well-posedness of this problem (up to a small technical condition) is equivalent to the fact that A is a generator of a C_0-semigroup $(S(t); t \geqslant 0)$ on X. So the question whether or not a given operator A generates a C_0-semigroup is intimately connected with the applications to the Cauchy problem and hence with the differential law of the evolution of a physical system.

Let us now restrict ourselves to a case which is important in applications to physics, namely $X = L^p := L^p(M, \mathcal{M}, \mu)$ for some measure space $(M, \mathcal{M}, \mu)$ and $p \in [1, \infty)$. We will think of $(S(t); t \geqslant 0)$ resp. (1) to describe the evolution of a "free" physical system.

Let $q : M \mapsto \mathbb{C}$ be measurable. By $q^{(n)}$ we denote the truncation of q, i.e. $q^{(n)} = (|q| \wedge n)\operatorname{sgn} q$. In the following we will use the same symbol for the function q and its associated multiplication operator and we will think of q as a "complex absorption rate", i.e., formally we want to solve the problem

$$\frac{d}{dt}u(t) = (A - q)u(t) \ (t \geqslant 0), \quad u(0) = f. \tag{2}$$

Important traditional examples are the heat equation in $L^p(\mathbb{R}^d)$ with an absorption-excitation rate $V : M \mapsto \mathbb{R}$ where $A = \Delta_p$ (with Δ_p the maximal Laplace operator in $L^p(\mathbb{R}^d)$ and $q = V$ (cf. [V1]); and the non-relativistic Schrödinger equation in $L^2(\mathbb{R}^d)$ for a spinless particle moving in $\mathbb{R}^d$ under the influence of the potential $V : M \mapsto \mathbb{R}$ where up to a constant $A = i\Delta_2$ and $q = iV$.

Below we give statements of the main results and some ideas of proofs. For the detailed exposition see [LM].

The first question, which arises, is how to associate a natural generator A_q with the formal expression '$A - q$' for highly singular potentials q. The following (traditional) procedure offers itself as very natural and intuitive: Since $q^{(n)}$ is a bounded linear operator in $X = L^p$, by standard perturbation theorems $A - q^{(n)}$ is the generator of a C_0-semigroup denoted by $(e^{t(A-q^{(n)})}; t \geqslant 0)$. If q corresponds to

1991 *Mathematics Subject Classification.* 47D03, 47D06, 47A55.

Research of the first author was supported by the Alexander von Humboldt Foundation and of the second author by Deutsche Forschungsgemeinschaft

some "real" physical perturbation, then with respect to the continuous dependence on the parameters (here q), which must hold for every realistic physical model, for all $t \geqslant 0$ the limit

$$S_q(t) := s - \lim_{n \to \infty} e^{t(A - q^{(n)})} \tag{3}$$

should exist and define a C_0-semigroup. Whenever this is true we will denote the generator of the semigroup $S_q(t); t \geqslant 0$ by A_q.

Obviously this is true if q is bounded and then $A_q = A - q$. In general one has $A_q \supset A - q$ where $A - q$ is the operator sum. Therefore, A_q should be considered as the natural realization of the formal sum "$A - q$". There are important cases, when A_q exists and $D(A) \cap D(q) = \{0\}$.

Let us now assume additionally, that there exists a positive C_0-semigroup $U(\cdot) := (U(t); t \geqslant 0)$ which dominates the given semigroup $S(\cdot)$, i.e.

$$|S(t)f| \leqslant U(t)|f| \quad (f \in L^p, \ t \geqslant 0). \tag{4}$$

Our main goal now is to show, that under the additional assumption (4) for a wide class of potentials q the limit (3) exists for all $t \geqslant 0$ and defines a C_0-semigroup $S_q(\cdot)$, which is independent of the approximating sequence, in a sense to be made precise.

This program was realized by Voigt for positive semigroups and real potentials q with the main application to Schrödinger semigroups (see [V1], [V2]).. Our aim is to extend his results in two directions. First, we relax the assumption on the initial semigroup demanding only that it should be dominated by a positive C_0-semigroup $(U(t); t \geqslant 0)$. This opens a new range of applications of the theory. In particular, it applies to the Schrödinger operator with magnetic field since the semigroup generated by this operator is dominated by the semigroup generated by the Laplacian. Second, instead of real-valued potentials as perturbations we consider singular complex valued potentials.

The study of Schrödinger operators with complex potentials was started by Nelson [N], and then the results were extended by Kato [K1], Brezis, Kato [BK], Devinatz [De] and Bivar-Weinholtz and Lapidus [BWL]. Recently perturbations of regular Dirichlet forms by complex L^1_{loc}-potentials were studied in [LSt] using the Feynman–Kac formula. Here we treat the corresponding problem in a more abstract context.

In order to describe the results we need to introduce some important definitions (cf. [V1], [V2]).

Definition 1.

(a) If V is bounded below then $0 \leqslant U_{V^{(n+1)}}(t) \leqslant U_{V^{(n)}}(t)$, $(n \geqslant -\inf V, \ t \geqslant 0)$ and therefore the limit

$$U_V(t) := s - \lim U_{V^{(n)}}(t) \tag{5}$$

exists for all $t \geqslant 0$. V is called $U(\cdot)$-admissible if $U_V(\cdot)$ is a C_0-semigroup. Then $U_V(\cdot) = s - \lim_{n \to \infty} U_{V^{(n)}}(\cdot)$.

(b) If V is bounded above then $U_{V^{(n+1)}}(t) \geqslant U_{V^{(n)}}(t)$, $(n \geqslant \sup V, \ t \geqslant 0)$. V is called $U(\cdot)$-admissible if the limit (5) exists for all $t \geqslant 0$, and $U_V(\cdot)$ is a C_0-semigroup.

(c) V is called $U(\cdot)$-admissible if there are potentials $V_\pm \geqslant 0$ such that $V = V_+ - V_-$ and V_+ resp. $-V_-$ are $U(\cdot)$-admissible.

Definition 2. $V \geqslant 0$ is called $U(\cdot)$-*regular* if V is $U(\cdot)$-admissible, and $U(\cdot) = (U_V)_{-V}(\cdot)(= s - \lim_{n\to\infty} U_{V-V^{(n)}}(\cdot))$.

Admissibility and regularity of a potential..

Let us illustrate the notions of admissibility and regularity of a potential in the case of $U(t) = e^{t\Delta_\Omega^D}$ ($\Omega \subset \mathbb{R}^d$ is an open set, Δ_Ω^D is the corresponding Dirichlet Laplacian).

First, for $V \geqslant 0$ the notions of admissibility and regularity are p-**independent**. For $V \leqslant 0$, in contrast, admissibility depends on p.

So for $V \geqslant 0$ one chooses L^2-space for the characterization.

Theorem (Voigt,'86).

1) $V \geqslant 0$ *is* $e^{t\Delta_\Omega^D}$-*admissible if and only if*

$$Q(\Delta_\Omega^D) \cap Q(V) \text{ is dense inb } L^2(\Omega).$$

Here $Q(B) = D(B^{1/2})$ *for a nonnegative selfadjoint operator* B.
In particular, if $V \in L^1_{loc}(\Omega \setminus S)$, $\text{meas} S = 0$ *then* V *is* $e^{t\Delta_\Omega^D}$-*admissible.*

2) $V \geqslant 0$ *is* $e^{t\Delta_\Omega^D}$-*regular if and only if*

$$Q(\Delta_\Omega^D) \cap Q(V) \text{ is a form core of } \Delta_\Omega^D.$$

In particular, if $V \in L^1_{loc}(\Omega \setminus S)$, *where* S *is a closed set with* $\text{cap} S = 0$, *then* V *is* $e^{t\Delta_\Omega^D}$-*regular.*

Note that Stollmann and Voigt constructed in [StV] an example of a regular potential which is nowhere in L^1. So L^1_{loc}-conditions are not adequate in this context.

One can also give a characterization of regularity in terms of Wiener measure.

. $V \geqslant 0$ *is* $e^{t\Delta_\Omega^D}$-*regular if and only if for all* $t > 0$ *the integral* $\int_0^t V(\mathbf{b}(s))ds$ *is finite for* $\mathbf{P}_x$-*a.e.* $\mathbf{b}$ *and a.e.* x *where* $\mathbf{b}$ *is the standard Brownian motion and* $\mathbf{P}_x$ *is the Wiener measure.*

Now looking at the Feynman–Kac formula with a complex-valued potential (which clearly holds if the imaginary part is bounded) one gets the intuition that the semigroup can be defined if the absolute value of the imaginary part of the potential is $U(\cdot)$-regular.

This observation finds its rigorous formulation in the next proposition.

Proposition 1. *Let* $Re\ q$ *be* $U(\cdot)$-*admissible and* $|Im\ q|$ *be* $U(\cdot)$-*regular. Then for all* $t \geqslant 0$ *the limit (3) exists and* $S_q(\cdot)$ *thus defined is a* C_0-*semigroup.*

Theorem 1. *(General dominated convergence theorem for complex potentials) Let* $q, q_n : M \mapsto \mathbb{C}(n \in \mathbb{N})$ *be such that* $q_n \longrightarrow q$ *a.e. Suppose that* $Re\ q$ *is* $U(\cdot)$-*admissible and there exist* $V_\pm, W_+ \geqslant 0$, *where* $-V_-$ *is* $U(\cdot)$-*admissible and* V_+, W_+ *are* $U(\cdot)$-*regular, such that* $-V_- \leqslant Re\ q_n \leqslant Re\ q + V_+$ *and* $|W_n| \leqslant W_+(n \in \mathbb{N})$. *Then the assumptions of Proposition 1 are fulfilled for* q *and* $q_n(n \in \mathbb{N})$ *and moreover*

$$S_q(t) = s - \lim_{n\to\infty} S_{q_n}(t)\ (t \geqslant 0). \tag{6}$$

Corollary. *(Dominated convergence theorem) Let $q, q_n : M \mapsto \mathbb{C}(n \in \mathbb{N})$ be such that $q_n \longrightarrow q$ a.e. Suppose that there exist $V_\pm, W_+ \geqslant 0$, where $-V_-$ is $U(\cdot)$-admissible and V_+, W_+ are $U(\cdot)$-regular, such that $-V_- \leqslant Re\ q_n \leqslant V_+$ and $|W_n| \leqslant W_+(n \in \mathbb{N})$. Then the assumptions of Proposition 1 are fulfilled for q and $q_n(n \in \mathbb{N})$ and moreover*

$$S_q(t) = s - \lim_{n \to \infty} S_{q_n}(t)\ (t \geqslant 0).$$

Remark. The assumptions of Theorem 1 are fulfilled particularly if $Re\ q$ is $U(\cdot)$-admissible, $|Im\ q|$ is $U(\cdot)$-regular, $-(Re\ q)^- \leqslant Re\ q_n \leqslant (Re\ q)^+$ and $|Im\ q_n| \leqslant |Im\ q|$, where $(Re\ q)^\pm$ denotes the positive resp. negative part of $Re\ q$, i.e. $(Re\ q)^\pm = (\pm \Re\ q) \vee 0$. This follows from the fact that $U(\cdot)$-admissibility of $Re\ q$ implies that $-(Re\ q)^-$ is $U(\cdot)$-admissible and hence $(Re\ q)^-$ is $U(\cdot)$-regular (!). Therefore one can take $V_+ := (Re\ q)^-$ and $W_+ := |Im\ q|$.

Next we explain the main observation which links the theory of positive semigroups with the theory of dominated semigroups. It is contained in the following proposition.

Proposition 2. *Let $q_1 = V_1 + iW_1, q_2 = V_2 + iW_2$, $q_1, q_2 \in L^\infty$, $V_1 \leqslant V_2$. Then for all $f \in L^p, t \geqslant 0$ the following inequalities hold*
(a) $|S_{q_2}(t)f| \leqslant U_{V_1}(t)|f|$,
(b) $|S_{q_1}(t)f - S_{q_2}(t)f| \leqslant U_{V_1 - |W_1 - W_2|}(t)|f| - U_{V_2}(t)|f|$.

The proof of (a) is based on the Trotter product formula. (b) follows from (a) and the Duhamel formula.

Remark. Proposition 2(b) shows that the difference of the perturbed dominated semigroups reacts in an appropriate way to changes of the potentials.

We give some comments to the proofs of Proposition 1 and Theorem 1. The crucial point is that $S_q(\cdot)$ can be defined by an iteration procedure, namely $S_q(t) = ((S_{(Re\ q)^+})_{-(Re\ q)^-})_{iIm\ q}(t)$. Now looking at Proposition 2 one can see that if $(Re\ q)^+$ is $U(\cdot)$-admissible then $S_{(Re\ q)^+}(\cdot)$ exists and will be dominated by the positive C_0-semigroup $U_{(Re\ q)^+}(\cdot)$. Repeating this kind of argument one shows the existence of the C_0-semigroup $(S_{(Re\ q)^+})_{-(Re\ q)^-}(\cdot)$ which will be dominated by the semigroup $(U_{(Re\ q)^+})_{-(Re\ q)^-}(\cdot)$. To handle the purely imaginary case one needs an additional perturbation argument.

Schrödinger semigroups with singular magnetic field and complex potentials.

Here we consider Schrödinger operators with singular magnetic field and singular complex potential. We construct a C_0-semigroup whose generator corresponds to the formal differential expression

$$(i\nabla + b)^2 + q,\ q = V + iW.$$

Let $\Omega \subset \mathbb{R}^d$ be an arbitrary open set.

The notions of admissibility and regularity will be considered further on with respect to the semigroup $e^{t\Delta_\Omega^D}$.

Let $b : \Omega \mapsto \mathbb{R}^d$ be measurable and such that $b^2 = \sum_{j=1}^d b_j^2$ is $U(\cdot)$-admissible. Define the sesquilinear form

$$\langle (i\nabla + b)u, (i\nabla + b)v \rangle = \sum_{j=1}^d \int_\Omega (i\partial_j + b_j)u\overline{(i\partial_j + b_j)v}dx\ \ (u, v \in \mathcal{D}(\nabla) \cap \mathcal{D}(|b|)).$$

The form is closable as a sum of positive closable forms. The closure will be denoted by $h^D(b)$ and the associated operator by $H^D(b)$. This operator will be called Schrödinger operator with magnetic field.

Our next goal is to prove the domination property. The next result generalizes the results of T.Kato and B.Simon.

. *Let b^2 be $U(\cdot)$-admissible. Then*

$$|e^{-tH^D(b)}f| \leqslant e^{-tH_0^D}|f|.$$

Our proof is based on a domination criterion by Ouhabaz [Ou].

The next result is an approximation theorem for the Schrödinger with magnetic field.

Theorem 2. *Let $g : \Omega \mapsto \mathbb{R}^+$, g^2 be $U(\cdot)$-regular and $b, b_n : \Omega \mapsto \mathbb{R}^d$ ($n \in \mathbb{N}$) such that $b_n \to b$ a.e. as $n \to \infty$ and $|b_n| \leqslant g$ ($n \in \mathbb{N}$). Let q and q_n satisfy the conditions of Theorem 1. Then*

$$e^{-tH^D(b)_q} = s - \lim_{n \to \infty} e^{-tH^D(b_n)_{q_n}}$$

uniformly for t in bounded subsets of $[0, \infty)$.

In the next theorem we reduce our generality to the case of L^1_{loc}-potentials and L^2_{loc}-magnetic fields.

Theorem 3. *Let $q, q_n \in L^1_{loc}(\Omega)$, $b, b_n \in [L^2_{loc}(\Omega)]^d$ ($n \in \mathbb{N}$). Let*

$$\text{for every } \phi \in C_0^\infty(\Omega) \quad \|(q_n - q)\phi\|_1 + \|(b_n - b)\phi\|_2 \to 0 \ (n \to \infty)$$

and in form sense

$$V_k^-, V^- \leqslant \beta H_0^D + c(\beta) \quad (k \in \mathbb{N}) \ \text{for some} \ \beta \in (0, 1), \ c(\beta) \geqslant 0. \tag{7}$$

Then

$$e^{-tH^D(b)_q} = s - \lim_{n \to \infty} e^{-tH^D(b_n)_{q_n}}$$

uniformly for t in bounded subsets of $[0, \infty)$.

Remarks.1. It is well known that $-V_k^-, -V^-$ satisfing (7) are $U(\cdot)$-admissible.

Nevertheless one cannot deduce this theorem from Theorem 2 since in general the sequence (q_n) does not satisfy the conditions of the general dominated convergence theorem.

2. Note that if $0 \leqslant V \in L^1_{loc}(\Omega)$ then V is $U(\cdot)$-regular.

3. In case $\operatorname{Im} q = 0$ the operator $H^D(b)_q$ is the form sum $H^D(b) \dot{+} q$. For $b = 0$ and $\operatorname{Im} q = 0$ Theorem 3 coincides with the corresponding approximation result of Brüning and Gestezy [BG].

References

[BWL]   A. Bivar-Weinholtz and M.L. Lapidus, *Product formula for resolvents of normal operators and the modified Feynman integral*, Proc. AMS **110** (1990), 449–460.

[BK]   H.-Brezis and T. Kato, *Remarks on the Schrödinger operator with singular complex potentials*, J.Math.Pure Appl. **58** (1979), 137–151.

[BG]   E. Brüning and F. Gesztesy, *Continuity of wave and scattering operators with respect to interactions*, J.Math.Phys. **24** (1983), 1516–1528.

[De]   A. Devinatz, *Schrödinger operators with singular potentials*, J.Operator Theory **4** (1980), 25–35.

[K]   T. Kato, *Remarks on Schrödinger operators with vector potentials*, Integr. Equat. Operator Theory **1** (1978), 103–113.

[K1]   T. Kato, *On some Schrödinger operators with a singular complex potential*, Ann. Scuola Norm. Sup. Pisa **5** (1978), 105–114.

[LM]   V. Liskevich and A. Manavi, *Dominated semigroups with singular complex potentials*, J. Funct. Anal. **151** (1997), 281–305.

[LSt]   V.A. Liskevich and P. Stollmann, *Schrödinger operators with singular complex potentials as generators: existence and stability*, Semigroup Forum, to appear.

[N]   E. Nelson, *Feynman integrals and the Schrödinger equation*, J.Math.Phys. **5** (1964), 332–343.

[Ou]   E.-M. Ouhabaz, *Invariance of closed convex sets and domination criteria for semigroups*, Potential Anal. **5** (1996), 611–625.

[PSe]   M.A. Perelmuter and Yu. A. Semenov, *On decoupling of finite singularities in the scattering theory for the Schrödinger operators with a magnetic field*, J.Math.Phys. **22(3)** (1981), 521–533.

[StV]   P. Stollmann and J. Voigt, *A regular potential which is nowhere in L_1*, Letters Math.Phys. **9** (1985), 227–230.

[V1]   J. Voigt, *Absorption semigroups, their generators and Schrödinger semigroups*, J. Funct. Anal. **67** (1986), 167–205.

[V2]   J. Voigt, *Absorption semigroups*, J. Operator Theory **20** (1988), 117–131.

SCHOOL OF MATHEMATICS, UNIVERSITY OF BRISTOL, BRISTOL BS8 1TW, UK

INSTITUT FÜR ANALYSIS, FACHRICHTUNG MATHEMATIK, TU DRESDEN, 01062 DRESDEN, GERMANY

Contemporary Mathematics
Volume **221**, 1999

ON SOME NONLINEAR DISPERSIVE EQUATIONS

Mi Ai Park

ABSTRACT. We consider the initial value problem for the Rosenau equation. The global existence, uniqueness and the smoothing effect in the time variable for the initial value problem are invesgated.

1. Introduction

We consider two fifth order differential equations:

$$(1) \qquad u_t + u_{xxxxt} + \phi(u)_x = 0 \qquad \text{for } x \in \mathbf{R},\, t > 0$$

$$(2) \qquad u_t + u_{xxxxx} + \psi(u)_x = 0 \qquad \text{for } t, x \in \mathbf{R}.$$

Equation (1), the Rosenau equation, models approximately the dynamics of certain large discrete systems, when $\phi(u) = u + u^2$ ([6]). We may regard (1) as a variant of the Benjamin-Bona-Mahony equation

$$(3) \qquad u_t - u_{xxt} + \phi(u)_x = 0 \qquad \text{for } x \in \mathbf{R}.$$

Equation (2) is not only a mathematical model for the fifth order dispersive term instead of the third one of the KdV equation, but is important in the magneto-acoustic wave propagating at an angle to an external magnetic field, the gravitational wave under the effect of surface tension, and the wave in a nonlinear LC circuit with mutual inductance between neighboring inductors, when $\psi(u) = u^2$ ([8]). The global (in time) existence and uniqueness of the initial-value problem for (2) have been investigated ([2]).

We consider the initial-value problem for the Rosenau equation

$$(4) \qquad u_t + u_{xxxxt} + \phi(u)_x = 0 \quad \text{for } x \in \mathbf{R}, t > 0$$

$$(5) \qquad u(x,0) = f(x) \qquad \text{for } x \in \mathbf{R}$$

It has been proved that the initial-value problem (4)-(5) has a unique global solution for $\phi \in C^2(\mathbf{R})$, $f \in H_0^4(\mathbf{R})$ and that the solution is C^∞ if $\phi \in C^\infty(\mathbf{R})$ and $f \in C^\infty(\mathbf{R})$ ([3]). We have generalized the results to the higher dimensional cases ([4]).

1991 Mathematics Subject Classification. Primary 35A05,35D05; Secondary 35Q53.

The author is partially supported by the GARC-KOSEF at Seoul National Univ.

Now we consider the initial-value problem

(6) $u_t + u_{xxxxt} + \epsilon u_{xxxx} + \phi(u)_x = 0 \quad$ for $x \in \mathbf{R}, t > 0$

(7) $u(x,0) = f(x) \qquad$ for $x \in \mathbf{R}$.

We will show the initial-value problem (6)-(7) has a unique solution in $C(0, \infty, H^s)$ for $\epsilon > 0$ and for $f \in H^s$, where $s \geq 2$. We will also prove the smoothing effect in the time variable. It will be proved that for each $k > 0$, $\partial_t^k u$ lies in $C(0, T; H^s)$ for each $T > 0$ and $s \geq 2$.

All the above results carry over in the limit as ϵ tends to zero, but with $\epsilon = 0$, $\partial_t^k u \in C(0, T : H^{s+1})$ for $k > 0$ and $T > 0$. Therefore we will have the existence and uniqueness for the initial-value problem (4)-(5) in $C(0, \infty; H^s)$ where $f \in H^s$ for $s \geq 2$, which improves the earlier result. We will also have the smoothing effect in the time variable for the initial-value problem (4)-(5).

This is an analogue of the Amick-Bona-Schonbek's result ([1]) on decay of solutions of some nonlinear wave equations.

2. Notation

If X is any Banach space its norm will generally be denoted $|| \cdot ||_X$, except for the special abbreviations mentioned below. If $\Omega \subset \mathbf{R}^n$ is a Lebesgue measurable set, $L_p(\Omega)$ is the Banach space of the pth power integrable functions normed by

$$|f|_{L_p(\Omega)} = \{ \int_\Omega |f(x)|^p dx \}^{\frac{1}{p}}.$$

If Ω is understood from the context, we shall write L_p for $L_p(\Omega)$ and $|f|_p$ for the norm of $f \in L_p(\Omega)$. The usual modification will be presumed if $p = \infty$. If $f \in L_p(\Omega)$ and its distributional derivatives up to order k also lie in $L_p(\Omega)$, we write $f \in W_p^k(\Omega)$. This class of functions is a Banach space with norm

$$||f||_{W_p^k(\Omega)} = \sum_{|\alpha| \leq k} |\partial^\alpha f|_{L_p(\Omega)},$$

where the usual multi-index notation is being employed. The case $p = 2$ deserves the special notation $H^k(\Omega)$. If Ω is understood, then H^k will stand for $H^k(\Omega)$, and the norm of f in $H^k(\Omega)$ will be abbreviated to

$$||f||_k = ||f||_{H^k(\Omega)}.$$

If $f \in L_1(\mathbf{R})$, its Fourier transform is defined as

$$\hat{f}(y) = (\mathcal{F}f)(y) = \int_{-\infty}^{\infty} f(x) e^{-ixy} dx.$$

If $f \in L_2(\mathbf{R})$, then $\mathcal{F}f \in L_2(\mathbf{R})$ and $\mathcal{F}$ extends by continuity to a Hilbert-space isomorphism of $L_2(\mathbf{R})$ onto itself. If k is a positive integer and $f \in H^k(\mathbf{R})$, then

$$||f||_k^2 = \int_{-\infty}^{\infty} (1 + y^2 + \cdots + y^{2k}) |\hat{f}(y)|^2 dy.$$

These elementary facts may be found in Yosida's text([7]). If I is a closed interval in $\mathbf{R}$, say $I = [a, b]$ where $a = -\infty$ or $b = +\infty$ is allowed, and X is a Banach space, let $C_b(a, b; X) = C_b(I; X)$ denote the bounded continuous mappings $u : I \to X$. This is again a Banach space with the norm

$$||u||_{C_b(I;X)} = \sup_{t \in I} ||u(t)||_X.$$

If I is bounded, the subscript b, for bounded, will be dropped. Similarly, if $1 \le p \le \infty$ and I and X are as above, $L_p(I; X)$ is the collection of measurable functions $u : I \to X$ such that

$$||u||_{L_p(I;X)} = \{\int_I ||u(s)||_X^p ds\}^{\frac{1}{p}} < \infty.$$

3. Main Theorem

We prove existence, uniqueness, regularity and continuous dependence for the initial-value problem (4)-(5).

LEMMA. *Let $f \in H^2$. Then there is a unique solution u of (6)-(7). This solution satisfies*

$$||u(\cdot, t)||_2 \le ||f||_2.$$

For the proof, see Park [5].

THEOREM. *Let $f \in H^s$, where $s \ge 2$. Then there exists a unique function in $C_b(0, \infty; H^2)$ which also lies in the class $C(0, T; H^s)$ for each $T > 0$ and is such that u solves the initial value problem (4)-(5). For each $k > 0$, $\partial_t^k u$ also lies in $C(0, T; H^s)$ for each $T > 0$. The mapping that associates to f in H^s the solution u of (4) with initial value f is continuous from H^s to $C^k(0, T; H^s)$, for $k \ge 0$. If $f \in W_1^k$, then u and all of its temporal derivatives lie in $C(0, T; W_1^k)$, for all $T > 0$.*

Proof. Let u be a solution of (4) and let the Green's function

$$\frac{1}{8}\{(\sqrt{2} - \sqrt{2}i)e^{-(\frac{\sqrt{2}}{2} - \frac{\sqrt{2}}{2i})|z|} + (\sqrt{2} + \sqrt{2}i)e^{-(\frac{\sqrt{2}}{2} + \frac{\sqrt{2}}{2i})|z|}\}$$

for $1 + \partial_x^4$ on the whole line be denoted by $K(z)$. Evaluate (4) for u at the point (y, t), multiply the result by $K(x - y)$, and then integrate with respect to y over $\mathbf{R}$. We obtain

$$\int_{-\infty}^{\infty} K(x - y)(u_t + u_{yyyyt} + \epsilon u_{yyyy})dy = -\int_{-\infty}^{\infty} K(x - y)\phi(u)_y dy,$$

and consequently

$$(8) \quad u_t(x, t) = -\epsilon u(x, t) - \epsilon \int_{-\infty}^{\infty} K(x - y)u(y, t)dy - \int_{-\infty}^{\infty} K'(x - y)\phi(u)dy$$

where $K'(z) = \frac{1}{4} sgn(z)\{ie^{-(\frac{\sqrt{2}}{2} - \frac{\sqrt{2}}{2})|z|} - ie^{-(\frac{\sqrt{2}}{2} + \frac{\sqrt{2}}{2}i)|z|}\}.$

Equation (8) may be viewed as an ordinary differential equation of the form

$$\dot{u} = -\epsilon u + g,$$

where the dot denotes differentiation with respect to the time variable. This interpretation of (8) allows one to deduce readily the formula (i.e. the integral equation)

$$u(x,t) = e^{-\epsilon t} f(x) + \int_0^t e^{\epsilon(s-t)} \{ \epsilon \int_{-\infty}^{\infty} K(x-y) u(y,s) dy$$

$$- \int_{-\infty}^{\infty} K'(x-y)\phi(u) dy \} ds$$

(9) $$= f_0(x,t) + A(u)(x,t) \equiv B(u)(x,t)$$

for functions u which are sufficiently smooth.

It is easy to see that if T is sufficiently small, then B is a contraction mapping of a ball centered at zero in $C(0,T;H^2)$. Hence (9) possesses a solution in $C(0,T;H^2)$, at least for T small. It is immediate from (9) that if $f \in H^s$ for $s > 2$, then $u \in C(0,T;H^s)$. Moreover, the time interval T for which B is known to be contractive depends inversely on $\|f\|_2$ It follows that if $\|u(\cdot,t)\|_2$ is shown to be bounded on finite time intervals, then this contraction-mapping argument may be succesfully iterated to produce a solution of (9) defined for all $t \geq 0$.

More precisely, suppose u is defined in $C(0,T_1;H^2)$ as a solution of (9). Then one can use $u(\cdot,T_1)$ as initial data and use the corresponding reduction (4) to (9) to extend u as a solution of (4) over the interval $[0,T_1+\Delta T]$. The positive quantity ΔT depends inversely on $\|u(\cdot,T_1)\|_2$. One continues this procedure inductively and makes use of the a priori information about $\|u(\cdot,t)\|_2$ to conclude that u may be extended to any bounded interval in a finite number of such steps. By the above lemma, the H^2-norm of a solution of (4) is actually bounded for all time, and hence that the above argument is effective in producing global solutions of the initial-value problem for (4).

One way to establish the L^1 results in the statement of the theorem is to mimic the argument just outlined relative to the function class H^2. It turns out that using both the L^1 and H^2 norms simultaniously is effective. Define

$$X_T = C(0,T;H^2) \cap C(0,T;L^1)$$

with the norm

$$\||u\||_T = \|u\|_{C(0,T;H^2)} + \|u\|_{C(0,T;L^1)}.$$

For $u,v \in X_T$, let $a > 0$, $\|u(\cdot,s)\|_2 \leq a$, $\|v(\cdot,s)\|_2 \leq a$. Then

$$|A(u)(\cdot,t) - A(v)(\cdot,t)|_1$$

$$\leq t\{|K|_1 + |K'|_1 \sup_{0 \leq s \leq t} \|\phi'\|_{L_\infty[-\tilde{a},\tilde{a}]}\} \times \sup_{0 \leq s \leq t} |u(\cdot,s) - v(\cdot,s)|_1$$

$$\leq tC\|u - v\|_{C(0,t;L_1)},$$

whenever $||u||_\infty, ||v||_\infty \leq \tilde{a}$. Hence, for any $T > 0$,

(10) $$||A(u) - A(v)||_{C(0,T;L_1)} \leq TC||u - v||_{C(0,T;L_1)}$$

where

(11) $$C = C(||u||_{C(0,T;H^2)}, ||v||_{C(0,T;H^2)}),$$

and A is defined by (9).

Exactly the same estimate as (10) holds in $C(0,T;H^2)$, and the constant has the dependence depicted in (11). It follows that

$$|||A(u) - A(v)|||_T \leq TC|||u - v|||_T,$$

where

$$C = C(|||u|||_T, |||v|||_T).$$

We conclude that B is contractive in a ball of radius R centered at zero in X_T provided R is large enough and T is small enough. The desired result will follow if it can be shown that, for a solution u of (9), $|||u|||_T$ is uniformly bounded as T ranges over any bounded interval. Since

(12) $$||u||_{C(0,T;H^2)} \leq ||f||_2$$

for all $T > 0$ by the lemma, it suffices to bound the L_1-norm of u on bounded time intervals. But from (9),

(13) $$|u(\cdot,t)|_1 \leq |f|_1 + \int_0^t C|u(\cdot,s)|_1 ds$$

where C is as in (11), except that C is now known to be bounded independently of t from (12). Gronwall's lemma applies to (13) and gives the desired result. Now we consider the further spatial and temporal regularity of u. Notice that the mapping A defined in (9) carries $C(0,T;H^s) \cap C(0,T;W_1^s)$ into $C(0,T;H^{s+1}) \cap C(0,T;W_1^{s+1})$ for any $s \geq 0$. This fact follows easily from the properties of the kernel K and formula (9). A simple induction shows that u is as smooth spatially as the initial datum f. From this and (8) it is then deduced that u_t lies in $C(0,T;H^s) \cap C(0,T;W_1^s)$. For $m > 1$, another easy induction demostrates that

$$\partial_t^m u = \epsilon \partial_t^{m-1} - \epsilon \int_\infty^\infty K(x - y)\partial_t^{m-1} u(y,t)dy$$

$$- \int_{-\infty}^\infty K'(x - y)\partial_t^{m-1}[\phi(u)]dy,$$

and consequently the final result in the statement of the theorem is seen to be valid.

Remark. We can establish the similar results for the initial value problem (2).

REFERENCES

1. C.J.Amick,J.L.Bona and M.E.Schonbek, *Decay solutions of some nonlinear wave equations,* J. Diff. Eqns. **81** (1989), 1-55.
2. Y.Choi, *Ph.D.thesis,* Univ. of Chicago, 1994.
3. M.A.Park, *On the Rosenau equation,* Matem. Aplic. e Comput. **9** (1990), 145-152.
4. M.A.Park, *On the Rosenau equation in multidimensional space,* Nonlinear Anal. TMA **21** (1993), 77-85.
5. M.A.Park, *On nonlinear dispersive equations,* Diff. Int. Equ. **9** (1996), 1331-1335.
6. Ph.Rosenau, *Dynamics of dense discrete systems,* Prog. Theoretical Phys. **79** (1988), 1028-1042.
7. K.Yosida, *Functional Analysis,* Springer-Verlag, Berlin, 1974.
8. K.Yoshimura and S.Watanabe, *Chaotic behavior of nonlinear evolution equation with fifth order dispersion,* J. Phys. Soc. Jpn. (1982), 3028-3035.

Department of Mathematics, Soongsil University, Seoul 156-743, Korea
E-mail address: mapark@math.soongsil.ac.kr.

Contemporary Mathematics
Volume **221**, 1999

On perturbation theory for linear elliptic and parabolic operators; the method of Nash

by

Yu. A. Semenov

0. Introduction.

It is known that the heat equation $\partial_t u = \Delta u$ in $\mathbb{R}^d$, $d \geq 1$, has

the explicit heat kernel

$$(0.1) \quad p_0(t, x, y) = (4\pi t)^{-d/2} \exp(-|x - y|^2/4t) \qquad (t > 0, \ x, y \in \mathbb{R}^d)$$

so that the semigroup

$$e^{t\Delta} f(x) := \langle p_0(t, x, \cdot) f(\cdot) \rangle = \int_{\mathbb{R}^d} p_0(t, x, z) f(z) dz$$

delivers the unique solution for a suitable initial data $f(x)$. Based

upon (0.1) one can recover all main inequalities of the theory of

Sobolev spaces, in particular, the Sobolev inequality

$$\langle |f|^{2d/(d-2)} \rangle^{(d-2)/d} \leq c_s \langle (\nabla f)^2 \rangle, \qquad (d \geq 3)$$

and the Nash inequality

$$(0.2) \qquad \langle f^2 \rangle^{1+2/d} \leq c_N \langle (\nabla f)^2 \rangle \langle |f| \rangle^{4/d}, \qquad (d \geq 1)$$

which are valid for all (real) $f \in C_0^\infty(\mathbb{R}^d)$.

Amongst semigroups closely related to $e^{t\Delta}$ the Hermite semi-

group plays a central role in the discussing theory. Its (minus) gen-

erator $\mathcal{H}_s := -\Delta + |o - x|^2/16s^2 - d/4s$, $s > 0$, has the discrete

---

1991 *Mathematics Subject Classification.* Primary 35K10; Secondary 35K15.

spectrum $\mathrm{Spect}(\mathcal{H}_s) = \{0, 1/2s, 2/2s, 3/2s, \ldots\}$, the corresponding

eigenfunctions $\phi_i(x) = \phi_i(s, o, x)$, $i = 0, 1, 2, \ldots$, are the Hermitian

polynomials so that $\langle \phi_i, \phi_j \rangle = \delta_{ij}$, $\phi_0(s, o, x) = \sqrt{p_0(s, o, x)}$, and the

Poincaré type inequality holds

$$(0.3) \qquad \left\langle \phi_0^2 \left(f - \langle \phi_0^2 f \rangle \right)^2 \right\rangle \leq 2s \langle \phi_0^2 (\nabla f)^2 \rangle.$$

We consider now the heat equation $\partial_t u + Au = 0$ with $A = -\nabla \cdot a \cdot \nabla$, where the symmetric smooth matrix $a(x)$: $\mathbb{R}^d \mapsto \mathbb{R}^d \otimes \mathbb{R}^d$

satisfies the uniform ellipticity condition:

$$(\mathbf{H_u}) \qquad\qquad \sigma I \leq a(x) \leq \zeta I \qquad (x \in \mathbb{R}^d)$$

for some constants $o < \sigma \leq \zeta < \infty$.

One of the most significant results for this equation is the Gaussian upper and lower bounds on the heat kernel $p(t, x, y)$: *There are constants $0 < c_i < \infty$, $i = 1, 2, 3, 4$, depending only on d, σ and ζ such that for all $t > 0$, $x, y \in \mathbb{R}^d$*

$$(0.4) \qquad\qquad p(t, x, y) \leq c_1 p_0(c_2 t, x, y)$$

$$(0.5) \qquad\qquad p(t, x, y) \geq c_3 p_0(c_4 t, x, y)$$

This result is essentially due to John Nash [N] who treated even a more general time-dependent equation

$$\partial_t u + A(t)u = 0, \quad A(t) = -\nabla \cdot a(t, x) \cdot \nabla.$$

His proof rests on three principal cornerstones: *(0.2), (0.3)* and a *geometric condition* which in the case of the Euclidean structure turns into the equality $\langle \exp(-b|x - \cdot|^2) \rangle = C(d) b^{-d/2}$ ($\forall b > 0$, $x \in \mathbb{R}^d$).

In order to obtain (0.4), (0.5) Nash introduced an "entropy"

$$Q(t) \equiv Q(t, x) := -\langle p \log p \rangle \equiv - \int_{\mathbb{R}^d} p(t, x, y) \log p(t, x, y) dy$$

and a "moment" $M_1(t) := \langle |x - \cdot| p(t, x, \cdot) \rangle$. Using (0.2) and the geometric condition he derived the two-sided estimate

$$(0.6) \qquad \widetilde{Q}(t) \mp C_1 \leq Q(t) \leq \widetilde{Q}(t) + C_2 \qquad (t > 0, \ x \in \mathbb{R}^d)$$

where $\widetilde{Q}(t) := \frac{d}{2} \log t$ and positive constants C_i depend on d, σ and ζ only.

Our first goal is to show that (0.6) yields (0.4) in the following sharp form initially discovered by E.B. Davies [D]:

$$(0.4s) \qquad p(t, x, y) \leq c \cdot (\varepsilon t)^{-d/2} \exp \left\{ -\frac{D^2(x, y)}{4(1 + \varepsilon)t} \right\}$$

$$(\varepsilon \in]0, 1[, \ t > 0, \ x, y \in \mathbb{R}^d)$$

where $c = c(d, \sigma, \zeta)$ and $D(x, y) := \sup\{\psi(x) - \psi(y); \ \psi \in C_0^\infty(\mathbb{R}^d), \ \nabla \psi \cdot a \cdot \nabla \psi \leq 1\}$.

Actually, we will see that Nash's method of deriving (0.6) can be generalized in such a way that it becomes possible for the first time to construct the L^1-theory for a fairly wide class of semigroups corresponding to the operator $-\nabla \cdot a \cdot \nabla + b \cdot \nabla - \nabla \cdot \hat{b} + V$ with measurable unbounded leading and lower coefficients $a, b, \hat{b}, V$.

Armed with (0.6) it is also easy to obtain (0.5) again along the way indicated by John Nash [N]. In fact, the main step in proving (0.5) is to obtain the *off-diagonal* lower bound

$$(0.7) \qquad p(t, x, y) \geq K t^{-d/2} \qquad (t > 0, \ x, y \in B_{\sqrt{t}}(o))$$

for any $o \in \mathbb{R}^d$ with a generic constant $K(d, \sigma, \zeta)$. (Note that the RHS
of (0.6) implies the *on-diagonal* lower bound $p(t, x, x) \geq 2^{d/2} e^{-C_2 t - d/2}$
$(t > 0, x \in \mathbb{R}^d)$). Let $p_0(\cdot) := p_0(\delta t, o, \cdot)$ with any $\delta > \zeta$. One has

$$p(2t, x, y) = \langle p(t, x, \cdot) p(t, y, \cdot) \rangle$$

$$\geq \left\langle \frac{p_0(\cdot)}{\|p_0\|_\infty} p(t, x, \cdot) p(t, y, \cdot) \right\rangle$$

$$= (4\pi\delta t)^{d/2} \langle p_0(\delta t, o, \cdot) p(t, x, \cdot) p(t, y, \cdot) \rangle$$

so that, by Jensen's inequality,

$$\log p(2t, x, y) \geq \frac{d}{2} \log 4\pi\delta + \widetilde{Q}(t) + \Theta_x(t) + \Theta_y(t)$$

where we set $\Theta_z(t) = \langle p_0(\delta t, o, \cdot) \log p(t, z, \cdot) \rangle$.

The main estimate (0.7) follows from the bound

(0.8) $\Theta_z(t) \geq -\widetilde{Q}(t) - c \qquad \left(z \in B_{\sqrt{t}}(o) \right)$

with some generic constant c.

In order to obtain (0.8) introduce Nash's G-function

$$G(\tau) := \langle p_0(\delta t, o, \cdot) \log p(\tau, z, \cdot) \rangle \qquad (t/2 \leq \tau \leq t)$$

and note that $G(t) = \Theta_z(t)$. Now,

$$\frac{d}{d\tau} G(\tau) \equiv G'(\tau) = \left\langle \frac{p_0}{p} p' \right\rangle = -\left\langle \frac{p_0}{p} Ap \right\rangle$$

$$= -\left\langle \frac{\nabla p_0}{p_0} \cdot a p_0 \cdot \frac{\nabla p}{p} \right\rangle + \left\langle \frac{\nabla p}{p} \cdot a p_0 \cdot \frac{\nabla p}{p} \right\rangle$$

$$\geq -\frac{1}{2} J_0 + \frac{1}{2} J$$

where we set $J_0 = \left\langle \frac{\nabla p_0}{p_0} \cdot a p_0 \cdot \frac{\nabla p_0}{p_0} \right\rangle$ and $J = \left\langle \frac{\nabla p}{p} \cdot a p_0 \cdot \frac{\nabla p}{p} \right\rangle$.

Put $\hat{\phi}_0(\cdot) = \phi_0(\delta t, o, \cdot)$ so that $\left(-\Delta. + \frac{|o - \cdot|^2}{(4\delta t)^2} - \frac{d}{4\delta t}\right)\hat{\phi}_0(\cdot) = 0$ and $\langle(\nabla\hat{\phi}_0)^2\rangle \le \frac{d}{4\delta t}$. Thus, recalling that $\phi_0 = \sqrt{p_0}$, one has

$$J_0 \le \zeta\left\langle\frac{(\nabla p_0)^2}{p_0}\right\rangle = 4\zeta\langle(\nabla\hat{\phi}_0)^2\rangle$$

$$\le \frac{\zeta}{\delta}\frac{d}{t} = \frac{2\zeta}{\delta}\widetilde{Q}'(t).$$

In order to estimate J (from below) use (0.3), obtaining

$$J \ge \sigma\langle p_0(\nabla\log p)^2\rangle \ge \frac{\sigma}{2\delta t}\langle p_0(\log p - G(\tau))^2\rangle.$$

Thus

$$G'(\tau) \ge -\frac{\zeta}{\delta}\widetilde{Q}'(t) + \frac{\sigma}{4\delta t}\langle p_0(\log p - G(\tau))^2\rangle.$$

Now, set $\Psi(\cdot) = \log p(\tau, z, \cdot)$ and recall that by our choice $\delta > \zeta$. Then the last inequality takes the form

$$(0.9) \qquad \left(G(\tau) + \widetilde{Q}(\tau)\right)' \ge \frac{\sigma}{4\delta t}\langle p_0(\Psi - G(\tau))^2\rangle.$$

where we have used the inequality $\widetilde{Q}'(\tau) \ge \widetilde{Q}'(t)$ if $\tau \le t$.

Next, observe that, due to (0.4s) and the triangle inequality,

$$(0.10) \qquad p_0(\delta t, o, \cdot) \ge k \cdot p(\tau, z, \cdot), \qquad z \in B_{\sqrt{t}}(o), \ \tau \in [t/2, t],$$

with some generic positive constant k and any fixed $\delta > \zeta$.

Let us return to (0.9). By (0.10) and the Cauchy-Bunjakovskii inequality,

$$\langle p_0(\Psi - G(\tau))^2\rangle \ge k\langle p|\Psi - G(\tau)|\rangle^2$$

so

$$(0.11) \qquad \left(G(\tau) + \widetilde{Q}(\tau)\right)' \ge \frac{\sigma k}{4\delta t}\langle p|\Psi - G(\tau)|\rangle^2$$

and

$$\langle p | \Psi - G(\tau)| \rangle \geq \langle p\big(\Psi - G(\tau)\big)\rangle$$

$$= \langle p \log p \rangle - G(\tau)$$

$$= -Q(\tau) - G(\tau)$$

so, by the RHS of (0.6),

(0.12) $$\langle p | \Psi - G(\tau)| \rangle \geq -G(\tau) - \widetilde{Q}(\tau) - C_2 \, .$$

Now, if $-G(\tau) - \widetilde{Q}(\tau) - 2C_2 \geq 0$ for all $\tau \in [t/2, t]$ then (0.11) and (0.12) imply

$$\big(G(\tau) + \widetilde{Q}(\tau)\big)' \geq \frac{\sigma k}{4\delta t}\big(-G(\tau) - \widetilde{Q}(\tau) - C_2\big)^2$$

$$\geq \frac{\sigma k}{16\delta t}\big(-G(\tau) - \widetilde{Q}(\tau)\big)^2$$

and hence

$$G(t) \geq -C - \widetilde{Q}(t), \qquad C = 32\delta/\sigma k \, ,$$

after integration over $\tau \in [t/2, t]$.

If $-G(\tau) - \widetilde{Q}(\tau) - 2C_2 \leq 0$ for some $\tau \in [t/2, t]$ then, by (0.9),

$$G(t) + \widetilde{Q}(t) \geq G(\tau) + \widetilde{Q}(\tau) \geq -2C_2 \, .$$

The latter means that (0.8) and therefore (0.7) is established.

The above proof of (0.7) is self-contained, rather transparent, clear and *uses nothing but basic results and ideas set forth in* [N]. Exploiting essentially the same ideas we can obtain (0.5) appealing only to (0.3) and the Nash moment bound (NMB): $\langle |x - \cdot| p(t, x, \cdot)\rangle \leq c\sqrt{t}$, excluding any usage of (0.4) or (0.4s). The Nash algorithm for derivation of the Gaussian lower bound (0.5) now reads as follows.

$$(0.2) \Leftrightarrow (\text{NMB}) \Leftrightarrow (0.6), \quad (\text{NMB}) + (0.3) \Rightarrow (0.5).$$

Appealing only to (0.6) (we call it Nash's entropy estimate and denote by (NEE)) one can sharpen Gaussian lower bound (0.5). Indeed, the reproductive property of $p(t, x, y)$, Jensen's inequality, Nash's initial estimate $p(t, x, y) \leq C_N t^{-d/2}$ and (0.5) combined yield the inequality

$$\log p(t+\theta t, x, y) \geq \log \frac{c_3}{C_N} - \frac{d}{2} \log \theta - \frac{1}{c_4 \theta t} M_2(t, x) - Z(t) \qquad (\theta > 0)$$

where $M_2(t, x) = \langle |x - \cdot|^2 p(t, x, \cdot) \rangle$ and
$Z(t) = -\langle p(t, x, \cdot) \log p(t, \cdot - x + y, x) \rangle$.

By (NEE) or (0.4), $M_2(t, x) \leq kt$ with $k(d, \sigma, \zeta) < \infty$. Define the function $Z(s)$ putting

$$Z(s) = -\langle p(s, x, \cdot) \log p(s, \cdot + h(s), x) \rangle, \qquad 0 \leq s \leq t,$$

where

$$h(s) = \begin{cases} 0, & 0 \leq s \leq \delta t \\ \frac{1}{1-\delta} \left(\frac{s}{t} - \delta \right) (y - x), & \delta t \leq s \leq t \end{cases} \qquad (0 < \delta < 1).$$

By straightforward evaluation,

$$\frac{d}{ds} Z(s) \leq \left(1 + \frac{1}{\varepsilon} \right) \frac{\zeta}{\sigma} \frac{d}{ds} Q(s) + \frac{1+\varepsilon}{(1-\delta)^2} \frac{|x-y|^2}{4\sigma t^2}$$

or

$$Z(t) \leq \frac{d}{2} \log t + \frac{1+\varepsilon}{1-\delta} \cdot \frac{|x-y|^2}{4\sigma t} + K \qquad (\varepsilon > 0)$$

after integration over s from δt to t, and then using (NEE), where

$$K = C_2 + \left(1 + \frac{1}{\varepsilon} \right) \frac{\zeta}{\sigma} \left[C_1 + C_2 - \frac{d}{2} \log \delta \right] + \frac{d}{2} \log \delta.$$

Therefore, $\log p(t+\theta t, x, y) \geq -\frac{d}{2} \log \theta t - \frac{1+\varepsilon}{1-\delta} \cdot \frac{|x-y|^2}{4\sigma t} - K - \log \frac{c_N}{c_3} - \frac{k}{c_4\theta}$ so that we arrived at the *sharpened* Gaussian lower bound

$$(sLGB) \qquad c(\theta)t^{-d/2} \exp\left[-(1+\theta)\frac{|x-y|^2}{4\sigma t}\right] \leq p(t,x,y),$$

$t, \theta > 0, \; x, y \in \mathbb{R}^d$.

To understand the relation between (NEE) and the Gaussian upper bound (0.4s) let us estimate the quantity

$$I(s,x) = \int_0^s \|b(\cdot) \cdot \nabla.p(t,x,\cdot)\|_1 \, dt, \qquad s > 0, \; x \in \mathbb{R}^d,$$

where $b: \mathbb{R}^d \to \mathbb{R}^d$ satisfies the condition $\|b \cdot a^{-1} \cdot b\|_\infty = U_1 < \infty$. One has

$$\begin{aligned}
I(s,x) &\leq \int_0^s \langle pb \cdot a^{-1} \cdot b \rangle^{1/2} \left\langle \nabla p \cdot \frac{a}{p} \cdot \nabla p \right\rangle^{1/2} dt \\
&\leq U_1^{1/2} \int_0^s (Q'(t))^{1/2} dt \\
&\leq \frac{1}{2}(U_1 d)^{1/2} \int_0^s \left(d^{-1}Q'(t)t^{1/2} + t^{-1/2}\right) dt .
\end{aligned}$$

By (NEE),

$$\int_0^s Q'(t)\sqrt{t}\, dt = Q(s)\sqrt{s} - \int_0^s Q(t)d\sqrt{t} \leq (C_1 + C_2 + d)\sqrt{s}$$

so that

$$I(s,x) \leq \hat{c}\sqrt{U_1 s} \quad \hat{c} = (C_1 + C_2 + 3d)/2\sqrt{d}.$$

Due to Fubini's theorem the above estimate for $I(s,x)$ leads to the bound

$$\int_0^s \|b \cdot \nabla e^{-tA_1} f\|_1 \, dt \leq \beta(s)\|f\|_1, \; f \in L^1(\mathbb{R}^d), \; \text{where } \beta(s) = \hat{c}\sqrt{U_1 s}.$$

Therefore, by the elementary perturbation theory,

$$\| \exp[-t(A_1 + b \cdot \nabla)] \|_{1 \to 1} \leq 2 \exp(\omega U_1 t), \qquad \omega = 4\hat{c}^2 \log 2.$$

The latter, combined with an inequality of Nash and the equality $\exp[-t(A_1 + b \cdot \nabla)]1 = 1$, allows one to reiterate the argument and derive the bounds

$$\| e^{-t\mathcal{L}_1} \|_{1 \to 1} \leq c_1 \exp(c_2 U_1 t),$$

$$[*] \quad \| e^{-t\mathcal{L}_1} \|_{1 \to \infty} \leq c_3 (\varepsilon t)^{-d/2} \exp\left[(1 + \varepsilon) U_1 t\right], \qquad t > 0, \ \varepsilon \in]0, 1[$$

where $\mathcal{L}_1 = A_1 + b \cdot \nabla + \nabla \cdot b - b \cdot a^{-1} \cdot b$, $D(\mathcal{L}_1) = D(A_1)$, $c_i = c_i(d, \sigma, \zeta)$. It is a simple matter now to complete the derivation of (UGB) by employing Davies's original idea [D]. Indeed, putting $b = a \cdot \alpha$, $\alpha \in \mathbb{R}^d$, $\phi(x) = x \cdot \alpha$, one obtains from [*]

$$e^{-x \cdot \alpha} p(t, x, y) e^{y \cdot \alpha} \leq \| e^{-\phi} e^{-tA_1} e^{\phi} \|_{1 \to \infty} \leq \text{R.H.S. of } [*].$$

Finally, put $\alpha = (y - x)/2(1 + \varepsilon)\zeta t$ and use the inequality $U_1 \leq \alpha^2 \zeta$.

We have just demonstrated a unique role of (NEE) in proving the deepest results of the theory of second order divergence form operators. In the main body of this paper we establish sharpened lower and upper bounds on the weak heat kernels corresponding to general elliptic operators. To formulate the results let us introduce a new class of drifts and recall the definition of the Kato classes K_{d+1} and K_d.

We say that a locally square integrable function f belongs to the Nash class N_2 if

$$\lim_{\delta \to \infty} n(\delta, f) = 0$$

where $n(\delta, f) = \text{ess sup}_{x \in \mathbb{R}^d} \int_0^\infty e^{-\delta t} \left(e^{t\Delta} |f|^2(x) \right)^{1/2} t^{-1/2} dt$.

It is said that a locally integrable function f belongs to the Kato class K_{d+1} if

$$\lim_{\delta \to \infty} k(d+1; \delta, f) = 0$$

where $k(d+1; \delta, f) = \text{ess sup}_{x \in \mathbb{R}^d} \int_0^\infty e^{-\delta t} e^{t\Delta} |f|(x) t^{-1/2} dt$.

It is said that a locally integrable function f belongs to the Kato class K_d if

$$\lim_{\delta \to \infty} k(d; \delta, f) = 0$$

where $k(d; \delta, f) = \text{ess sup}_{x \in \mathbb{R}^d} \int_0^\infty e^{-\delta t} e^{t\Delta} |f|(x) dt$.

Note that

$$\left[L^q(\mathbb{R}^d) + L^\infty(\mathbb{R}^d), \ q > d, \ d \geq 2 \right] \subset N_2 \subset K_{d+1}$$

(the last inclusion follows from the inequality $e^{t\Delta}|f| \leq (e^{t\Delta}|f|^2)^{1/2}$).

Our first theorem contains a refinement and generalization of Aronson's upper and lower bounds [A] and an extension of Davies's upper bound [D].

Theorem 0.1. *Suppose that (i) a measurable symmetric matrix-valued function $a(\cdot)\colon \mathbb{R}^d \mapsto \mathbb{R}^d \otimes \mathbb{R}^d$ satisfies $(\mathbf{H_u})$, (ii) $b, \hat{b}\colon \mathbb{R}^d \to \mathbb{R}^d$, $b, \hat{b} \in N_2$, $b^2, \hat{b}^2, V \in K_d$. Then the form sum $\mathcal{L}_2 \supset -\nabla \cdot a \cdot \nabla + b \cdot \nabla - \nabla \cdot \hat{b} + V$ is defined and the heat kernel $r(t, x, y)$ for $e^{-t\mathcal{L}_2}$ enjoys the bounds*

$$c_\varepsilon t^{-d/2} \exp\left[-\frac{(1 + \varepsilon + \delta)|x - y|^2}{4\sigma t} - \chi_\delta t \right]$$

$$\leq r(t, x, y) \leq c(\varepsilon t)^{-d/2} \exp\left[-\frac{1}{1 + \varepsilon + \delta} \frac{D^2(x, y)}{4t} + \omega_\delta t \right]$$

which are valid for all $t > 0$, $x, y \in \mathbb{R}^d$, *all* $\varepsilon, \delta \in]0, 1[$ *and some positive*

constants c, c_ε, ω_δ *and* χ_δ.

Our next result seems to be new even when $V = 0$.

Theorem 0.2. *Suppose that (i) a measurable symmetric matrix-valued function* $a(\cdot) \colon \mathbb{R}^d \to \mathbb{R}^d \otimes \mathbb{R}^d$ *satisfies* $(\mathbf{H_u})$, *(ii)* $b \colon \mathbb{R}^d \to \mathbb{R}^d$, $b \in N_2$, $V = V^+ - V^-$, $V^\pm \geq 0$, $V^+ \in L^1_{\mathrm{loc}}(\mathbb{R}^d)$, $V^- \in K_d$. *Then* $b \cdot \nabla$, V^- *are small operator perturbations of* A_1, $\Lambda_1 := A_1 + b \cdot \nabla + V$ *of domain* $D(A_1) \cap D(V_1^+)$ *generates a* C_0-*semigroup on* $L^1(\mathbb{R}^d)$.

We remark that the assumption "$b \in N_2$" does not guarantee the form sum $A_2 \dotplus b \cdot \nabla$ to be defined. Nevertheless we will show that $T_1^t := \exp\left[- t(A_1 + b \cdot \nabla) \right]$ is a contraction on L^∞, i.e. $\|T_1^t f\|_\infty \leq \|f\|_\infty$, $f \in L^1(\mathbb{R}^d) \cap L^\infty(\mathbb{R}^d)$, $t \geq 0$, and hence extends by continuity to a C_0-semigroup on $L^p(\mathbb{R}^d)$ for all $1 < p < \infty$. Moreover, for each $\varepsilon > 0$ there is $C_\varepsilon < \infty$ such that

$$\|T_1^t\|_{1 \to \infty} \leq C_\varepsilon t^{-(d+\varepsilon)/2} \qquad (0 < t \leq 1).$$

The following result is closely related to Theorem 0.2.

Theorem 0.3. *If* $b \in K_{d+1}$ *then* $\Lambda_1 := -\Delta + b \cdot \nabla$ *of domain* $(1 - \Delta)^{-1} L^1(\mathbb{R}^d)$ *generates a* C_0-*semigroup in* $L^1(\mathbb{R}^d)$; *its integral kernel admits the bound*

$$0 \leq e^{-t\Lambda_1}(x, y) \leq c_\varepsilon t^{-d/2} \exp\left[- \frac{|x - y|^2}{(1 + \varepsilon)4t} + t\omega_\varepsilon \right],$$

$$t > 0, \ x, y \in \mathbb{R}^d, \ \varepsilon \in]0, 1].$$

We comment that without further hypotheses on b the expression $-\Delta + b \cdot \nabla$ cannot be defined on $C_0^\infty(\mathbb{R}^d)$ as an operator in $L^p(\mathbb{R}^d)$ for any $p > 1$ and it cannot be defined in $L^2(\mathbb{R}^d)$ as the form sum, but as the (minus) generator of

$$T_p^t f(x) := \int_{\mathbb{R}^d} e^{-t\Lambda_1}(x,y) f(y)\,dy\,, \qquad f \in L^p(\mathbb{R}^d)\,.$$

So far we have discussed the uniformly elliptic case. We consider now the semigroup theory for $\mathcal{L}(a,b,\hat{b},V) = -\nabla \cdot a \cdot \nabla + b \cdot \nabla - \nabla \cdot \hat{b} + V$ assuming that

$(\mathbf{H_1})$
$$\sigma I \le a(\cdot) \text{ for some } \sigma > 0\,,$$
$$a(\cdot) = a(\cdot)^* : \ \mathbb{R}^d \to \mathbb{R}^d \otimes \mathbb{R}^d\,, \quad a(\cdot) \in \left[L^1_{\mathrm{loc}}(\mathbb{R}^d)\right]^{d \times d}\,.$$

The hypothesis assures the quadratic form
$$E(u,v) := \langle \nabla \bar{u} \cdot a \cdot \nabla v \rangle$$
$$= \int_{\mathbb{R}^d} \sum_{i,j=1}^{d} a_{ij}(x) \frac{\partial \bar{u}(x)}{\partial x_i} \frac{\partial v(x)}{\partial x_j}\,, \qquad u,v \in C_0^1(\mathbb{R}^d)$$

to be closable in $L^2(\mathbb{R}^d)$. Let A denote the operator associated with the closure of E. It follows from the Beurling-Deny theory that there exist positivity preserving contraction consistent C_0-semigroups on $L^p = L^p(\mathbb{R}^d)$, $1 \le p < \infty$, with generators $-A_p$ and $A_2 = A$. One may now treat $\mathcal{L}(a,b) = -\nabla \cdot a \cdot \nabla + b \cdot \nabla$ regarding $b \cdot \nabla$ as a small perturbation of A either in the sense of quadratic forms in L^2 or operators in L^p for appropriate p's.

The main result of applying the quadratic form method to $\mathcal{L}(a,b)$ can be formulated as follows. Let us define the class $PK_\beta(A)$ of A-form bounded potentials by
$$PK_\beta(A)$$

$$= \{W \in L^1_{\mathrm{loc}}(\mathbb{R}^d); |\langle f, Wf \rangle| \le \beta E(f,f) + c(\beta)\|f\|_2^2\,, \quad f \in C_0^1(\mathbb{R}^d)\}$$

where β, $c(\beta) \in \mathbb{R}^1$ and $\langle f, g \rangle = \langle \bar{f}g \rangle := \int_{\mathbb{R}^d} \bar{f}g\, dx$.

If $b \cdot a^{-1} \cdot b \in PK_\beta(A)$ for some $\beta < 1$ then

$$|\langle \nabla f \cdot bf \rangle| \leq \sqrt{\beta} E(f, f) + \frac{c(\beta)}{2\sqrt{\beta}} \|f\|_2^2, \qquad f \in C_0^1(\mathbb{R}^d)$$

and hence, according to the KLMN-Theorem [K, Ch. 6, §1], $A + b \cdot \nabla$ is defined as the quasi m-sectorial operator associated with the closure of the form

$$t(f, g) = E(f, g) + \langle \nabla \bar{f} \cdot bg \rangle, \qquad D(t) = D(E).$$

The hypothesis $(\mathbf{H_1})$ and "$b \cdot a^{-1} \cdot b \in PK_\beta(A)$, $\beta < 1$" imply the existence of positivity preserving L^∞-contraction consistent C_0-semigroups $e^{-t\Lambda_p}$, $t \geq 0$, $2/2 - \sqrt{\beta} \leq p < \infty$, such that $\Lambda_2 = A + b \cdot \nabla$ (see Appendix, Th.A.1).

The above approach is applicable to the general operator $\mathcal{L}(a, b, \hat{b}, V)$ and yields, in particular, the following. Let $V = V^+ - V^-$, $0 \leq V^\pm \in L_{\text{loc}}^1(\mathbb{R}^d)$. If

$$b \cdot a^{-1} \cdot b, \; \hat{b} \cdot a^{-1} \cdot \hat{b}, \; V^- \in PK_1(\beta A + V^+) \text{ for all } \beta > 0$$

then there are quasi-contraction consistent C_0-semigroups on L^p, $1 < p < \infty$, with generators $-\mathcal{L}_p$ such that $\mathcal{L}_2 = A + b \cdot \nabla \dot{-} \nabla \cdot \hat{b} + V$ and

$$\|e^{-t\mathcal{L}_p}\|_{p \to q} \leq C_{p,q} e^{t\omega_{p,q}} t^{-d(p-q)/2pq}, \qquad 1 < p < q < \infty, \; t > 0.$$

The related results in $L^1(\mathbb{R}^d)$ were unknown even under rather restricted assumptions on the coefficients (e.g. $a(\cdot) = a^1(\cdot) + a^2(\cdot)$, $a_{ij}^1(\cdot) \in L^\infty(\mathbb{R}^d)$, $a_{ij}^2(\cdot) \in \bigcap_{r < \infty} L_{\text{com}}^r(\mathbb{R}^d)$; $b_i(\cdot) \in L_{\text{com}}^\infty(\mathbb{R}^d)$, $\hat{b} = 0$,

$V = 0$). The main obstacles in proving them are the following: 1) $T^t := e^{-t(A+b\cdot\nabla)}$, $t > 0$, is no longer a *(quasi) contraction* on L^1 because of the presence of $b(\cdot)$; 2) its heat kernel does not admit any upper bound *leading to* $\|T^t f\|_1 \leq ce^{t\omega}\|f\|_1$, $f \in L^1 \cap L^2$, because of the presence of $a^2(\cdot)$.

Given constants $c_0 < \infty$, $0 < d_1 \leq d$, $L > 0$ and $v > 0$, define the subclass $\mathcal{T}$ of $[C^1(\mathbb{R}^d)]^d$ by

$$\mathcal{T} = \mathcal{T}(c_0, d_1, L, v)$$

$$:= \left\{ \psi \in [C^1(\mathbb{R}^d)]^d; \sup_{x\in\mathbb{R}^d} \int \exp\left(-k|\psi(x) - \psi(y)|\right) dy \leq c_0 k^{-d_1} \right.$$

$$\left. \text{for all } k \geq 1, \text{ and } |\psi(x) - \psi(y)| \geq L|x - y|^v \text{ for all large } |x - y| \right\}.$$

Given $\psi \in \mathcal{T}$, we define

$$\Gamma(\psi) = \Gamma(\psi, y) := \text{ess sup}_{x\in\mathbb{R}^d} \Psi(x, y) \cdot a(y) \cdot \Psi(x, y),$$

$$\Psi(x, y) = \nabla_y |\psi(x) - \psi(y)|.$$

We develop the L^1-theory of $\mathcal{L}(a, b, \hat{b}, V)$ imposing the following hypothesis on $a(\cdot)$.

There is a $\psi \in \mathcal{T}$ such that $\Gamma(\psi) \leq \Gamma_1 + \Gamma_2$,

(**H$_2$**) Γ_1 is a positive finite number, $0 \leq \Gamma_2 \in L^q(\mathbb{R}^d)$ for some

$q > d/2$, $d \geq 3$.

For instance $a(\cdot) = a^*(\cdot)$ satisfies (**H$_1$**), (**H$_2$**) with $v = 1$, $d_1 = d$, if

$$\sigma I \leq a(\cdot) \in \left[L^\infty(\mathbb{R}^d) + L^q(\mathbb{R}^d)\right]^{d\times d} \text{ for some } q > d/2.$$

Given $x \in \mathbb{R}^d$, $\varepsilon > 0$ and ψ from (**H$_2$**), we define the function

$$\mathbb{R}^d \ni \cdot \to \xi(\cdot) = \eta\left(\varepsilon|\psi(x) - \psi(\cdot)|\right)$$

where

$$\eta(s) = \begin{cases} 1, & 0 \le s \le 1 \\ \left(1 - \frac{s-1}{n_0}\right)^{n_0}, & 1 \le s \le n_0 + 1 \\ 0, & n_0 + 1 \le s, \quad (n_0 > 2). \end{cases}$$

Let $p(t, x, y)$ be the heat kernel for e^{-tA}, $t > 0$. Put $p_\theta = \theta + p$, $\theta = \varepsilon^{d/v}$ and introduce

$$Q(t) = Q(t, x) = Q(t, x; \varepsilon)$$

$$:= -\langle \xi p \log p_\theta \rangle = -\langle \xi(\cdot) p(t, x, \cdot) \log p_\theta(t, x, \cdot) \rangle,$$

$$\mathcal{N}(t) = \mathcal{N}(t, x) = \mathcal{N}(t, x; \varepsilon) := \left\langle \xi \nabla p \cdot \frac{a}{p_\theta} \cdot \nabla p \right\rangle.$$

We obtain the following inequalities lying at the heart of the method.

There are finite constants c_1, c_2, c_N such that for all $t \in]0, 1]$, $\varepsilon \in]0, 1/2]$ and a.e. $x \in \mathbb{R}^d$

$$\mathcal{N}(t) \le \left(1 + \sqrt{\varepsilon}\right) Q'(t) + \varepsilon c_1 + \varepsilon c_2 t^{-d/2q} (\log t)^2,$$

$$Q(t) \ge \langle p(t, x, \cdot) \rangle \left(\widetilde{Q}(t) - c_N \right), \qquad \widetilde{Q}(t) := \frac{d}{2} \log t.$$

There is a constant $t_0 \in]0, 1]$ such that

$$Q(t) \le \langle p(t, x, \cdot) \rangle \left(\frac{d_1}{d} \hat{\gamma} \widetilde{Q}(t) + C_N(\hat{\gamma}) \right)$$

for all $t \in]0, t_0]$, $\varepsilon \in]0, 1/2]$, $\hat{\gamma} \in \left]0, 1 - \frac{d}{2q}\right[$, a.e. $x \in \mathbb{R}^d$ and some $C_N(\hat{\gamma}) < \infty$.

The above are powerful inequalities. We will use them to obtain the estimate

$$[**] \quad \text{ess sup}_{x \in \mathbb{R}^d} \int_0^s \|b(\cdot) \cdot \nabla . p(t, x, \cdot)\|_1 dt \le \beta(s), \qquad s \in]0, t_0]$$

with $\beta(s) \to 0$ as $s \to 0$, assuming $(\mathbf{H_1})$, $(\mathbf{H_2})$ and

$(\mathbf{H_3'})$ $b \cdot a^{-1} \cdot b \in L^\infty(\mathbb{R}^d) + L^q(\mathbb{R}^d)$, $q > d/2$.

Define the operators B, $\widehat{B}$ in $L^1(\mathbb{R}^d)$ by $B = B(b) := b \cdot \nabla$, $\widehat{B} := B(\hat{b})$,

$$D(B) = \{f \in L^1(\mathbb{R}^d); \nabla f \in L^1_{\text{loc}}(\mathbb{R}^d) \text{ and } b \cdot \nabla f \in L^1(\mathbb{R}^d)\}$$

where ∇ is calculated in the weak sense.

Armed with [**] we obtain our first main result for the non-uniformly elliptic case.

Theorem 0.4. *Hypotheses* $(\mathbf{H_1})$, $(\mathbf{H_2})$, $(\mathbf{H_3'})$ *imply that*

1) B *is a Rellich-Kato perturbation of* A_1 *with relative bound zero (i.e.* $D(B) \supset D(A_1)$ *and* $\|Bf\|_1 \leq \vartheta \|A_1 f\|_1 + k_\vartheta \|f\|_1$, $f \in D(A_1)$, *for all* $\vartheta > 0$ *and* $k_\vartheta < \infty$).

2) $\int_0^s \|Be^{-tA_1}f\|_1 dt \leq \beta(s)\|f\|_1$ $(f \in D(A_1)$, $s \in]0, t_0])$ *with* $\beta(s)$ *from* [**].

Due to [M], *the latter means that* $A_1 + B$ *of domain* $D(A_1)$ *generates a* C_0-*semigroup* T_1^t *on* $L^1(\mathbb{R}^d)$ *consistent with* $T^t = \exp[-t(A + b \cdot \nabla)]$, *and the bound*

$[***]$
$$\|T_1^t\|_{1 \to 1} \leq (1 - \beta(s))^{-1} \exp(t\omega), \qquad \omega = -s^{-1} \log(1 - \beta(s)), \ t > 0,$$

holds for all $s \in]0, t_0]$ *such that* $\beta(s) < 1$.

Next, due to [***] and Nash's inequality, we literally reiterate the above argument and obtain our second main result.

Theorem 0.5. *Hypotheses* $(\mathbf{H_1})$, $(\mathbf{H_2})$ *and*

$(\mathbf{H_3})$ $b \cdot a^{-1} \cdot b$, $\hat{b} \cdot a^{-1} \cdot \hat{b}$, $V^- \in L^\infty(\mathbb{R}^d) + L^q(\mathbb{R}^d)$ *for some* $q > d/2$

imply that

1) B is a Rellich-Kato perturbation of $(A \dot{-} \nabla \cdot \hat{b})_1$ with relative bound zero.

2) $\int_0^{\hat{s}} e^{-t\omega} \| B e^{-t(A \dot{-} \nabla \cdot \hat{b})_1} f \|_1 dt \leq \widetilde{\beta}(\hat{s}) \| f \|_1$, $f \in L^1(\mathbb{R}^d)$ for all small enough $\hat{s} > 0$, $\widetilde{\beta}(\hat{s}) \to 0$ as $\hat{s} \to 0$, and ω entering [***].

3) $\mathcal{L}_1 = (A \dot{-} \nabla \cdot \hat{b})_1 + B - V^-$ of domain $D\big((A \dot{-} \nabla \cdot \hat{b})_1\big)$ generates a C_0-semigroup on $L^1(\mathbb{R}^d)$.

4) The heat kernel $r(t,x,y)$ of $e^{-t\mathcal{L}_1}$, $t > 0$, has the properties:

$$0 \leq r(t,x,y) \leq ct^{-d/2} e^{\tilde{\omega} t},$$

$$\text{ess sup}_{x \in \mathbb{R}^d} \langle r(t,\cdot,x) \rangle \leq ce^{\tilde{\omega} t},$$

$$\text{ess sup}_{x \in \mathbb{R}^d} \langle r(t,x,\cdot) \rangle \leq ce^{\tilde{\omega} t} . cr$$

Thus, under rather mild assumptions on the coefficients we prove that $\hat{b} \cdot \nabla$ and $b \cdot \nabla$ are small operator perturbations of A_1 and $(A \dot{-} \nabla \cdot \hat{b})_1$ respectively, and that standard tools of perturbation theory for semigroups are applicable to $\mathcal{L}(a, b, \hat{b}, V)$. Of course, the possibility of carrying out this very natural program rests entirely on the startlingly deep Nash's ideas.

Though some fragments of the Nash method had been understood for a long time it was only displayed in [FS] that ideas set forth in [N] plus a Gaussian upper bound could be used to obtain a Gaussian lower bound proceeding from first principles. A Nash type proof of a Gaussian upper bound was found in [Fa]. The fact that Gaussian upper and lower bounds yield both Moser's parabolic Harnack inequality and Nash's local Hölder continuity of the weak solutions has been known since the seventies (see [L]).

The approach we develop here is based upon the following key observations: First (NEE) allows one to obtain sharpened heat kernel bounds from above and below under rather general assumptions on the lower term coefficients; second, some generalized (NEE)'s deliver a powerful tool for employing the beautiful Kato-Miyadera-Rellich theory in order to build up the L^1-theory of second order divergence form equations with measurable unbounded leading and lower coefficients.

The paper is arranged in six sections and an appendix. In section 1 we outline an elementary approach to the perturbation theory which is based upon DuHamel's formula, recall Miyadera's theorem, and establish some facts about semigroup heat kernels satisfying Nash's initial estimate. We also prove Theorem 0.4, but postpone the proof of crucial estimates to section 2.

Being the key part of the paper section 2 contains a proof of some generalization of (NEE) for the case with unbounded leading coefficients. We have to work with a specially cut-off entropy and other functions because of the lack of information on the decay at infinity of the heat kernels under consideration.

In sections 3 and 4 we derive the upper and lower bounds on the heat kernels in the case of bounded coefficients. Once again, the advantage of our approach to the subject is revealed in section 5 where

Theorems 0.1 and 0.2 are proved. In the last section we introduce a purely L^1-method of deriving the upper bound in the case when the notion of the weak solution cannot be applied.

1.   Some basic facts.

Let $\mathcal{B}$ be a Banach space. For a (linear) operator A in $\mathcal{B}$ we denote by $D(A)$ its domain. $\mathcal{L}(\mathcal{B})$ denotes the set of all bounded operators in $\mathcal{B}$.

A one-parameter strongly continuous semigroup on $\mathcal{B}$, or simply a C_0-semigroup is a family $(P^t,\ t \geq 0)$ in $\mathcal{L}(\mathcal{B})$ satisfying

$$P^0 = 1, \quad P^{t+s} = P^t P^s \qquad (t, s \geq 0)$$

$[0, \infty[\ni t \mapsto P^t f,\ f \in \mathcal{B}\mathrm{f}$, is continuous in $\mathcal{B}$.

These properties imply the bound $\|P^t\| \leq Ne^{\omega t}$, $t \geq 0$. Define A by $Af = \lim\limits_{t \to 0+} \frac{f - P^t f}{t}$, when this limit exists in $\mathcal{B}$; $-A$ is the (infinitesimal) generator of P^t.

$\forall f \in D(A)$ $f_t = P^t f$ is strongly differentiable and $\frac{d}{dt} f_t = -A f_t$. P^t is said to be (quasi-bounded) holomorphic if for all $t > 0$

$$P^t \mathcal{B} \subset D(A) \quad \text{and} \quad \|AP^t\| \leq Nt^{-1}e^{\omega t}\,.$$

Let $T^t = e^{-tT}$ and $T_0^t = e^{-tT_0}$ be C_0-semigroups, $\|T_0^t\| \leq Ne^{\omega t}$ $(t \geq 0)$. Assume that there exists a dense subspace $\mathcal{D}$ of $\mathcal{B}$, such that $T_0^t \mathcal{D} \subset \mathcal{D} \subset D(T_0) \cap D(T)$ $(t > 0)$, and for all $f \in \mathcal{D}$ the function $[0, \infty[\ni s \to T^{t-s}T_0^s f$ is strongly continuously differentiable.

$$\frac{d}{ds} T^{t-s}T_0^s f = T^{t-s}(T - T_0)T_0^s f\,,$$

so integrating this inequality over s from 0 to t gives the DuHamel formula

$$T^t f = T_0^t f + \int_0^t T^{t-s} B T_0^s f \, ds, \qquad f \in \mathcal{D},$$

where we set $B = T_0 - T$.

Let us suppose that for some $s \in\,]0, \infty]$ and $\beta = \beta(s) < 1$

$$\int_0^s e^{-\omega t} \| B T_0^t f \| \, dt \le \beta \| f \|, \qquad f \in \mathcal{D}.$$

Then iteration of the DuHamel formula yields the equality

$$T^t = \sum_{n=0}^{\infty} F_n^t, \qquad t \in [0, s]$$

where we set $F_0^t = T_0^t$, $F_{n+1}^t = \int_0^t F_n^{t-\tau} B T_0^\tau \, d\tau$. In fact, since

$$\| F_{n+1}^t f \| \le \int_0^t \| F_n^{t-\tau} \| \| B T_0^\tau f \| \, d\tau$$

$$\le \beta^n N e^{\omega t} \int_0^t e^{-\omega \tau} \| B T_0^\tau f \| \, d\tau$$

$$\le \beta^{n+1} N e^{\omega t} \| f \|$$

by induction, we discover that

$$\| T^t \| \le \sum_{n=0}^{\infty} \| F_n^t \| \le \frac{N}{1 - \beta} e^{\omega t}.$$

If $s < \infty$ and $t > s$, choose $n \in \mathbb{N}$ such that $s(n-1) \le t \le ns$. Then

$$\| T^t \| \le \| T^{t/n} \|^n \le \left(\frac{N}{1 - \beta} \right)^n e^{\omega t} \le \frac{N}{1 - \beta} \left(\frac{N}{1 - \beta} \right)^{t/s} e^{\omega t}$$

or

$$\| T^t \| \le \frac{N}{1 - \beta} \exp \left[\left(\omega + \frac{1}{s} \log \frac{N}{1 - \beta} \right) t \right], \qquad t \ge 0.$$

The same argument also gives the following.

Suppose that for some $\lambda \in \mathbb{R}^1$ and $\hat{\beta} = \hat{\beta}(\lambda) < 1$

$$\int_0^\infty e^{-(\lambda+\omega)t} \|BT_0^t f\| dt \leq \hat{\beta} \|f\|, \qquad f \in \mathcal{D}.$$

Then

$$\|T^t\| \leq \frac{N}{1-\hat{\beta}} e^{(\omega+\lambda)t}, \qquad t \geq 0.$$

The above is *sufficient* for the derivation of the lower and upper bounds on the heat kernel of the semigroup corresponding to $\mathcal{L}(a, b, \hat{b}, V)$. For other use we need more sophisticated results of I. Mijadera and J. Voigt. Namely, the DuHamel formula can be used *to define* a perturbed semigroup.

Theorem ([M], [V]). *Let $-T_0$ be the generator of a C_0-semigroup in $\mathcal{B}$, $\|e^{-tT_0}\| \leq N e^{\omega t}$ ($t \geq 0$). Let B be a linear operator in $\mathcal{B}$. Suppose, that*

1) B is a Rellich-Kato perturbation of T_0 (i.e. $D(B) \supset D(T_0)$ and there are finite constants $\delta > 0$, $c(\delta)$ such that $\|Bf\| \leq \delta\|T_0 f\| + c(\delta)\|f\|$, $f \in D(T_0)$).

2) $\int_0^s e^{-\omega t}\|Be^{-tT_0}f\| dt \leq \beta\|f\|$, $f \in D(T_0)$ for some constants $s > 0$ and $\beta < 1$. Then $-T_0 - B$ of domain $D(T_0)$ is the generator of a C_0-semigroup and

$$\|e^{-t(T_0+B)}\| \leq \frac{N}{1-\beta} \exp\left[\left(\omega + \frac{1}{s} \log \frac{N}{1-\beta}\right)t\right], \qquad t \geq 0.$$

Next, let $(M, \mathcal{M}, \mu)$ be a measure space with a positive measure μ. $L^p = L^p(M, \mu)$, $1 \leq p \leq \infty$, denote the Banach spaces of complex measurable functions with finite norms $\|f\|_p = \left(\int |f|^p d\mu\right)^{1/p}$, $1 \leq$

$p < \infty$, and $\|f\|_\infty = \inf\{\lambda; |\{x \in M; |f(x)| > \lambda\}| = 0\}$, where
$|\Omega| := \mu(\Omega)$, $\Omega \in \mathcal{M}$. By $\langle f, g \rangle$, $f \in L^p$, $g \in L^{p'}$, $p'p = p' + p$, we
denote the duality $\langle f, g \rangle = \int \bar{f} g d\mu = \langle \bar{f} g \rangle$.

We let $\|T\|_{p \to q}$ denote the operator norm of T in $\mathcal{L}(L^p, L^q)$.

Definition 1.1. Let H be an operator in L^2. We say that H satisfies
an inequality of Nash if there exist a number $d > 0$ and finite constants
C and ω such that

$$\|f\|_2^{2+4/d} \le C \operatorname{Re}\langle (H + \omega)f, f \rangle \|f\|_1^{4/d}, \qquad f \in D(H) \cap L^1.$$

Lemma 1.2. *Let T^t be a C_0-semigroup of quasi-contractions on L^2,
such that its generator $-\mathcal{L}$ and the adjoint operator $-\mathcal{L}^*$ satisfy an
inequality of Nash with the same constants $d > 0$, C and ω. Suppose
also that T^t is a quasi-bounded map on L^1 and L^∞, i.e.*

$$\|T^t f\|_{p_i} \le c_i e^{\omega, t} \|f\|_{p_i}, \qquad f \in L^1 \cap L^\infty, \ t \ge 0, \ p_1 = 1, \ p_2 = \infty.$$

*Let $T_p^t = (T^t \mid [L^p])^\sim_{L^p \to L^p}$. Then $(T_p^t, \ t \ge 0)$ is a C_0-semigroup for
all $p \in]1, \infty[$ and*

$$\|T_1^t f\|_\infty \le c(\varepsilon t)^{-d/2} e^{(\omega + \varepsilon\omega_1 + \varepsilon\omega_2)t} \|f\|_1, \qquad f \in L^1$$

*for all $t > 0$, $\varepsilon \in \left]0, \frac{1}{2}\right[$, where $c = \left(\frac{dC}{2}\right)^{d/2} (c_1 \cdot c_2)^2$. Moreover, if T_1^t
is strongly continuous then one can take $c = \left(\frac{dC}{2}\right)^{d/2} c_1 \cdot c_2$.*

Proof. Let $f \in D(\mathcal{L}) \cap L^1$ and set $f_t = e^{-\omega t} T^t f$ and $u(t) = \|f_t\|_2^2$. Then

$$-\frac{d}{dt} u(t) = 2 \operatorname{Re}\langle (\mathcal{L} + \omega)f_t, f_t \rangle \ge \frac{2}{C} u^{1+2/d} \|f_t\|_1^{-4/d}.$$

Therefore

$$u(t)^{-2/d} \geq u(t)^{-2/d} - u(0)^{-2/d}$$

$$\geq \frac{4}{dC} \int_0^t \|f_s\|_1^{-4/d} ds$$

$$\geq \frac{4}{dC} (c_1\|f\|_1)^{-4/d} \int_0^{\varepsilon t} e^{-4\omega_1 s/d} ds$$

$$\geq \frac{4}{dC} (c_1\|f\|_1)^{-4/d} e^{-4\varepsilon\omega_1 t/d} \varepsilon t, \qquad 0 < \varepsilon < 1,$$

or

$$\|f_t\|_2 \leq K_1 e^{\varepsilon\omega_1 t}(\varepsilon t)^{-d/4}\|f\|_1, \qquad K_1 = \left(\frac{dC}{4}\right)^{d/4} c_1.$$

Choose $f = \left(1 + \frac{\mathcal{L}}{n}\right)^{-1} g$, $g \in L^1 \cap L^2$, so that $f \in D(\mathcal{L}) \cap L^1$ and $\|f\|_1 \leq \frac{n}{n-\omega_1} c_1\|g\|_1$, $n > \omega_1 \vee 1$. One has

$$\|f_t\|_2 \leq K_1 \frac{n}{n - \omega_1} c_1 e^{\varepsilon\omega_1 t}(\varepsilon t)^{-d/4}\|g\|_1$$

and, since $\left(1 + \frac{\mathcal{L}}{n}\right)^{-1} \to 1$ strongly in L^2 as $n \to \infty$,

$$\|T^t e^{-\omega t} g\|_2 \leq K_1 c_1 e^{\varepsilon\omega_1 t}(\varepsilon t)^{-d/4}\|g\|_1, \qquad g \in L^1.$$

The same estimate holds for $(T^t)^* e^{-\omega t}$:

$$\|(T^t)^* e^{-\omega t} g\|_2 \leq K_2 c_2 e^{\varepsilon\omega_2 t}(\varepsilon t)^{-d/4}\|g\|_1, \qquad g \in L^1, \ 0 < \varepsilon < 1.$$

Now,

$$\|T^t\|_{1\to\infty} \leq \|T^{t/2}\|_{1\to 2} \cdot \|T^{t/2}\|_{2\to\infty} = \|T^{t/2}\|_{1\to 2} \cdot \|(T^{t/2})^*\|_{1\to 2}$$

and the Lemma is proved.

The Dunford-Pettis theorem tells us that T^t is an integral operator, $T^t f(x) = \int T^t(x,y)f(y)d\mu(y)$, and, for all $t > 0$, all $\varepsilon \in \left]0, \frac{1}{2}\right[$ and a.e. $(x,y) \in M \times M$,

$$|T^t(x,y)| \leq c(\varepsilon t)^{-d/2} e^{(\omega+\varepsilon\omega_1+\varepsilon\omega_2)t}.$$

Lemma 1.3. *Let $A \geq 0$ be a self-adjoint operator in L^2 such that*

$$\|e^{-tA}f\|_2 \leq ct^{-d/4}\|f\|_1, \qquad f \in L^1 \cap L^2,$$

for some constant c and all $t > 0$. Then A satisfies the Nash inequality

$$\|f\|_2^{2+4/d} \leq \mathcal{K}\langle A^{1/2}f, A^{1/2}f\rangle\|f\|_1^{4/d}, \qquad f \in D(A^{1/2}) \cap L^1$$

where $\mathcal{K} = \mathcal{K}(c, d) < \infty$.

Proof. Fix $f \in D(A^{1/2})$. Set $f_t = e^{-tA}f$. Since $\|f_t\|_2 \leq \|f\|_2$ and e^{-tA} commutes with $A^{1/2}$, one has

$$\begin{aligned}
\|f\|_2^2 &= \|f_t\|_2^2 - \int_0^t \frac{d}{ds}\|f_s\|_2^2 ds \\
&= \|f_t\|_2^2 + 2\int_0^t \langle e^{-sA}A^{1/2}f, e^{-sA}A^{1/2}f\rangle ds \\
&\leq c^2 t^{-d/2}\|f\|_1^2 + 2t\langle A^{1/2}f, A^{1/2}f\rangle.
\end{aligned}$$

Now put $t = \left(c^2\|f\|_1^2/\langle A^{1/2}f, A^{1/2}f\rangle d\right)^{\frac{2}{d+2}}$.

Remark 1.4. Let L be an operator in L^2 such that $D(L) \subset D(A^{1/2})$ and for some constants $k > 0$ and $\omega < \infty$

$$\mathrm{Re}\langle (L+\omega)f, f\rangle \geq k\langle A^{1/2}f, A^{1/2}f\rangle, \qquad f \in D(L)$$

where A satisfies the hypothesis of Lemma 1.3. Then L satisfies an inequality of Nash.

Lemma 1.5. *Let $T^t = e^{-tH}$, $t \geq 0$, be a C_0-semigroup on L^2. Suppose that*

1) $\|T^t f\|_{p_i} \leq c_i\|f\|_{p_i}$, $f \in L^1 \cap L^\infty$, *for $p_1 = 1$, $p_2 = \infty$ and some constants $c_i < \infty$.*

2) H is m-sectorial and satisfies an inequality of Nash
$\|f\|_2^{2+4/d} \le c\operatorname{Re}\langle Hf, f\rangle \|f\|_1^{4/d}$, $f \in D(H) \cap L^1$.

Then $q(t, x, y) := T^t(x, y)$ fulfills

(i) $\|q(t, x, \cdot)\|_1 \le c_2$,

(ii) $\overline{q(t, x, y)}$ as a function of y belongs to $D(H^*)$ and satisfies the
equation $\left[(\partial_t + H^*)\overline{q(t, x, \cdot)}\right](y) = 0$ for all $t > 0$ and a.e. $x \in M$.

(iii) $H^* e^{-tH^*} f(x) = \int_M \left[H^* \overline{q(t, y, \cdot)}\right](x) f(y) d\mu(y)$ for all $f \in L^2$.

(iv) $|q^{(n)}(t, x, y)| \le c_{(n)} t^{-n-\frac{d}{2}}$ for all $t > 0$, $n = 0, 1, 2, \ldots$, and
a.e. $x, y \in M$ where $q^{(n)} = \frac{d^n}{dt^n} q$.

Proof. H^* inherits the properties of H to be m-sectorial and
satisfy an inequality of Nash. Thus, (i) and (iv), for $n = 0$, follow
from Lemma 1.2. Since $\|T^t\|_{1\to 2} \le ct^{-d/4}$, $u(t, y) = q(t, x, y) \in L^2$ for
a.e. $x \in M$. Furthermore, HT^t has the integral kernel $K(t, x, y) \in L^2$
as a function of y. Indeed, $\|HT^t\|_{2\to\infty} \le \|T^{t/2}\|_{2\to\infty}\|HT^{t/2}\|_{2\to 2} \le Ct^{-d/4-1}$. To prove (ii) pick up $f \in D(H)$, $g \in L^2$. One has, using
the Fubini-Tonelli theorem,

$$\langle HT^t f, g\rangle = \langle Hf, (T^t)^* g\rangle = \int g(y)\langle Hf, \overline{q(t, y, \cdot)}\rangle d\mu(y),$$

$$\langle HT^t f, g\rangle = \int g(y)\langle K(t, y, \cdot), \overline{f(\cdot)}\rangle d\mu(y),$$

so that $\langle Hf, \overline{q(t, y, \cdot)}\rangle = \langle f, \overline{K(t, y, \cdot)}\rangle$. The latter means that for a.e.
$y \in M$

$$\overline{q(t, y, \cdot)} \in D(H^*) \quad \text{and} \quad \left[H^* \overline{q(t, y, \cdot)}\right](x) = \overline{K(t, y, x)}.$$

Since $K(t, y, x) = -\partial_t q(t, y, x)$, (ii) is proved. (iii) follows from (ii)
and the closeness of H^*. (iv), for $n = 1$ and hence for $n = 2, 3, \ldots$,
follows from (ii) and (iii).

From now on, let (M, μ) be $(\mathbb{R}^d, dx)$, $d \geq 3$. Throughout the rest of this section we will be assuming that the hypotheses $(\mathbf{H_1})$, $(\mathbf{H_2})$, $(\mathbf{H_3'})$ are fulfilled. Recall, that A denotes the operator associated with the closure of $E(u, v) = \langle \nabla \bar{u} \cdot a \cdot \nabla v \rangle$, $u, v \in C_0^1(\mathbb{R}^d)$, and $(e^{-tA}, t \geq 0)$ is a symmetric Markov semigroup.

Note that equalities $\|e^{t\Delta}\|_{1 \to 1} = 1$ and $\|e^{t\Delta}\|_{1 \to \infty} = (4\pi t)^{-d/2}$, combined with Lemma 1.3, $(\mathbf{H_1})$ and Remark 1.4, imply the Nash initial bound $\|e^{-tA}\|_{1 \to \infty} \leq C_N t^{-d/2}$ $(t > 0)$.

Let $p(t, x, y)$ denote the heat kernel for e^{-tA}. Due to Lemma 1.5, it possesses the properties:

$0 \leq p(t, x, y) = p(t, y, x) \leq C_N t^{-d/2}$;

$|p'(t, x, y)| \leq c_{(1)} t^{-d/2-1}$ for all $t > 0$ and a.e. $x, y, \in \mathbb{R}^d$;

$\langle p(t, x, \cdot) \rangle \leq 1$, for a.e. $x \in \mathbb{R}^d$;

$p(t, x, y)$ as a function of x belongs to $D(A)$, for all $t > 0$ and a.e. $y \in \mathbb{R}^d$, and solves the equation $(\partial_t + A)u(t, x) = 0$.

Let 1_n, $n = 1, 2, \ldots$, be the indicator of the set $\{|x| \leq n, |b(x)| \leq n\}$. Define $b_n = b1_n$ and $B_n = B(b_n)$ and set $f_t = e^{-tA_1} f$, $f = \operatorname{Re} f$. The following key result is proved in section 2 (see Lemma 2.6).

There exist $t_0 > 0$ and function $[0, t_0] \ni s \to \beta(s)$ with $\beta(s) \to 0$ as $s \to 0$ such that

$$\operatorname{ess\,sup}_{x \in \mathbb{R}^d} \int_0^s \|b(\cdot) \cdot \nabla.p(t, x, \cdot)\|_1 dt \leq \beta(s).$$

We will now use it to prove

Lemma 1.6. *1)* $\sup_n \int_0^s \|B_n f_t\|_1 dt \leq \beta(s)\|f\|_1$, $f \in L^1$, $s \in]0, t_0]$.

2) There is a function $[0, t_0] \ni s \to \gamma(s)$ *such that*

$$\int_0^s \|\varphi \nabla f_t\|_1 dt \le \gamma(s) \|f\|_1, \qquad f \in L^1, \ s \in]0, t_0]$$

where $\varphi(x) = \exp(-x^2)$.

First we prove

Proposition 1.7. *The functions* $]0, \infty[\in t \to B_n f_t, \varphi \nabla f_t$ *are* L^1-*strongly continuous for all* $f \in L^1$ *and* $n \ge 1$.

Proof. One has $\langle 1_n | b \cdot \nabla f_t | \rangle \le \langle 1_n b \cdot a^{-1} \cdot b \rangle^{1/2} \langle \nabla f_t \cdot a \cdot \nabla f_t \rangle^{1/2}$,

$$\langle \varphi | \nabla f_t | \rangle^2 \le \langle \varphi \rangle \sigma^{-1} \langle \nabla f_t \cdot a \cdot \nabla f_t \rangle,$$

$$\langle \nabla f_t \cdot a \cdot \nabla f_t \rangle = \langle f_t A f_t \rangle \le \|f_t\|_1 \|A_1 f_t\|_\infty$$

$$\le \|f\|_1 C_N (3/t)^{d/2+1} \|f\|_1.$$

If $t, \delta > 0$, then $\|B_n f_{t+\delta}\|_1 \le c(n, d, t) \|f_\delta\|_1$, $\|\varphi \nabla f_{t+\delta}\|_1$ $\le c(\sigma, d, t) \|f_\delta\|_1$ and hence

$$\|B_n(f_{t+\delta} - f_t)\|_1 \le c(n, d, t) \|f_\delta - f\|_1 \to 0, \quad \|\varphi \nabla(f_{t+\delta} - f_\delta)\|_1 \to 0$$

as $\delta \to 0$, which ends the proof.

Proposition 1.8. $B_n f_t(x) = \int [B_n p(t, \cdot, y)](x) d\mu_f(y)$, $f \in L^1$ *where* $d\mu_f(y) := f(y) dy$.

Proof. The estimate $\langle \nabla p \cdot a \cdot \nabla p \rangle = \langle p, -p' \rangle \le \langle p \rangle \|p'\|_\infty \le$ $C_{(1)} t^{-d/2-1}$, $t > 0$, and the Fubini-Tonelli theorem yield, for any

$R < \infty$,

$$\int_{|x|<R} \left[\int |\nabla_x p(t,x,y)| d\mu_{|f|}(y) \right] dx$$

$$= \int \left[\int_{|x|<R} |\nabla_x p(t,x,y)| dx \right] d\mu_{|f|}(y)$$

$$\leq \|f\|_1 \sigma^{-1/2} c_d R^{-d/2} \text{ess sup}_y \langle \nabla.p(t,\cdot,y) \cdot a(\cdot) \cdot \nabla.p(t,\cdot,y) \rangle^{1/2}$$

$$\leq \|f\|_1 \sigma^{-1/2} c_d R^{-d/2} C_{(1)}^{1/2} t^{-d/4-1/2}$$

so that $\left| \int \nabla_x p(t,x,y) d\mu_f(y) \right| < \infty$ for a.e. $x \in \mathbb{R}^d$.

The inequalities

$$\left\| \int \nabla.p(t,\cdot,y) d\mu_f(y) \right\|_2 \leq \int \|\nabla_x p\|_2 d\mu_{|f|} \leq \|f\|_1 \sigma^{-1/2} C_{(1)}^{1/2} t^{-d/4-1/2}$$

show that $\int \nabla_x p(t,x,y) d\mu_f(y) \in [L^2]^d$. Since the weak gradient is closed in L^2 and $f_t = \int p d\mu_f$, one has $[L^2]^d \ni \nabla f_t = \int \nabla_x p(t,x,y) d\mu_f(y)$. Since $|b_n| \in L^\infty$, one concludes that $L^2 \ni b_n \cdot \nabla f_t(x) = \int [b_n \cdot \nabla p(t,\cdot,y)](x) d\mu_f(y)$, and since $|b_n| \in L^\infty_{\text{com}}$, one has $L^1 \ni B_n f_t(x) = \int [B_n p(t,\cdot,y)] d\mu_f(y)$.

Remark 1.9. The same proof gives

$$[L^1]^d \ni \varphi \nabla f_t(x) = \int [\varphi \nabla p(t,\cdot,y)](x) d\mu_f(y), \qquad f \in L^1.$$

Proposition 1.10. *Given $n \geq 1$, $s \in]0, t_0]$, $f \in L^1$,*

$$\int_0^s \|B_n f_t\| dt \leq \|f\|_1 \text{ess sup}_y \int_0^s \int \left| [B_n p(t,\cdot,y)](x) \right| dx dt.$$

Proof. By the Fubini-Tonelli theorem and L^1-continuity of $B_n f_t$ (Prop. 1.7),

$$\int_0^s \|B_n f_t\|_1 dt = \left\| \int_0^s |B_n f_t| dt \right\|_1.$$

Now apply Proposition 1.8 and the Fubini-Tonelli theorem.

Remark 1.11. The same proof gives

$\forall s \in]0, t_0]$

$$\int_0^s \|\varphi \nabla f_t\|_1 dt \le \|f\|_1 \text{ ess sup}_y \int_0^s \int \left| [\varphi \nabla p(t, \cdot, y)](x) \right| dx dt.$$

The proof of Lemma 1.6. Apply Proposition 1.10, Remark 1.11 and Lemma 2.6 to complete the proof.

Given $\lambda > 0$ and $s \in]0, t_0]$, define operators $F_s, \Phi_s, \widehat{F}_\lambda, \widehat{\Phi}_\lambda$ by

$$F_s f := \int_0^s B_n f_t dt, \qquad \Phi_s f := \int_0^s \varphi \nabla f_t dt,$$

$$\widehat{F}_\lambda := \int_0^\infty e^{-\lambda t} B_n f_t dt, \qquad \widehat{\Phi}_\lambda f := \int_0^\infty e^{-\lambda t} \varphi \nabla f_t dt.$$

Proposition 1.12. $F_s, \Phi_s, \widehat{F}_\lambda, \widehat{\Phi}_\lambda \in \mathcal{L}(L^1); \quad \|F_s\|_{1 \to 1} \le \beta(s)$, $\|\Phi_s\|_{1 \to 1} \le \gamma(s), \|\widehat{F}_\lambda\|_{1 \to 1} \le \frac{e^\lambda}{e^\lambda - 1} \beta(s), \|\widehat{\Phi}_\lambda\|_{1 \to 1} \le \frac{e^\lambda}{e^\lambda - 1} \gamma(s)$.

Proof. The assertions about F_s, Φ_s follow from Lemma 1.6; the others follow then by Miyadera's argument [M] (see also [V]).

Lemma 1.13. $D(B_n) \supset D(A_1)$ and $B_n(\lambda + A_1)^{-1} = \widehat{F}_\lambda$ for all $n \ge 1$, $\lambda > 0$.

Proof. Since the operator $\varphi \nabla$ is closed in L^1, one has, using Lemma 1.6 and Proposition 1.12, $[L^1]^d \ni \widehat{\Phi}_\lambda f = \varphi \nabla (\lambda + A_1)^{-1} f$, and since $|b_n| \in L^\infty$, $\int_0^\infty e^{-\lambda t} \varphi b_n \nabla f_t dt = b_n \widehat{\Phi}_\lambda f$, $f \in L^1$. Thus $\int_0^\infty e^{-\lambda t} \varphi B_n f_t dt = \varphi B_n (\lambda + A_1)^{-1} f$. Once more, $\widehat{F}_\lambda f \in L^1$ and $\varphi \in L^\infty$ imply that $\int_0^\infty e^{-\lambda t} \varphi B_n f_t dt = \varphi \widehat{F}_\lambda f \in L^1$ and hence $\varphi \widehat{F}_\lambda f = \varphi B_n (\lambda + A_1)^{-1} f$. Now observe that $\varphi 1_n(y) \ge \delta_n 1_n(y)$ for some $\delta_n > 0$ and all $y \in \mathbb{R}^d$ so that $\widehat{F}_\lambda f = B_n (\lambda + A_1)^{-1} f \in L^1$.

Theorem 1.14. *Under hypotheses* $(\mathbf{H_1})$, $(\mathbf{H_2})$, $(\mathbf{H_3'})$ B *is a Rellich-Kato perturbation of* A_1 *with relative bound zero;* $A_1 + B$ *of domain*

$D(A_1)$ generates a C_0-semigroup and

$$\|e^{-t(A_1+B)}\|_{1\to 1} \le (1 - \beta(s))^{-1} \exp\left[-\frac{t}{s} \log(1 - \beta(s))\right], \qquad t \ge 0$$

for all $s > 0$ such that $\beta(s) < 1$.

Proof. Proposition 1.12, Lemma 1.13 and the monotone convergence theorem yield

$$\|b \cdot \nabla(\lambda + A_1)^{-1} f\|_1 = \lim_n \|1_n b \cdot \nabla(\lambda + A_1)^{-1} f\|_1 = \lim_n \|\widehat{F}_\lambda f\|_1$$

$$\le \frac{e^\lambda}{e^\lambda - 1} \beta(s)\|f\|_1 \qquad f \in L^1, \ \lambda > 0.$$

The strong L^1-continuity of $[0, \infty[\ni t \to Bf_t, \ f \in D(A_1)$, is now evident. The inequality

$$\int_0^s \|Bf_t\|_1 dt \le \beta(s)\|f\|_1, \qquad f \in D(A_1)$$

follows from Lemma 1.6 and the monotone convergence theorem. Now apply Miyadera's theorem to A_1, B and obtain the final assertion of Theorem 1.14.

Theorem 1.15. (i) Semigroups $T^t := \exp[-t(A_1 + B)]$ and $\widehat{T}^t := \exp[-t(A \dotplus b \cdot \nabla)]$ are consistent.

(ii) $T^t \in \mathcal{L}(L^1, L^\infty)$ with $\|T^t\|_{1\to\infty} \le ct^{-d/2}e^{\omega t}, \ t > 0$.

(iii) $0 \le q(t, x, y) := T^t(x, y) \le ct^{-d/2}e^{\omega t}, \ \langle q(t, x, \cdot)\rangle \le 1, \ \langle q(t, \cdot, x)\rangle \le c_1 e^{\omega t}$.

(iv) For any $f \in L^p, \ 1 \le p < \infty, \ \langle q(t, x, \cdot)f(\cdot)\rangle$ and $\langle q(t, \cdot, x)f(\cdot)\rangle$ determine C_0-semigroups T_p^t and $\check{T}_p^t$ respectively such that

$$T_1^t = T^t, \quad T_2^t = \widehat{T}^t, \quad \check{T}_2^t = \exp[-t(A\dot- \nabla \cdot b)] = (\widehat{T}^t)^*.$$

Proof. Since $[L^\infty + L^q,\ q > d/2,\ d \geq 3] \subset \bigcap_{0 < \beta} PK_\beta(A)$, one concludes that $(\mathbf{H_3'})$ implies $b \cdot a^{-1} \cdot b \in \bigcap_{0 < \beta} PK_\beta(A)$ so $A + b \cdot \nabla$, $A - \nabla \cdot b$ are well defined. Also, $A + b_n \cdot \nabla \to A + b \cdot \nabla$, $A_1 + B_n \to A_1 + B$ as $n \to \infty$ in the strong resolvent sense in L^2 and in L^1 respectively. The first convergence is proved in the Appendix (cf. [LS, Th. 4.3]), the second follows from Theorem 1.14 and the estimate

$$\|(\lambda + A_1 + B_n)^{-1} f - (\lambda + A_1 + B)^{-1} f\|_1$$

$$\leq c_\lambda \|(1 - 1_n) B (\lambda + A_1 + B)^{-1} f\|_1$$

which is valid for all $f \in L^1$, $n \geq 1$ and all large $\lambda > 0$. Thus, to prove (i) we need only state the equality

$$(\lambda + A + b_n \cdot \nabla)^{-1} f = (\lambda + A_1 + B_n)^{-1} f, \qquad f \in L^1 \cap L^2$$

for all $n \geq 1$ and all large $\lambda = \lambda(n)$. The latter follows from the estimates

$$\|B_n (\lambda + A_1)^{-1}\|_{1 \to 1} < 1, \qquad \|b_n \cdot \nabla (\lambda + A)^{-1}\|_{2 \to 2} < 1,$$

the equalities

$$(\lambda + A_1)^{-1} f = (\lambda + A)^{-1} f,$$

$$B_n (\lambda + A_1)^{-1} f = b_n \cdot \nabla (\lambda + A)^{-1} f, \qquad f \in L^1 \cap L^2,$$

and from the coincidence of the perturbation series for $(\lambda + A_1 + B_n)^{-1}$ and $(\lambda + A + b_n \cdot \nabla)^{-1}$ on $L^1 \cap L^2$.

Since $\widehat{T}^t$ is a contraction on L^∞ and $A + b \cdot \nabla$ is m-sectorial and satisfies an inequality of Nash, (ii) follows from (i), Theorem 1.14 and Lemma 1.2. "$\langle q(t, x, \cdot) \rangle \leq 1$" expresses the fact that $\widehat{T}^t$ and, due to (i), T^t are contractions on L^∞. The first inequality in (iii) follows from (ii), and the third from Theorem 1.14 by duality. Thus (iii) is proved and implies (iv).

2.   The method of Nash.

In this section we apply the Nash method to estimate from above the quantities

$$\int_0^s \langle |b \cdot \nabla.p(t,x,\cdot)| \rangle dt\,, \quad \int_0^s e^{-\omega t} \langle |\hat{b} \cdot \nabla.q(t,x,\cdot)| \rangle dt$$

where $p(\cdot)$, $q(\cdot)$ denote the heat kernels for e^{-tA}, $T^t = \exp[-t(A + b \cdot \nabla)]$ respectively. Throughout this section we will be assuming that hypotheses $(\mathbf{H_1})$, $(\mathbf{H_2})$, $(\mathbf{H_3})$ are fulfilled. The main objects we work with have already been defined in the Introduction. In particular, we will make use of the following properties of functions η, ξ:

$$-\eta' \le \eta^{1-m}\,, \quad m = 1/n_0\,, \quad \xi(\cdot) \in D(A^{1/2})\,,$$

$$\langle \xi(\cdot)^r \rangle \le c(d, n_0, L, v)\theta^{-1} \text{ for all } r \in]0, \infty[$$

where $\theta = \varepsilon^{d/v}$, $c(d, n_0, L, v) = \left(\frac{n_0+1}{L}\right)^{d/v} \cdot c_d$.

The last inequality follows from

$$\langle \xi(\cdot)^r \rangle \le \int_{\varepsilon|\psi(x)-\psi(y)|\le 1+n_0} dy \le \int_{\varepsilon L|x-y|^v \le 1+n_0} dy\,.$$

Lemma 2.1. *There are constants* $c_i(d, C_N, \Gamma_1, \|\Gamma_2\|_q, L, v)$, $i = 1, 2$, *such that*

$$\mathcal{N}(t) \le (1 + \sqrt{\varepsilon})Q'(t) + \varepsilon c_1 + \varepsilon c_2 t^{-d/2q}(\log t)^2$$

for all $t \in]0,1]$, $\varepsilon \in]0, 1/2]$ and a.e. $x \in \mathbb{R}^d$.

Proof. Writing down $Q(t) = -\langle \xi p \log p_\theta \rangle = -\Big\langle \xi p \log \Big(1 + \frac{p}{\theta} \Big) \Big\rangle - \langle \xi p \rangle \log \theta$ and using the properties of ξ and p listed before Lemma 1.6, one has

$$Q'(t) = -\Big\langle \xi p' \log \Big(1 + \frac{p}{\theta} \Big) \Big\rangle - \langle \xi p' \rangle \log \theta - \Big\langle \xi \frac{p}{p_\theta} p' \Big\rangle$$

$$= \Big\langle \xi A p, \log \Big(1 + \frac{p}{\theta} \Big) \Big\rangle + \langle \xi A p \rangle (\log \theta + 1) + \theta \Big\langle \frac{\xi}{p_\theta} A p \Big\rangle$$

$$= \Big\langle \xi \nabla p \cdot \frac{a}{p_\theta} \cdot \nabla p \Big\rangle + \langle \nabla \xi \cdot a \cdot \nabla p, \log p_\theta \rangle + \langle \nabla \xi \cdot a \cdot \nabla p \rangle$$

$$- \theta \Big\langle \nabla \xi \cdot \frac{a}{p_\theta} \cdot \nabla p \Big\rangle + \theta \Big\langle \xi \nabla p \cdot \frac{a}{p_\theta^2} \cdot \nabla p \Big\rangle$$

$$\geq \mathcal{N}(t) - \langle |\nabla \xi \cdot a \cdot \nabla p| (2 + |\log p_\theta|) \rangle .$$

The Nash initial bound yields

$$|\log p_\theta| \leq -\widetilde{Q}(t) - \log \theta + \log(1 + C_N) \quad \Big(t \in]0,1], \ \widetilde{Q}(t) := \frac{d}{2} \log t \Big)$$

so that for all $v_1 \in \Big] 0, \frac{1}{2} \Big]$

$$\langle |\nabla \xi \cdot a \cdot \nabla p| (2 + |\log p_\theta|) \rangle$$

$$\leq \big(-\widetilde{Q}(t) - \log \theta + 2 + C_N \big) \Big\langle \xi \nabla p \cdot \frac{a}{p_\theta} \cdot \nabla p \Big\rangle^{\frac{1}{2}} \Big\langle \nabla \xi \cdot \frac{a p_\theta}{\xi} \cdot \nabla \xi \Big\rangle^{\frac{1}{2}}$$

$$\leq v_1 \mathcal{N}(t) + \frac{1}{4 v_1} \big[-\widetilde{Q}(t) - \log \theta + 2 + C_n \big]^2 \Big\langle p_\theta \nabla \xi \cdot \frac{a}{\xi} \cdot \nabla \xi \Big\rangle .$$

The estimates

$$\nabla \xi \cdot \frac{a}{\xi} \cdot \nabla \xi \leq \varepsilon^2 \xi^{1-2m} \Gamma(\psi) \leq \varepsilon^2 \xi^{1-2m} (\Gamma_1 + \Gamma_2),$$

$$\Big\langle \nabla \xi \cdot \frac{a}{\xi} \cdot \nabla \xi \Big\rangle \leq \varepsilon^2 \Gamma_1 \langle \xi^{1-2m} \rangle + \varepsilon^2 \|\Gamma_2\|_q \langle \xi^{(1-2m)q'} \rangle^{1/q'}$$

$$\leq \varepsilon^2 (\Gamma_1 + \|\Gamma_2\|_q) c(d, n_0, L, v) \theta^{-1},$$

$$\Big\langle p \nabla \xi \cdot \frac{a}{\xi} \cdot \nabla \xi \Big\rangle \leq \varepsilon^2 (\Gamma_1 + \langle p \Gamma_2 \rangle) \leq \varepsilon^2 \big(\Gamma_1 + C_N^{1/q} \|\Gamma_2\|_q t^{-d/2q} \big)$$

yield

$$\left\langle p_\theta \nabla \xi \cdot \frac{a}{\xi} \cdot \nabla \xi \right\rangle$$

$$\leq \varepsilon^2 \left[(\Gamma_1 + \|\Gamma_2\|_q) c(d, n_0, L, v) + \Gamma_1 + C_N^{1/q} \|\Gamma_2\|_q t^{-d/2q} \right].$$

Thus, gathering the estimates above, one has

$$Q'(t) \geq (1 - v_1)\mathcal{N}(t)$$

$$- \frac{c}{4v_1} \varepsilon^2 \left[-\widetilde{Q}(t) - \log \theta + 2 + C_N \right]^2 \left[\Gamma_1 + \|\Gamma_2\|_q t^{-d/2q} \right]$$

where $c = C_N^{1/q} \vee 1 \vee c(d, n_0, L, v)$.

Now, let $v_1 = \sqrt{\varepsilon}/(1 + \sqrt{\varepsilon})$. This ends the proof of Lemma 2.1.

Lemma 2.2. *Given $\psi \in \mathcal{T}$, define the moment*

$$M(t) = M(t, x)cr := \langle \xi(\cdot) | \psi(x) - \psi(\cdot) | p(t, x, \cdot) \rangle$$

$$(t > 0, \ \varepsilon \in]0, 1/2], \ \text{a.e.} \ x \in \mathbb{R}^d).$$

There is a constant $C = C(d, n_0, L, v) \in]0, \infty[$ such that

$$M(t) \leq C \int_0^t \left(\Gamma_1^{1/2} + \|\Gamma_2\|_q^{1/2} s^{-d/4q} \right) \mathcal{N}(s)^{1/2} ds.$$

Proof. One has

$$M'(t) = \langle \xi(\cdot) | \psi(x) - \psi(\cdot) | p'(t, x, \cdot) \rangle$$

$$\leq \langle \xi | \nabla . | \psi(x) - \psi(\cdot) | \cdot a \cdot \nabla p | \rangle + \langle | \psi(x) - \psi(\cdot) | | \nabla \xi \cdot a \cdot \nabla p | \rangle$$

$$\leq \left\langle \xi \nabla p \cdot \frac{a}{p_\theta} \cdot \nabla p \right\rangle^{1/2} \langle \xi p_\theta \Gamma(\psi) \rangle^{1/2} - \varepsilon \langle | \psi(x) - \psi(\cdot) | \xi'(\cdot) | \nabla | \psi(x)$$

$$- \psi(\cdot) | \cdot a \cdot \nabla p | \rangle.$$

By the definition of ξ, $-\varepsilon | \psi(x) - \psi(\cdot) | \xi'(\cdot) \leq (1 + n_0)\xi(\cdot)^{1-m}$ so that

$$M'(t) \leq \mathcal{N}(t)^{\frac{1}{2}} \langle \xi p_\theta (\Gamma_1 + \Gamma_2) \rangle^{\frac{1}{2}}$$

$$+ (1 + n_0) \langle \xi^{1-m} | \nabla | \psi(x) - \psi(\cdot) | \cdot a \cdot \nabla p | \rangle.$$

Since

$$\langle \xi^{1-m} |\nabla| \psi(x) - \psi(\cdot)| \cdot a \cdot \nabla p| \rangle$$

$$\leq \left\langle \xi \nabla p \cdot \frac{a}{p_\theta} \cdot \nabla p \right\rangle^{1/2} \langle \xi^{1-2m} p_\theta \Gamma(\psi) \rangle^{1/2}$$

$$\leq \mathcal{N}(t)^{1/2} \left[\theta^{1/2} \langle \xi^{1-2m}(\Gamma_1 + \Gamma_2) \rangle^{1/2} + \langle p(\Gamma_1 + \Gamma_2) \rangle^{1/2} \right],$$

one has

$$M'(t) \leq \mathcal{N}(t)^{1/2} \left[(\theta \Gamma_1)^{1/2} \langle \xi \rangle^{1/2} + (1 + n_0)(\theta \Gamma_1)^{1/2} \langle \xi^{1-2m} \rangle^{1/2} \right.$$

$$\left. + \theta^{1/2} \langle \xi \Gamma_2 \rangle^{1/2} + (2 + n_0) \langle p(\Gamma_1 + \Gamma_2) \rangle^{1/2} \right]$$

or

$$M'(t) \leq C \cdot (\Gamma_1^{1/2} + \|\Gamma_2\|_q^{1/2} t^{-d/4q}) \mathcal{N}(t)^{1/2}$$

$$M(t) - M(0) \leq C \int_0^t (\Gamma_1^{1/2} + \|\Gamma_2\|_q^{1/2} s^{-d/4q}) \mathcal{N}(s)^{1/2} ds$$

after integration.

Finally, using the fact: $e^{-tA_1} \to I$ strongly as $t \searrow 0$, one easily concludes that $\lim\limits_{t \searrow 0} \operatorname{ess} \sup_{x \in \mathbb{R}^d} M(t, x) = 0$. Therefore, we can put $M(0) = 0$. The proof of Lemma 2.2 is completed.

Lemma 2.3. For all $t > 0$, $\varepsilon > 0$ and a.e. $x \in \mathbb{R}^d$,

$$\exp \frac{Q}{\tilde{d} \langle \xi p \rangle} \leq \left[\frac{ec_0}{\langle \xi p \rangle} \right]^{1/\tilde{d}} \frac{M}{\langle \xi p \rangle}$$

where $\tilde{d} = d_1$ if $M \leq \langle \xi p \rangle$ and $\tilde{d} = d/v$ if $M > \langle \xi p \rangle$.

Proof. First observe that for any real ν

$$p \log p_\theta + \nu p \geq -e^{-1-\nu}, \qquad p = p(t, x, y).$$

Put $\nu = c + k|\psi(x) - \psi(y)|$, where c, k are any constants, and integrate over space with respect to the measure $\xi(y) dy$, obtaining

$$-Q + c\langle \xi p \rangle + kM \geq -e^{-1-c} \int_{\mathbb{R}^d} \xi(y) e^{-k|\psi(x)-\psi(y)|} dy \geq -e^{-1-c} c_0 k^{-\tilde{d}}.$$

Now set $kM = \langle \xi p \rangle$ and $e^{-1-c}c_0 k^{-\tilde{d}} = \langle \xi p \rangle$. Then $Q/\langle \xi p \rangle \leq c + 2$ or

$$\exp \frac{Q}{\langle \xi p \rangle} \leq e^{c+2} = e\langle \xi p \rangle^{-1} c_0 k^{-\tilde{d}} = e c_0 \langle \xi p \rangle^{-1} \left(\frac{M}{\langle \xi p \rangle} \right)^{\tilde{d}}.$$

Lemma 2.4. *There are positive constants t_0, C_{t_0} such that*

$$\langle \xi p \rangle = \langle \xi(\cdot) p(t, x, \cdot) \rangle \geq C_{t_0}$$

for all $t \in [0, t_0]$, $\varepsilon \in]0, 1]$ and a.e. $x \in \mathbb{R}^d$.

Proof. Put $w(\cdot) = \chi(\cdot)\xi(\cdot)$, $\chi(\cdot) := \exp(-|\psi(x) - \psi(\cdot)|)$ and fix $\gamma = 2/n_0$ by the conditions $0 < \mu_0 = \frac{d}{n_0} \frac{3n_0 - 2}{n_0 - 2} < 1$ and $\mu_1 = \frac{d}{2} \left(\frac{1}{q} + \frac{2}{n_0} \right) < 1$. Using an idea in [S, pp. 337–338] we will prove that $\Theta(t) := -\langle wp(1 + p_\theta^{-\gamma}) \rangle$ satisfies the inequality

$$\Theta'(\tau) \leq -(1 + \tau^{-\mu_0})\Theta(\tau) + c \cdot (1 + \tau^{-\mu_1}), \qquad 0 < \tau \leq 1$$

for some constant c. Integrating the inequality over τ from 0 to t and using the bound $-\Theta(0) \geq 1$ yield

$$-\Theta(t) \geq \tilde{c} > 0 \text{ for all small } t, \text{ say } t \leq t_0, \text{ and all } \varepsilon \in]0, 1].$$

To end the proof of the Lemma it suffices then notice that

$$\langle wpp_\theta^{-\gamma} \rangle \leq \langle wp \rangle^{1-\gamma} \langle \chi \rangle^\gamma = \langle wp \rangle^{1-\gamma} c_0^\gamma,$$

$$\langle wp \rangle^{1-\gamma}(c_0^\gamma + \langle wp \rangle^\gamma) \geq \tilde{c}$$

and hence $\langle wp \rangle \geq [\tilde{c}/(1 + c_0^\gamma)]^{1/1-\gamma} =: C_{t_0}$.

Now, differentiating $\Theta(t)$ and using quadratic inequalities, one sees that

$$\Theta'(t) \leq \frac{1}{(1-\gamma)\gamma} \theta^{1-\gamma} \left\langle \nabla w \cdot \frac{a}{w} \cdot \nabla w \right\rangle + \frac{1+\gamma}{1-\gamma} \left\langle pp_\theta^{-\gamma} \nabla w \cdot \frac{a}{w} \cdot \nabla w \right\rangle$$

$$+ \frac{1}{2(1-\gamma)\gamma} \left\langle p^{1+\gamma} \nabla w \cdot \frac{a}{w} \cdot \nabla w \right\rangle.$$

One has

$$\nabla w \cdot \frac{a}{w} \cdot \nabla w \leq 2 \left(\chi \nabla \xi \cdot \frac{a}{\xi} \cdot \nabla \xi + \xi \nabla \chi \cdot \frac{a}{\chi} \cdot \nabla \chi \right),$$

$$\nabla \xi \cdot \frac{a}{\xi} \cdot \nabla \xi \leq \varepsilon^2 \xi^{1-\gamma} \Gamma(\psi), \quad \nabla \chi \cdot \frac{a}{\chi} \cdot \nabla \chi \leq \chi \Gamma(\psi),$$

so that $\nabla w \cdot \frac{a}{w} \cdot \nabla w \leq 4\chi \Gamma(\psi) \xi^{1-\gamma}$. The latter and the bound $p(t,x,y) \leq C_N t^{-d/2}$ yield

$$\left\langle \nabla w \cdot \frac{a}{w} \cdot \nabla w \right\rangle \leq 4 c_0^{1/q} (\Gamma_1 c_0^{1/q'} + \|\Gamma_2\|_q),$$

$$\left\langle p p_\theta^{-\gamma} \nabla w \cdot \frac{a}{w} \cdot \nabla w \right\rangle \leq 4 \Gamma_1 \langle \chi(\xi p)^{1-\gamma} \rangle + 4 \langle \chi p^{1-\gamma} \Gamma_2 \rangle$$

$$\leq \frac{2\gamma}{1-\gamma} \langle wp \rangle + c_0 (2\gamma)^\gamma (2\Gamma_1)^{1/\gamma} + 4 c_0^\gamma \|\Gamma_2\|_q C_N^{1/q} t^{-d/2q},$$

$$\left\langle p^{1+\gamma} \nabla w \cdot \frac{a}{w} \cdot \nabla w \right\rangle \leq 2(1-\gamma)\gamma \langle wp p_\theta^{-\gamma} \rangle t^{-\mu_0} + c_1 (1 + t^{-\mu_1}).$$

Finally, use the above estimates to obtain the claimed inequality.

We are now ready to prove the upper bound on $Q(t)$.

Lemma 2.5. *There is a constant $C_{\hat{\gamma}} < \infty$ such that*

$$Q(t) \leq \langle \xi p \rangle \left(\frac{d_1}{d} \hat{\gamma} \widetilde{Q}(t) + C_{\hat{\gamma}} \right)$$

for all $t \in]0, t_0]$, $\varepsilon \in \left]0, \frac{1}{2}\right]$, $\hat{\gamma} \in \left]0, 1 - \frac{d}{2q}\right[$ and a.e. $x \in \mathbb{R}^d$.

Proof. First of all, the bound $p(t,x,y) \leq C_N t^{-d/2}$ yields the lower bound on $Q(t)$:

$$Q(t) \geq \langle \xi p \rangle (\widetilde{Q}(t) - C_1), \qquad C_1 := \log(1 + C_N),$$

for all $t \in]0,1]$, $\varepsilon \in]0,1]$ and a.e. $x \in \mathbb{R}^d$.

By Lemma 2.2,

$$M(t) \leq C(\Gamma_1^{1/2} + \|\Gamma_2\|_q^{1/2}) \int_0^t \left[s^{-d/2q} \mathcal{N}(s) \right]^{1/2} ds \qquad (t \in]0,1]).$$

Using the inequality $2ab \le a^2 + b^2$ $(a, b \in \mathbb{R}^1)$, one has

$$\int_0^1 \left[s^{-d/2q} \mathcal{N}(s) \right]^{1/2} ds \le \frac{\sqrt{\tilde{d}}}{2} \left(\frac{1}{\mu} t^\mu + \frac{1}{\tilde{d}} \int_0^t s^\mu \mathcal{N}(s) ds \right)$$

where $\mu := \frac{1}{2} \left(1 - \frac{d}{2q} \right)$.

By Lemma 2.1,

$$\int_0^t s^\mu \mathcal{N}(s) ds \le (1 + \sqrt{\varepsilon}) \int_0^t Q'(s) s^\mu ds$$

$$+ \varepsilon c_1 \frac{1}{1 + \mu} t^{1+\mu} + \varepsilon c_2 \int_0^t s^{\mu - \frac{d}{2q}} (\log s)^2 ds.$$

Next, define $R(t)$ by

$$\langle \xi p \rangle \tilde{d} R(t) := Q(t) - \langle \xi p \rangle \big(\widetilde{Q}(t) - C_1 \big).$$

Since $R(t) \ge 0$, one estimates $I := \int_0^t Q'(s) s^\mu ds$ as follows.

$$I = \tilde{d} \int_0^t \big[\langle \xi p \rangle R(s) \big]' s^\mu ds + \int_0^t \big[\langle \xi p \rangle \big(\widetilde{Q}(s) - C_1 \big) \big]' s^\mu ds$$

$$\le \tilde{d} \langle \xi p \rangle R(t) t^\mu + \langle \xi p \rangle \big(\widetilde{Q}(t) - C_1 \big) t^\mu - \int_0^t \big(\widetilde{Q}(s) - C_1 \big) ds^\mu$$

$$\le \left[\tilde{d} R(t) - \widetilde{Q}(t) + C_1 + \frac{d}{2\mu} \right] t^\mu.$$

Thus

$$M(t) \le k \left[R(t) - \frac{d}{2\tilde{d}} \log t + 1 \right], \quad t \in \,]0, 1], \quad k = k(\mu, \Gamma_1, \|\Gamma_2\|_q, C_1),$$

By Lemmas 2.3 and 2.4,

$$e^{R(t)} \le k_1 t^{\mu - \frac{d}{2d}} \left[R(t) - \frac{d}{2\tilde{d}} \log t + 1 \right], \qquad t \in \,]0, t_0].$$

Setting $R_1(t) = R(t) + \left(\frac{d}{2\tilde{d}} - \frac{\hat{\gamma}}{2} \right) \log t$, one has

$$e^{R_1(t)} \le k_1 t^{\mu - \frac{\hat{\gamma}}{2}} \left[R_1(t) + \left(\frac{\hat{\gamma}}{2} - \frac{d}{\tilde{d}} \right) \log t + 1 \right].$$

The latter clearly shows that $\sup\limits_{t\in]0,t_0]} R_1(t) =: \hat{c} < \infty$. Thus

$$
\begin{aligned}
Q(t) &= \langle \xi p \rangle \left(\widetilde{Q}(t) - C_1 + \tilde{d}R(t) \right) \\
&= \langle \xi p \rangle \left(\widetilde{Q}(t) - C_1 + \tilde{d}R_1(t) - \tilde{d} \left(\frac{d}{2\tilde{d}} - \frac{\hat{\gamma}}{2} \right) \log t \right) \\
&\le \langle \xi p \rangle \left(\frac{\tilde{d}}{d} \hat{\gamma} \widetilde{Q}(t) - C_1 + \tilde{d}\hat{c} \right) .
\end{aligned}
$$

Finally, without loss one may suppose that $d/v \ge d_1$ so $\tilde{d} \ge d_1$. Setting $C_{\hat{\gamma}} = \hat{c}d/v$, one obtains the desired.

Lemma 2.6. *There is a constant $k < \infty$ such that for all $s \in]0, t_0]$ and a.e. $x \in \mathbb{R}^d$*

$$
\int_0^s \langle |b \cdot \nabla p(t,x,\cdot)| \rangle dt \le k \left[(C_{t_0} - 1)\widetilde{Q}(s) + C_1 + C_{\hat{\gamma}} + \frac{d+2}{2\mu} \right] s^\mu .
$$

Proof. *Step 1.* It follows from the definition of ξ and the monotone convergence theorem that

$$
\int_0^s \langle |b \cdot \nabla p| \rangle dt = \lim_{\varepsilon \to 0} \int_0^s \langle \xi(\cdot)|b \cdot \nabla p(t,x,\cdot)| \rangle dt .
$$

Step 2. By $(\mathbf{H_3'})$, $b \cdot a^{-1} \cdot b \le U_1 + U_2$, where $U_1 = \|U_1\|_\infty$ and $0 \le U_2 \in L^q(\mathbb{R}^d)$. One has, using Step 1 and Lemma 2.1,

$$
\begin{aligned}
\int_0^s \langle |b \cdot \nabla p| \rangle dt &\le \lim_{\varepsilon \to 0} \int_0^s \langle \xi p_\theta b \cdot a^{-1} \cdot b \rangle^{1/2} \mathcal{N}(t)^{1/2} dt \\
&\le c(U_1^{1/2} + \|U_2\|_q^{1/2}) \lim_{\varepsilon \to 0} \int_0^s [t^{-d/2q} \mathcal{N}(t)]^{1/2} dt \\
&\le c(U_1^{1/2} + \|U_2\|_q^{1/2}) \left(\mu^{-1} s^\mu + \lim_{\varepsilon \to 0} \int_0^s Q'(t) t^\mu dt \right) .
\end{aligned}
$$

Step 3. By Lemmas 2.5 and 2.4,

$$
\int_0^s Q'(t)t^\mu dt \le Q(s)s^\mu - \int_0^s \langle \xi p \rangle \big(\widetilde{Q}(t) - C_1\big)dt^\mu
$$

$$
\le Q(s)s^\mu + C_1 s^\mu - \int_0^s \widetilde{Q}(t)dt^\mu
$$

$$
= [Q(s) - \widetilde{Q}(s)]s^\mu + C_1 s^\mu + \frac{d}{2}\int_0^s t^{\mu-1}dt
$$

$$
\le \left[(C_{t_0} - 1)\widetilde{Q}(s) + C_1 + C_{\hat\gamma} + \frac{d}{2\mu}\right]s^\mu .
$$

As soon as Lemma 2.6 is proved one can apply Theorem 1.15 and slightly modified versions of Lemmas 2.1–2.6 to obtain the bound on $\int_0^s e^{-\omega t}\langle |\hat b \cdot \nabla q(t,x,\cdot)|\rangle dt$ which, in its turn, leads to Theorem 0.5. We leave the details to the reader but specify them in the uniformly elliptic case (see Section 4).

Remark 2.7. (1) Theorem 0.5 combined with the well known *localization* procedure (see [D, p. 62]) allows us to consider the theory for $\mathcal{L}(a,b,\hat b,V)$ in $L^p(\Omega)$, $1 \le p < \infty$, where $\Omega \subset \mathbb{R}^d$ is an open set. (2) Theorem 0.5 is a basis for proving the L^p-spectral independence of A_p: Assume along with $(\mathbf{H_1})$, $(\mathbf{H_2})$ that there is a ψ entering $(\mathbf{H_2})$ such that $|\psi(x) - \psi(y)| \le \ell$ for some $\ell < \infty$ and all $|x - y| \le 1$. Then the spectrum of A_p is p-independent, $1 \le p < \infty$.

3. The upper bound for the off-diagonal "inner entropy" and the lower bound for the heat kernel.

Throughout this section we will be assuming that $a(\cdot)$ is a smooth, bounded matrix valued function on $\mathbb{R}^d$, $d \ge 2$, satisfying the uniform ellipticity condition $(\mathbf{H_u})$. Thus $-A = \nabla \cdot a \cdot \nabla$ of domain $H^2(\mathbb{R}^d)$ is a symmetric Markov generator in $L^2(\mathbb{R}^d)$; the heat kernel p for e^{-tA} is a smooth positive function on $\mathbb{R}^1_+ \times \mathbb{R}^d \times \mathbb{R}^d$, $\langle p(t, x, \cdot) \rangle = 1$, p satisfies a Gaussian lower bound, p and $t|p'(t, x, y)|$ satisfy Gaussian upper bounds (with the constants which may depend on the smoothness of $a(\cdot)$), $u(\cdot) = p(t, x, \cdot) \in H^2(\mathbb{R}^d)$ for all $t > 0$, $x \in \mathbb{R}^d$. The listed properties of p show that the functions

$$Q(t) := -\langle p(t, x, \cdot) \log p(t, x, \cdot) \rangle, \quad Q'(t), \quad M(t) := \langle |x - \cdot| p(t, x, \cdot) \rangle$$

are well defined and

$$Q'(t) = -\langle p'(t, x, \cdot) \log p(t, x, \cdot) \rangle.$$

To see that $\mathcal{N}(t) := \left\langle \nabla p \cdot \frac{a}{p} \cdot \nabla p \right\rangle$ is finite and equals to $Q'(t)$ let us define the functions

$$Q_\nu(t) := -\langle p \log p_\nu \rangle, \quad p_\nu = p + \nu, \quad \nu > 0.$$

One has

$$Q'_\nu(t) = -\langle p' \log p_\nu \rangle - \left\langle \frac{p}{p_\nu} p' \right\rangle = \left\langle \nabla p \cdot \frac{a}{p_\nu} \cdot \nabla p \right\rangle + \nu \left\langle \nabla p \cdot \frac{a}{p_\nu^2} \cdot \nabla p \right\rangle$$

after integration by parts. Since $Q'_\nu \geq \left\langle \nabla p \cdot \frac{a}{p_\nu} \cdot \nabla p \right\rangle$ and $\lim_{\nu \to 0} Q'_\nu = -\langle p' \log p \rangle$, one concludes that

$$\lim_{\nu \to 0} \left\langle \nabla p \cdot \frac{ap}{p_\nu^2} \cdot \nabla p \right\rangle = \lim_{\nu \to 0} \left\langle \nabla p \cdot \frac{a}{p_\nu} \cdot \nabla p \right\rangle = \mathcal{N}(t) \leq Q'(t).$$

In particular, $\lim_{\nu \to 0} \nu \left\langle \nabla p \cdot \frac{a}{p_\nu^2} \cdot \nabla p \right\rangle = 0$ and hence $\mathcal{N}(t) = Q'(t)$.

We write down the two-sided estimate on the entropy $Q(t)$:

$$-C_1 \leq Q(t) - \widetilde{Q}(t) \leq C_2 \qquad (t > 0, \ x \in \mathbb{R}^d)$$

where $C_2 = -C_1 + d \sup_{t>0} R(t)$. $R(t)$ fulfills the inequality

$$e^{R(t)} \leq k[R(t) + 2], \quad k = \frac{1}{2}\sqrt{d}\zeta(c_0 e^{1+C_1})^{1/d} \quad c_0 = \int_{\mathbb{R}^d} e^{-|x|} dx.$$

Let us define an *off-diagonal inner entropy*

$$Z(t, x, y) := -\langle p(t, x, \cdot) \log p(t, \cdot + y - x, x) \rangle \qquad (t > 0, \ x, y \in \mathbb{R}^d).$$

Theorem 3.1. *For all* $t > 0$, $\delta, \varepsilon \in {]}0, 1{[}$, $x, y \in \mathbb{R}^d$,

$$(i) \ Z(t, x, y) \leq Q(\delta t, x) + \left(1 + \tfrac{1}{\varepsilon}\right) \tfrac{\delta}{\sigma}[Q(t, x) - Q(\delta t, x)]$$

$$+ \tfrac{1+\varepsilon}{1-\delta} \cdot \tfrac{|x-y|^2}{4\sigma t}.$$

(ii) $(LGB) \Rightarrow (sLGB)$.

Proof. (i) $\Rightarrow$ (ii) has already been proved in the Introduction. To prove (i) we use the function $Z(s)$ which has also been defined in the Introduction. Note that $Z(\delta t) = Q(\delta t, x)$ and the function $p_h(s, z) := p(s, z + h(s), x)$ solves the equation

$$\left(\partial_s - \nabla_z \cdot a(z + h(s)) \cdot \nabla_z + b(s) \cdot \nabla_z\right) u(s, z) = 0, \quad 0 < s < t, \ z \in \mathbb{R}^d,$$

where $b(s) := h'(s)$. Thus, letting $a_h(x) = a(x + h)$, one has

$$(3.1) \quad \begin{aligned} Z'(s) = &\langle \nabla p \cdot a \cdot \nabla \log p_h \rangle + \langle \nabla p \cdot a_h \cdot \nabla \log p_h \rangle \\ &- \langle \nabla \log p_h \cdot a_h p \cdot \nabla \log p_h \rangle + \langle pb \cdot \nabla \log p_h \rangle. \end{aligned}$$

Putting $I = \left\langle \nabla p \cdot \frac{a}{p} \cdot \nabla p \right\rangle$, $I_h = \langle \nabla \log p_h \cdot a_h p \cdot \nabla \log p_h \rangle$, one has

$$\langle \nabla p \cdot a \cdot \nabla \log p_h \rangle \leq \frac{1}{4\varepsilon_1} I + \varepsilon_1 \langle \nabla \log p_h \cdot ap \cdot \nabla \log p_h \rangle$$

$$\leq \frac{1}{4\varepsilon_1} I + \varepsilon_1 \frac{\zeta}{\sigma} I_h \qquad \left(0 < \varepsilon_1 < \frac{\sigma}{\zeta}\right),$$

$$\langle \nabla p \cdot a_h \cdot \nabla \log p_h \rangle \leq \frac{1}{4\varepsilon_2} \left\langle \nabla p \cdot \frac{a_h}{p} \cdot \nabla p \right\rangle + \varepsilon_2 I_h$$

$$\leq \frac{1}{4\varepsilon_2} \cdot \frac{\zeta}{\sigma} I + \varepsilon_2 I_h, \qquad \varepsilon_2 = \varepsilon_1 \frac{\zeta}{\sigma},$$

$$\langle pb \cdot \nabla \log p_h \rangle \leq \frac{1+\varepsilon}{4} \langle pb \cdot a_h^{-1} \cdot b \rangle + \frac{1}{1+\varepsilon} I_h, \qquad 0 < \varepsilon < 1.$$

Substituting these estimates into (3.1) gives

$$Z'(s) \leq \left(-1 + 2\frac{\zeta}{\sigma}\varepsilon_1 + \frac{1}{1+\varepsilon}\right) I_h + \frac{1}{2\varepsilon_1} I + \frac{1+\varepsilon}{(1-\delta)^2} \cdot \frac{|x-y|^2}{4\sigma t^2}$$

or

$$Z'(s) \leq \left(1 + \frac{1}{\varepsilon}\right) \frac{\zeta}{\sigma} Q'(s,x) + \frac{1+\varepsilon}{(1-\delta)^2} \cdot \frac{|x-y|^2}{4\sigma t^2} \; .$$

Integrating the last inequality over s from δt to t yields (i).

Remark 3.2. The above proof also works for time dependent matrices $a(t, x)$. In this connection see [NS]. There is a close relationship between the idea behind Theorem 3.1 (i) and the so-called lemma on *the tilted cylinder* (see [Kr, pp. 103–105]).

4.   The upper and lower bounds, I.

In this section we show how (NEE), combined with results of sections 1–3, yields the upper and lower bounds on the heat kernel of $\exp[-t\mathcal{L}_2(a, b, \hat{b}, V)]$ with bounded measurable coefficients.

Let first the coefficients be smooth bounded functions on a $\mathbb{R}^d$, $d \geq 2$, and $(\mathbf{H_u})$ be fulfilled. Then, due to the properties of $p(t, x, y)$ listed in section 3, most of the questions concerning the justification of the estimates

$$\int_0^s \|B_1 e^{-tA_1} f\|_1 dt \leq \|b \cdot a^{-1} \cdot b\|_\infty^{1/2} \int_0^s Q'(t)^{1/2} dt \|f\|_1$$

$$\leq \beta(s)\|f\|_1 \qquad f \in D(A_1),$$

(4.1)

$$\|\exp[-t(A_1 + B_1)]\|_{1\to 1} \leq (1 - \beta(s))^{-1} \exp\left[-\frac{t}{s} \log\left(1 - \beta(s)\right)\right],$$

$$\beta(s) := \|b \cdot a^{-1} \cdot b\|_{\infty}^{1/2}(C_1 + C_2 + 3d)\sqrt{s/2d}, \qquad \beta(s) < 1, \ t > 0,$$

turn out to be rather simple (the elementary perturbation theory is only needed).

Lemma 4.1. *Given* $\alpha \in \mathbb{R}^1$, $\phi \in C_b^1(\mathbb{R}^d)$ *with* $\nabla\phi \cdot a \cdot \nabla\phi \leq 1$, *define* $b := \alpha a \cdot \nabla\phi$. *Then*

i) $\int_0^s \|b \cdot \nabla e^{-tA_1} f\|_1 dt \leq \beta(s)\|f\|_1, \qquad f \in L^1$

where $\beta(s) = k|\alpha|\sqrt{s}$, $k = \sqrt{(C_1 + C_2 + 3d)/2d}$.

ii) $\|e^{-tH_1}\|_{1\to 1} \leq 2e^{\delta\alpha^2 t}, \qquad \delta = 4k^2, \ t \geq 0.$

iii) $\|e^{-tH}\|_{2\to 2} \leq e^{\frac{1}{4}\alpha^2 t}.$

iv) $\|e^{-tH_1}\|_{1\to\infty} \leq c_\varepsilon t^{-d/2} e^{\delta_1\alpha^2 t}, \quad t > 0, \ \delta_1 = \frac{1}{4} + \varepsilon, \ 0 < \varepsilon < 1,$

where $H = A + b \cdot \nabla$ *and* $H_1 = A_1 + b \cdot \nabla$.

Proof. By the preceding we have i). Taking $s = (2k|\alpha|)^{-2}$ in i) so $\beta(s) = \frac{1}{2}$, we obtain ii) by (4.1). iii) follows from the inequality

$$\text{Re}\langle Hf, f\rangle = \langle Af, f\rangle + \langle b \cdot \nabla f, f\rangle \geq -\frac{1}{4}\alpha^2\langle f, f\rangle,$$

and iv) from ii), iii), the inequality $\text{Re}\langle Hf, f\rangle \geq \frac{1}{2}\langle A^{1/2}f, A^{1/2}f\rangle - \frac{1}{2}\alpha^2\langle f, f\rangle$ and Lemma 1.2.

As a result, the heat kernel $q(t, x, y)$ for e^{-tH}, $t > 0$, has the properties: it is smooth, admits a Gaussian lower bound, q and $t|q'(t, x, y)|$ admit Gaussian upper bounds (with entering constants that may depend on smoothness of $a(\cdot)$), $\langle q(t, x, \cdot)\rangle = 1$,

$$0 < q(t, x, y) \leq C_\varepsilon t^{-d/2} e^{(1+\varepsilon)\frac{1}{4}\alpha^2 t}$$

for all $0 < \varepsilon < 1$, $t > 0$, $x, y \in \mathbb{R}^d$, $\alpha \in \mathbb{R}^1$, where C_ε depends on C_1, C_2, d and $0 < \varepsilon < 1$ only.

Let us now define functions $Q(t) = Q(t, y)$, $N(t) = N(t, y)$, $M(t) = M(t, y)$ by

$$Q(t) := -\langle q(t, y, \cdot) \log q(t, y, \cdot) \rangle,$$

$$N(t) := \left\langle \nabla.q(t, y, \cdot) \cdot \frac{a(\cdot)}{q(t, y, \cdot)} \cdot \nabla.q(t, y, \cdot) \right\rangle,$$

$$M(t) := \langle |y - \cdot| q(t, y, \cdot) \rangle.$$

One has

$$Q'(t) = N(t) + \langle b(\cdot) \cdot a^{-1}(\cdot) \cdot \nabla q(t, y, \cdot) \rangle$$

$$\geq N(t) - N(t)^{1/2} (e^{-tH} b \cdot a^{-1} \cdot b)^{1/2}$$

$$\geq \frac{1}{2} N(t) - \frac{1}{2} \alpha^2.$$

Repeating the proof of (NEE) for $p(t, x, y)$ yields

Corollary 4.2. *In the hypotheses of Lemma 4.1, for all $t > 0$, $y \in \mathbb{R}^d$, $\alpha \in \mathbb{R}^1$,*

$$\tilde{Q}(t) - \omega t - C_1 \leq Q(t, y) \leq \tilde{Q}(t) + \omega t + C_3, \qquad \omega = \delta \alpha^2,$$

where C_3 and δ depend on C_1, ζ and d only.

$$\int_0^s e^{-t\lambda} \|b \cdot \nabla e^{-t(A \dot{+} \nabla \cdot b)_1} f\|_1 dt$$

$$\leq \|b \cdot a^{-1} \cdot b\|_\infty^{1/2} \int_0^s e^{-t\lambda} (2Q'(t) + \omega)^{1/2} dt \|f\|_1,$$

A simple calculation, using Corollary 4.2, shows that

$$\int_0^s e^{-t\lambda} \|b \cdot \nabla e^{-(A \dot{+} \nabla \cdot b)_1} f\|_1 dt \leq \beta(s, \lambda) \|f\|_1$$

$$= \frac{c|\alpha|}{\sqrt{\lambda}} \left(1 + \frac{\omega}{\lambda}\right) \|f\|_1, \quad f \in L^1(\mathbb{R}^d),$$

where $c = c(d, C_1, C_3)$.

Letting $\lambda = \rho\omega$, one has $\beta(s, \lambda) = \frac{c}{\sqrt{\rho\delta}} \left(1 + \frac{1}{\rho}\right)$.

Fix ρ by the condition $\beta(s, \lambda) = 1/2$. Then

$$\| \exp\{-t\mathcal{L}_1(a, b, -b, 0)\} \|_{1 \to 1} \leq 2 \exp[(1 + \rho)\omega t],$$

$$\| \exp\{-t\mathcal{L}_1(a, b, -b, -b \cdot a^{-1} \cdot b)\} \|_{1 \to 1} \leq 2 \exp\{[1 + (1 + \rho)\delta]\alpha^2 t\}.$$

Since

$$\mathrm{Re}\langle \mathcal{L}_2(a, b, -b, -b \cdot a^{-1} \cdot b)f, f \rangle = \mathrm{Re}\langle \mathcal{L}_2(a, 0, 0, -b \cdot a^{-1} \cdot b)f, f \rangle$$

$$\geq \langle (A - \alpha^2)f, f \rangle,$$

one can apply Lemma 1.2 and obtain the bound

$$(4.2) \quad \| \exp[-t\mathcal{L}_2(a, b, -b, -b \cdot a^{-1} \cdot b)] \|_{1 \to \infty} \leq c(\varepsilon t)^{-d/2} e^{(1+\varepsilon)\alpha^2 t},$$

for all $t > 0$, $\alpha \in \mathbb{R}^1$, $0 < \varepsilon < 1$ and some constant $c = c(C_i, d)$, $i = 1, 2$.

Let us now define the operators S_α, $\alpha \in \mathbb{R}^1$, putting $S_\alpha f(x) = e^{\alpha\phi(x)} f(x)$, $\phi \in C_b^1(\mathbb{R}^d)$. Then it is easily seen that

$$S_{-\alpha} e^{-tA} S_\alpha f = \exp[-t\mathcal{L}_2(a, b, -b, -b \cdot a^{-1} \cdot b)]f, \qquad f \in L^1(\mathbb{R}^d).$$

Therefore, by (4.2)

$$e^{-tA}(x, y) \leq c(\varepsilon t)^{-d/2} e^{\alpha(\phi(x) - \phi(y)) + (1+\varepsilon)\alpha^2 t}.$$

Putting $\alpha = \frac{1}{(1+\varepsilon)2} (\phi(y) - \phi(x))$, one obtains from the last inequality the Davies upper bound

$$(\mathrm{UDB}) \qquad e^{-tA}(x, y) \leq c(t\varepsilon)^{-d/2} \exp\left(-\frac{D^2(x, y)}{(1 + \varepsilon)4t} \right)$$

$$(t > 0, \ 1 > \varepsilon > 0, \ x, y \in \mathbb{R}^d).$$

Armed with (UDB) it is easy to prove it without any smoothness assumptions on the coefficients. Indeed let a be a non-smooth matrix

satisfying all the other assumptions of Lemma 4.1. It is a well known
fact that if $\{a_n\}_1^\infty$ is any sequence of smooth matrices satisfying the
assumptions of Lemma 4.1 and such that $a_n(\cdot) \to a(\cdot)$ pointwise (a.e.)
then $e^{-tA(a_n)} \to e^{-tA}$ strongly in L^2 and, hence, due to (UGB),
strongly in L^1. In particular, $e^{-tA}1 = 1$. Given $f, g \in L^1 \cap L^\infty$, one
has

$$\left| \langle e^{-\alpha\phi} e^{-tA} e^{\alpha\phi} f, g \rangle \right| = \lim_n \left| \langle e^{-\alpha\phi} e^{-tA(a_n)} e^{\alpha\phi} f, g \rangle \right|$$

$$\leq c(\varepsilon t)^{-d/2} e^{(1+\varepsilon)\alpha^2 t} \|f\|_1 \|g\|_1 .$$

Therefore $\|e^{-\alpha\phi} e^{-tA} e^{\alpha\phi}\|_{1\to\infty} \leq c(\varepsilon t)^{-d/2} e^{(1+\varepsilon)\alpha^2 t}$ so the claim easily
follows.

By (LGB) $e^{-tA} f = \lim_n e^{-tA(a_n)} f \geq c e^{\mu t \Delta} f$, $0 \leq f \in L^1 \cap L^\infty$,
so one concludes that the same lower bound holds for the heat kernel
without any smoothness of $a(\cdot)$. We also recall that (LGB)+(UGB)
for $p_{a_n}(t, x, y) = e^{-tA(a_n)}(x, y)$ imply $p_{a_n}(t, x, y) \to p_a(t, x, y)$ bound-
edly and uniformly on compacts. The latter gives an alternative proof
of (LGB) and (UDB) for $p_a(t, x, y)$.

Further, let $\{b_m\}_1^\infty \subset C_b(\mathbb{R}^d, \mathbb{R}^d)$ be such that $\sup_m \|b_m\|_\infty < \infty$
and $b_m \to b$ pointwise a.e. Then slight modifications of Lemma 4.1
and Corollary 4.2 yield

$$e^{-t\Lambda_1(b_m)}(x, y) \leq c \exp\left(c \sup_m \|b_m \cdot a^{-1} \cdot b_m\|_\infty t - \frac{D^2(x, y)}{4ct} \right),$$

where $c = c(C_i, d)$ and $\Lambda_1(b) = A_1 + b \cdot \nabla$. Again, this bound
and (LGB) for $e^{-t\Lambda_1(b_m)}$ (see Theorem 5.3 below) yield $e^{-t\Lambda_1(b_m)} \to$
$e^{-t\Lambda_1(b)}$ strongly in L^1 (so $e^{-t\Lambda_1}1 = 1$), and

$$c \exp\left(-\frac{t}{c} \|b \cdot a^{-1} \cdot b\|_\infty - \frac{|x - y|^2}{ct} \right) \leq e^{-t\Lambda_1(b)}(x, y),$$

for all $t > 0$, $x, y \in \mathbb{R}^d$ and some $c = c(\sigma, \zeta, d) > 0$.

One can now repeat all steps leading to the two-sided bound on the diagonal entropy function corresponding to $\Lambda(b)$, and hence derive (UDB) for the uniformly elliptic operator $\mathcal{L}_2(a, b, \hat{b}, V)$ with measurable bounded coefficients.

5. The upper and lower bounds, II.

To consider more general classes of the coefficients we need the following:

Lemma 5.1. *Let $a\colon \mathbb{R}^d \to \mathbb{R}^d \otimes \mathbb{R}^d$ be a measurable function satisfying the uniform ellipticity condition $(\mathbf{H_u})$. If a measurable drift $b\colon \mathbb{R}^d \to \mathbb{R}^d$ satisfies the assumption*

$$\lim_{\lambda \to \infty} \left\| \int_0^\infty e^{-\lambda t}(e^{t\Delta}b \cdot a^{-1} \cdot b)^{1/2}t^{-1/2}dt \right\|_\infty = 0$$

or, equivalently, if $b \in N_2$, then

$$\int_0^\infty e^{-\lambda t}\|b \cdot \nabla e^{-tA_1}f\|_1 dt \le \hat{\beta}(\lambda)\|f\|_1$$

$$= c_\varepsilon \cdot n\left(\frac{\lambda(1+\varepsilon)}{2\zeta}, \sqrt{b \cdot a^{-1} \cdot b}\right)\|f\|_1, \quad f \in L^1$$

where (recall) $n(\lambda, h) := \left\| \int_0^\infty e^{-\lambda t}(e^{t\Delta}|h|^2)^{1/2}t^{-1/2}dt \right\|_\infty$ and $c_\varepsilon = c_\varepsilon(\sigma, \zeta, d)$, $\varepsilon \in]0, 1]$.

Proof. Let 1_n be the indicator of $\{x \in \mathbb{R}^d; |x| \le n$ and $b \cdot a^{-1} \cdot b(x) \le n\}$. Set $g = b \cdot a^{-1} \cdot b 1_n$. One has

$$\int_0^\infty e^{-\lambda t}\|b \cdot \nabla e^{-tA_1}f\|_1 dt$$

$$\le \lim_{n \to \infty} \left\| \int_0^\infty e^{-\lambda t}(e^{-tA}g)^{1/2}(Q'(t))^{1/2}dt \right\|_\infty \|f\|_1.$$

Using (UGB) yields

$$\int_0^\infty e^{-\lambda t}(e^{-tA}g)^{1/2}\big(Q'(t)\big)^{1/2}dt$$

$$\le c_\varepsilon^{1/2}\int_0^\infty e^{-\lambda t}(e^{\nu t\Delta}g)^{1/2}\big(Q'(t)\big)^{1/2}dt$$

where $\nu = (1+\varepsilon)^{-1}\zeta$.

Letting $\eta(t) = (e^{t\Delta}g)^{1/2}$, one has

$$\int_0^\infty e^{-\lambda t}\eta(\nu t)\big(Q'(t)\big)^{1/2}dt \le \frac{1}{2}\int_0^\infty e^{-\lambda t}\eta(\nu t)\left(\frac{1}{\sqrt{t}}+\sqrt{t}Q'(t)\right)dt$$

$$= \frac{1}{2\sqrt{\nu}}n\left(\frac{\lambda}{\nu},\sqrt{g}\right) + \frac{1}{2}I\,,$$

$$I := -\int_0^\infty Q(t)d\big(e^{-\lambda t}\sqrt{t}\eta(\nu t)\big) = -\frac{1}{2}\int_0^\infty e^{-\lambda t}Q(t)\eta(\nu t)\frac{dt}{\sqrt{t}}$$

$$+ \lambda\int_0^\infty e^{-\lambda t}Q(t)\eta(\nu t)\sqrt{t}dt - \int_0^\infty e^{-\lambda t}Q(t)\eta'(\nu t)dt\,.$$

Since $\eta'(\nu t) := \frac{d}{dt}\eta(\nu t) = \varphi^+ - \varphi^-$, $\varphi^\pm \ge 0$, $\varphi^- = \frac{d}{4t}\eta(\nu t)$, one has

$$I \le \int_0^\infty e^{-\lambda t}\widetilde{Q}'(t)\eta(\nu t)\sqrt{t}dt + \frac{C_1}{2\sqrt{\nu}}n\left(\frac{\lambda}{\nu},\sqrt{g}\right)$$

$$+ \lambda C_2\int_0^\infty e^{-\lambda t}\eta(\nu t)\sqrt{t}dt$$

$$+ C_2\int_0^\infty e^{-\lambda t}\sqrt{t}\varphi^-dt + C_1\int_0^\infty e^{-\lambda t}\sqrt{t}\varphi^+dt$$

$$= \frac{d}{2\sqrt{\nu}}\cdot\left[1 + \frac{C_1+C_2}{2}\right]n\left(\frac{\lambda}{\nu},\sqrt{g}\right)$$

$$+ (C_1+C_2)\lambda\int_0^\infty e^{-\lambda t}\eta(\nu t)\sqrt{t}dt$$

$$\le \frac{d}{4\sqrt{\nu}}(2+C_1+C_2)n\left(\frac{\lambda}{\nu},\sqrt{g}\right) + \frac{2(C_1+C_2)}{\sqrt{\nu}}n\left(\frac{\lambda}{2\nu},\sqrt{g}\right)\,.$$

Thus,

$$\int_0^\infty e^{-\lambda t}\|b\cdot\nabla e^{-tA_1}f\|_1 dt$$

$$\le (c_\varepsilon\nu^{-1})^{\frac{1}{2}}\left\{\big[2^{-1}+(8)^{-1}(2+C_1+C_2)d\big]n\left(\frac{\lambda}{\nu},\sqrt{b\cdot a^{-1}\cdot b}\right)\right.$$

$$\left. + (C_1+C_2)n\left(\frac{\lambda}{2\nu},\sqrt{b\cdot a^{-1}\cdot b}\right)\right\}\|f\|_1\,.$$

This proves Lemma 5.1.

Lemma 5.1 leads to the proof of Theorem 0.2 and to the following bound on $T^t = \exp\{-t[A_1 + (\hat{b} - \alpha a \cdot \nabla\phi) \cdot \nabla]\}$, $t \geq 0$, for $\sqrt{\hat{b} \cdot a^{-1} \cdot \hat{b}} \in N_2$, which holds for all large $\lambda > 0$, $\delta > 1$ (the optimal choice of λ depends on the behaviour of $n(\mu, \sqrt{\hat{b} \cdot a^{-1} \cdot \hat{b}})$ at $\mu = \infty$ only):

$$\|T^t\|_{1 \to 1} \leq 2e^{(\lambda + \omega)t}, \qquad \omega = \delta\alpha^2.$$

Thus,

$$\widetilde{Q}(t) - \frac{1}{2}(\lambda + \omega)t - \widetilde{C}_1 \leq Q(t, y) \leq \widetilde{Q}(t) + \frac{1}{2}(\lambda + \omega)t + \widetilde{C}_2$$

where $Q(t, y)$ corresponds to $q(t, y, x) = T^t(y, x)$, for $\sqrt{\hat{b} \cdot a^{-1} \cdot \hat{b}} \in N_2 \cap L^\infty$.

Lemma 5.2. *If* $\sqrt{\hat{b} \cdot a^{-1} \cdot \hat{b}} \in N_2$, $\hat{b} \cdot a^{-1} \cdot \hat{b} \in K_d \cap L^\infty$ *then there are constants* c^0, λ_0 *depending on the constants from* (UGB), *for* e^{-tA}, *and on the behaviour of the Kato norm* $k(d; \mu, \hat{b} \cdot a^{-1} \cdot \hat{b})$ *as* $\mu \to \infty$ *only such that, uniformly on* $\alpha \in \mathbb{R}^1$,

$$\int_0^\infty e^{-t(\lambda_1 + \omega_1)} \left\| [\alpha\nabla\phi \cdot a \cdot \nabla \pm \alpha\nabla\phi \cdot \hat{b} + \alpha^2 \nabla\phi \cdot a \cdot \nabla\phi](T^t)^* f \right\|_1 dt$$

$$\leq \widetilde{\beta}(\lambda_1 - \lambda_0, \delta_1)\|f\|_1,$$

$$\widetilde{\beta}(\lambda_1 - \lambda_0, \delta_1) = (1 + \widetilde{C}_1 + \widetilde{C}_2)c_d \frac{1}{\sqrt{\delta_1}}$$

$$+ k\big(d; 2(\lambda_1 - \lambda_0)\nu^{-1}, \hat{b} \cdot a^{-1} \cdot \hat{b}\big) \frac{c^0}{\nu\sqrt{\delta_1 - 1}},$$

$\nu = (1 + \varepsilon)^{-1}\zeta$, $\omega_1 = \delta_1\alpha^2$, $\delta_1 \geq \delta$, $\lambda_1 \geq \lambda \vee \lambda_0 > 0$.

Proof. One has

$$\int_0^\infty e^{-(\lambda_1 + \omega_1)t} \|\alpha^2 \nabla\phi \cdot a \cdot \nabla\phi(T^t)^* f\|_1 dt$$

$$\leq \alpha^2 \int_0^\infty e^{-t(\lambda_1 + \omega_1)} dt \|f\|_1 \leq \delta_1^{-1}\|f\|_1,$$

$$\int_0^\infty e^{-(\lambda_1+\omega_1)t}\|\alpha\nabla\phi\cdot a\cdot\nabla(T^t)^*f\|_1\,dt$$

$$\le |\alpha|\left\|\int_0^\infty e^{-(\lambda_1+\omega_1)t}\left(2Q'(t)+\frac{1}{2}(\lambda+\omega)\right)^{1/2}dt\right\|_\infty \|f\|_1$$

$$\le (1+\widetilde{C}_1+\widetilde{C}_2)c_d\delta_1^{-1/2}\|f\|_1\,,$$

$$\int_0^\infty -(\lambda_1+\omega_1)t\|\alpha\nabla\phi\cdot\hat{b}(T^t)^*f\|_1\,dt$$

$$\le |\alpha|\left\|\int_0^\infty e^{-(\lambda_1+\omega_1)t}T^t\sqrt{\hat{b}\cdot a^{-1}\cdot\hat{b}}\,dt\right\|_\infty\|f\|_1\,.$$

To estimate the last term we use the perturbation theory or the Cameron-Martin formula, obtaining

$$T^t|f|\le e^{\alpha^2 t}\left(e^{-t(A-2\hat{b}\cdot a^{-1}\cdot\hat{b})}f^2\right)^{1/2}\,.$$

Using the Kac formula and (UGB) for e^{-tA} yield

$$e^{-t(A-2\hat{b}\cdot a^{-1}\cdot\hat{b})}|f|\le (c^0)^2 e^{2\lambda_0 t}e^{t\nu\Delta}|f|\,.$$

Thus,

$$|\alpha|\left\|\int_0^\infty e^{-(\lambda_1+\omega_1)t}T^t\sqrt{\hat{b}\cdot a^{-1}\cdot\hat{b}}\,dt\right\|_\infty$$

$$\le |\alpha|\left(\int_0^\infty e^{-(\delta_1-1)\alpha^2 t}dt\right)^{\frac{1}{2}}$$

$$c^0\left(\left\|\int_0^\infty e^{-2(\lambda_1-\lambda_0)t}e^{t\nu\Delta}\hat{b}\cdot a^{-1}\cdot\hat{b}\,dt\right\|_\infty\right)^{\frac{1}{2}}\,.$$

Lemma 5.2 easily follows.

Due to Lemma 5.2, we conclude that

$$e^{-t(A+\hat{b}\cdot\nabla)}(x,y)\le ct^{-d/2}e^{\lambda_2 t-\frac{|x-y|^2}{ct}}$$

where $c=c(\sigma,\zeta,d)>1,\lambda_2$ depends on $n(\lambda,\sqrt{\hat{b}\cdot a^{-1}\cdot\hat{b}})$ and $k(d;\lambda_1,\hat{b}\cdot a^{-1}\cdot\hat{b})$. Similarly, we have

$$T^t(x,y)\le ct^{-d/2}e^{-\frac{|x-y|^2}{4ct}}e^{(\lambda_2+\alpha^2)t}\,.$$

The latter allows one to estimate the quantity $T^t h$, $h \equiv b \cdot a^{-1} \cdot b$, as follows:

$$T^t h \le c e^{(\lambda_2 + \alpha^2)t} e^{tc\Delta} h \,.$$

Therefore, proceeding as in the proof of Lemma 5.2, we have, letting $f_t = (T^t)^* f$,

$$\int_0^\infty e^{-(\lambda_1 + \omega_1)t} \left(\|(b + \alpha a \cdot \nabla\phi) \cdot \nabla f_t\|_1 \right.$$
$$+ \|\alpha\nabla\phi \cdot (b - \hat{b}) f_t\|_1 + \|\alpha^2 \nabla\phi \cdot a \cdot \nabla\phi f_t\|_1) dt \le \tilde{\beta}(\lambda_1 - \lambda_0, \delta_1) \|f\|_1$$

where λ_0 depends on the behaviour of $n(\mu, \sqrt{\hat{b} \cdot a^{-1} \cdot \hat{b}})$, $k(d; \mu, \hat{b} \cdot a^{-1} \cdot \hat{b})$ and $k(d; \mu, b \cdot a^{-1} \cdot b)$ at $\mu = \infty$.

Thus

$$\left\| S_{-\alpha} \exp[-t\mathcal{L}_1(a, b, -\hat{b}, 0)] S_\alpha \right\|_{1 \to 1} \le c \exp[(\lambda + \delta\alpha^2)t]$$

for some $c = c(\sigma, \zeta, d)$ and λ, δ, depending on the behaviour of $n(\mu, \sqrt{h})$, $k(d; \mu, h)$, $h = b \cdot a^{-1} \cdot b$ or $\hat{b} \cdot a^{-1} \cdot \hat{b}$, at $\mu = \infty$.

The last bound allows us to obtain the estimate

$$\left\| S_{-\alpha} \exp[-t\mathcal{L}_1(a, b, -\hat{b}, 0)] S_\alpha \right\|_{1 \to \infty} \le c_\lambda(\varepsilon t)^{-d/2} e^{(\lambda + (1+\varepsilon)\alpha^2)t} \,,$$

and complete the derivation of the upper bound

(5.1)

$$\exp[-t\mathcal{L}_2(a, b, -\hat{b}, 0)](x, y) \le c_\lambda(\varepsilon t)^{-d/2} \exp[\lambda t - D^2(x, y)/4(1 + \varepsilon)t]$$

$$(t > 0, \ 0 < \varepsilon < 1)$$

for $a(\cdot)$ satisfying the uniform ellipticity condition, $\sqrt{b \cdot a^{-1} \cdot b}$, $\sqrt{\hat{b} \cdot a^{-1} \cdot \hat{b}} \in N_2$, $b \cdot a^{-1} \cdot b + \hat{b} \cdot a^{-1} \cdot \hat{b} \in K_d$, $\hat{b} \in L^\infty$.

Now it is easy to remove the restriction "$\hat{b} \in L^\infty$". Indeed, putting $\hat{b}_n = 1_n \hat{b}$ where 1_n denotes the indicator of the set $\{x \in \mathbb{R}^d; |\hat{b}(x)| \leq n\}$, one has, for $0 \leq W \in K_d$,

$$\exp\left\{-t\mathcal{L}_2\left(a, b, -\hat{b}_n, -(W \wedge n)\right)\right\} \to \exp\left\{-t\mathcal{L}_2(a, b, -\hat{b}, -W)\right\}$$

strongly in L^2 as $n \to \infty$. By (5.1), the convergence also holds in L^1 and the claim easily follows. In particular, we have partly proved Theorem 0.1 (the upper bound).

Theorem 5.3. *Let $a, b, \hat{b}, V$ be measurable functions on $\mathbb{R}^d$, a satisfying the uniform ellipticity condition, $b, \hat{b} \in N_2$ and $0 \leq V$, $b^2, \hat{b}^2 \in K_d$. Then there are constants $c > 0$ and $\lambda < \infty$ such that for all $t > 0$, $x, y \in \mathbb{R}^d$,*

$$ct^{-d/2} \exp\left\{-\lambda t - \frac{|x - y|^2}{ct}\right\} \leq \exp[-t\mathcal{L}_2(a, b, \hat{b}, V)](x, y).$$

Proof. Letting $p, q_b, q_{b,V}$ and r denote the heat kernels to the semigroups generated by $\frac{1}{2}A$, $\frac{1}{2}A + b \cdot \nabla$, $\frac{1}{2}A + b \cdot \nabla + V$ and $\frac{1}{2}A \dot{-} \frac{1}{\theta-2}(b \cdot a^{-1} \cdot b + V)$, $\theta > 2$, respectively, one has

$$p(t, x, y) \leq \left(q_{b,V}(t, x, y)\right)^{1/\theta} \left(q_{-b}(t, x, y)\right)^{1/\theta} \left(r(t, x, y)\right)^{1-2/\theta}$$

by the Cameron-Martin-Kac formula and Hölder's inequality. Now apply (LGB) to p and (UGB) to q_{-b} and r. The case $\hat{b} = 0$ is proved. The general case follows from this particular one by taking $\frac{1}{2}A - \nabla \cdot \hat{b}$ instead of $\frac{1}{2}A$ and using the same argument:

$$\hat{p}(t, x, y) \leq \left(\hat{q}_{b,V}(t, x, y)\right)^{1/\theta} \left(\hat{q}_{-b}(t, x, y)\right)^{1/\theta} \left(\hat{r}(t, x, y)\right)^{1-2/\theta}.$$

6.   The upper bound, III. $-\Delta + b \cdot \nabla$, $b \in K_{d+1}$.

To begin with let us briefly discuss some consequences of the definition of the Kato class K_{d+1}. If b: $\mathbb{R}^d \to \mathbb{R}^d$ then $b \in K_{d+1}$ means that $\||b|(\lambda - \Delta)^{-1/2}\|_{1 \to 1} = \|(\lambda - \Delta)^{-1/2}|b|\|_\infty \to 0$ as $\lambda \to \infty$. By interpolation,

$$\||b|^{1/p}(\lambda - \Delta)^{-1/2}|b|^{1/p'}\|_{p \to p} \le \|(\lambda - \Delta)^{-1/2}|b|\|_\infty , \qquad p' = p/p - 1 ,$$

for all $1 \le p \le \infty$. In particular, $|b| \in PK_\beta\big((-\Delta)^{1/2}\big)$ for all $\beta > 0$. Moreover,

$$\|(\lambda - \Delta)^{-1/4}b \cdot \nabla(\lambda - \Delta)^{-3/4}\|_{2 \to 2}$$
$$\le \|(\lambda - \Delta)^{-1/2}|b|\|_\infty^{1/2} \left\| \frac{b}{\sqrt{|b|}} \cdot \nabla(\lambda - \Delta)^{-3/4} \right\|_{2 \to 2} \to 0 ,$$

$\|b \cdot \nabla(\lambda - \Delta)^{-1}\|_{1 \to 1} \to 0$ as $\lambda \to \infty$, by the explicit formula for $(\lambda - \Delta)^{-\delta}$, $\delta > 0$. The latter means that if we define $H_1 = H_1(b)$, letting $H_1 = A_1 + b \cdot \nabla$, $D(H_1) = D(A_1)$, $A_p = -\Delta$, $D(A_p) = (1 - \Delta)^{-1}L^p(\mathbb{R}^d)$, $1 \le p < \infty$, then according to Hille's perturbation theorem (see [Ka, Ch. 9, §2.2]) H_1 generates a holomorphic semigroup on $L^1(\mathbb{R}^d)$. Moreover,

$$(\lambda + H_1)^{-1} = (\lambda - \Delta)^{-1}\big(1 + b \cdot \nabla(\lambda - \Delta)^{-1}\big)^{-1}, \qquad \text{for all large } \lambda > 0 .$$

Let 1_n be the indicator of $\{x \in \mathbb{R}^d; \ |b(x)| \le n\}$. The estimate

$$\big\|\big(\lambda + H_1(b_n)\big)^{-1}f - \big(\lambda + H_1(b)\big)^{-1}f\big\|_1$$
$$\le c(\lambda)\big\|(1 - 1_n)b \cdot \nabla\big(\lambda + H_1(b)\big)^{-1}f\big\|_1 , \quad f \in L^1$$

(in which a sufficiently large $\lambda > 0$ is chosen to be independent of $n = 1, 2, \dots,$) shows that $e^{-tH_1(b_n)} \to e^{-tH_1(b)}$ strongly in $L^1(\mathbb{R}^d)$ as

$n \to \infty$ and, since $e^{-tH_1(b_n)}$ is a contraction on L^∞, one concludes that e^{-tH_1}, $t \geq 0$, is a contraction on L^∞. Therefore there is a family of consistent L^∞-contraction C_0-semigroups $\{e^{-tH_p}, 1 \leq p < \infty\}$.

To establish the $L^1 \to L^\infty$ property of e^{-tH_1} one can prove first the $L^p \to L^q$ property of e^{-tH_p} for *some* $1 \leq p < q \leq \infty$, and then use the extrapolation (see [VSC, p. 9]).

To prove the $L^p \to L^q$ property we use the following representation of $(\lambda + H_2)^{-1}$:

(6.1)
$$(\lambda + H_2)^{-1} f$$

$$= (\lambda - \Delta)^{-3/4}\big(1 + (\lambda - \Delta)^{-1/4} b \cdot \nabla (\lambda - \Delta)^{-3/4}\big)^{-1}(\lambda - \Delta)^{-1/4} f,$$

$f \in L^2$, which is valid for all $\lambda > 0$ sufficiently large. Indeed, if $f \in L^1 \cap L^2$ then the R.H.S. of (6.1) coincides with the R.H.S. of $(\lambda + H_1)^{-1} f = (\lambda - \Delta)^{-1}\big(1 + b \cdot \nabla (\lambda - \Delta)^{-1}\big)^{-1} f$. Also, semigroups e^{-tH_p} are consistent so that $(\lambda + H_1)^{-1} f = (\lambda + H_2)^{-1} f$. (6.1) is proved. (6.1) shows that 1) e^{-tH_2} is holomorphic, and 2) $(\lambda + H_2)^{-1}$ extends by continuity to a map from L^p into L^q for all $2d/d+1 \leq p \leq 2$ and $2 \leq q \leq 2d/d - 3$, $d \geq 4$. (The case $d = 3$ reduces to the case $d \geq 4$.) Since e^{-tH_p} are consistent and holomorphic on L^p, $1 \leq p \leq 2$, one has, letting $p = 2d/d + 1$, $q = 2d/d - 3$,

$$\|e^{-t(\lambda + H_p)} f\|_q = \|(\lambda + H_2)^{-1}(\lambda + H_2) e^{-t(\lambda + H_2)} f\|_q$$

$$\leq c\|(\lambda + H_p) e^{-t(\lambda + H_p)} f\|_p \leq C t^{-1} \|f\|_p, \qquad f \in L^p \cap L^q.$$

Since $\frac{d}{2}\left(\frac{1}{p} - \frac{1}{q}\right) = 1$, one concludes that $\|e^{-tH_1}\|_{1 \to \infty} \leq c t^{-d/2} e^{t\lambda}$.

We now apply the above procedure to $T^t(\alpha) := S_\alpha e^{-tH_1} S_{-\alpha}$, $\alpha \in \mathbb{R}^d$, $d \geq 4$ where $S_\alpha f(x) := e^{\alpha \cdot x} f(x)$. Namely, we will prove that there are constants c_i, ℓ_i such that

$$\|T^t(\alpha)\|_{1 \to 1} \leq c_1 e^{\ell_1 (1+\alpha^2)t},$$

$$\|T^t(\alpha)\|_{\infty \to \infty} \leq c_2 e^{\ell_2 (1+\alpha^2)t},$$

$$\|T^t(\alpha)\|_{p \to q} \leq c_3 t^{-1} e^{\ell_3 (1+\alpha^2)t}$$

for all $\alpha \in \mathbb{R}^d$, $t > 0$. It is clear that these inequalities prove Theorem 0.3. Given $\delta > 1$, we define the operator Λ_0 acting in $L^1 = L^1(\mathbb{R}^d)$ by

$$\Lambda_0 = -\Delta + 2\alpha \cdot \nabla + (\delta - 1)\alpha^2, \quad D(\Lambda_0) = (1 - \Delta)^{-1} L^1$$

for any $\alpha \in \mathbb{R}^d$. Let $f_t = e^{-t\Lambda_0} f$. By straightforward calculation,

$$\int_0^\infty e^{-\lambda t} (\|b \cdot \nabla f_t\|_1 + \|\alpha \cdot b f_t\|_1) dt \leq \beta(\lambda) \|f\|_1, \qquad f \in L^1$$

where $\lambda > 0$; $\beta(\lambda)$ depends on $k(d+1; \lambda, b)$ only and tends to zero as $\lambda \to \infty$. The latter means that $H_1(b, \alpha) := \Lambda_0 + b \cdot \nabla - \alpha \cdot b$ of domain $D(\Lambda_0)$ generates a semigroup, say F^t, and there are constants M, λ_0, depending on the behaviour of $k(d + 1; \mu, b)$ at $\mu = \infty$, such that

$$\|F^t\|_{1 \to 1} \leq M e^{\lambda_0 t} \text{ for all } t > 0, \ \alpha \in \mathbb{R}^d, \ \delta > 1.$$

Set $\Sigma(\theta) = \left\{ \xi; |a \arg \xi| \leq \frac{1}{2}\pi + \theta, \ 0 < \theta < \frac{1}{2}\pi \right\}$.

Using the explicit formula for $(\xi - \Delta)^{-1}(x, y)$ and the equality

$$(6.2) \quad (\xi - \Delta + 2\alpha \cdot \nabla)^{-1}(x, y) = e^{\alpha \cdot (x-y)}(\xi + \alpha^2 - \Delta)^{-1}(x, y),$$

one has, letting $R_\lambda = (\lambda + \Lambda_0)^{-1}$,

$$(6.3) \qquad\qquad \|R_\lambda\|_{1 \to 1} \leq \frac{c_1(\theta_\delta)}{|\lambda + (\delta - 1)\alpha^2|}$$

$\left(\lambda = \lambda_0 + \xi, \ \lambda_0 > 0, \ \xi \in \Sigma(\theta_\delta), \ 0 < \theta_\delta < \theta\right)$. Set

$$(6.4) \qquad Gf := g * f, \ g(x) = |x|^{1-d} e^{-a|x|},$$

where $2a = 2a_{\lambda,\alpha} := \mathrm{Re}\sqrt{\lambda + \delta\alpha^2} - |\alpha|$. Using (6.2) and the inequality

$a_{\lambda,\alpha} \geq c_0 \cdot (\sqrt{|\lambda|} + |\alpha|)$ with $c_0 = c_0(\delta, \theta_\delta)$, $\lambda = \lambda_0 + \xi$, $\xi \in \Sigma(\theta_\delta)$, one

gets two crucial estimates

(6.5)

$$|\nabla R_\lambda f| \leq c_2 G|f|, \quad |R_\lambda f| \leq \frac{c_3}{a_{\lambda,\alpha}} G|f|, \quad c_i = c_i(\delta, \theta_\delta), \quad i = 2, 3.$$

Setting $\gamma_0 = b \cdot \nabla R_\lambda$, $\gamma_1 = -\alpha \cdot b R_\lambda$, one has

$$\|\gamma_0\|_{1\to1} \leq c_2 \|G|b|\|_\infty, \quad \|\gamma_1\|_{1\to1} \leq c_4 \|G|b|\|_\infty, \quad c_4 = c_1/c_0.$$

It follows from (6.3) and the definition of G that if $b \in K_{d+1}$ and $\lambda = \lambda_0 + \xi$, $\xi \in \Sigma(\theta_\delta)$, with large enough $\lambda_0 > 0$, then $(c_2 + c_4)\|G|b|\|_\infty < 1/2$ and

$$\|R_\lambda(1 + \gamma_0 - \gamma_1)^{-1}\|_{1\to1} \leq 2c_1/|\lambda + (\delta - 1)\alpha^2|.$$

The latter means that $H_1(b, \alpha)$ generates a holomorphic semigroup.

Next, for $\lambda = \lambda_0 + \xi$ as above, define

$$\Gamma_0 = R_\lambda^{1/4} b \cdot \nabla R_\lambda^{3/4},$$

$$\Gamma_1 = -R_\lambda^{1/4} \alpha \cdot b R_\lambda^{3/4} \text{ and}$$

$$\tilde{G}f = \tilde{g} * f,$$

$$\tilde{g}(x) = |x|^{\frac{1}{2}-d} e^{-a|x|}$$

where a is defined by (6.4). One has

$$|\nabla R_\lambda^{3/4} f| \leq c\tilde{G}|f|, \quad |R_\lambda^{1/4} b \cdot \nabla R_\lambda^{3/4} f| \leq c\tilde{G}(|b|\tilde{G}|f|)$$

so that

$$\|\Gamma_0\|_{2\to 2} \le c(\lambda_0),$$

$$\|\Gamma_1\|_{2\to 2} \le |\alpha|\|R_\lambda^{1/2}\|_{2\to 2} \left\| R_\lambda^{1/4} \frac{\alpha}{|\alpha|} \cdot b R_\lambda^{1/4} \right\|_{2\to 2}$$

$$\le c \left\| R_\lambda^{1/4} \frac{\alpha}{|\alpha|} \cdot b R_\lambda^{1/4} \right\|_{2\to 2} \le \tilde{c}(\lambda_0)$$

where $c(\lambda_0)$ and $\tilde{c}(\lambda_0)$ tend to zero as $\lambda_0 \to \infty$ uniformly on $\alpha \in \mathbb{R}^d$.

Let $\tilde{R}_\lambda := \left(\lambda + H_1(b,\alpha)\right)^{-1}$. Then the estimates above lead to the equality

$$\tilde{R}_\lambda f = R_\lambda^{3/4}(1 + \Gamma_0 - \Gamma_1)^{-1} R_\lambda^{1/4} f, \qquad f \in L^1 \cap L^2$$

which allows one to get the bound $\|\tilde{R}_\lambda\|_{2\to 2} \le C(\lambda_0)/|\lambda|$ and claim that $F^t = e^{-tH_1(b,\alpha)}$ is holomorphic as a map from L^2 into L^2, and since F^t is holomorphic on L^1, as a map from L^r into L^r for all $r \in [1, 2]$, by interpolation. Using the embeddings $R_\lambda^{1/4} \colon L^p \to L^2$, $R_\lambda^{3/4} \colon L^2 \to L^q$, $p = 2d/d+1$, $q = 2d/d - 3$ yields the bound

$$(6.6) \qquad \|F^t\|_{p\to q} \le ct^{-\frac{d}{2}\left(\frac{1}{p}-\frac{1}{q}\right)} e^{ct}, \qquad t > 0, \ \alpha \in \mathbb{R}^d,$$

for $p = 2d/d + 1$, $q = 2d/d - 3$.

To claim that (6.6) holds for $p = 1$, $q = \infty$, we need only to prove the bound

$$(6.7) \qquad \|F^t f\|_\infty \le ce^{ct}\|f\|_\infty, \qquad f \in L^1 \cap L^\infty, \ t \ge 0, \ \alpha \in \mathbb{R}^d.$$

To do this let us set

$$P_\lambda := (\lambda + \Lambda_0 + b \cdot \nabla)^{-1}, \qquad \lambda = \lambda_0 + \xi, \ \xi \in \Sigma(\theta_\delta),$$

and estimate $\|P_\lambda\|_{\infty\to\infty}$. Writing down the second K. Neumann series for P_λ as an operator in L^1: $P_\lambda = R_\lambda - R_\lambda b \cdot \nabla R_\lambda + R_\lambda b \cdot \nabla R_\lambda b \cdot \nabla R_\lambda - \ldots$ and using (6.5), one has, for $f \in L^1 \cap L^\infty$, $\|f\|_\infty = 1$,

$$|R_\lambda \underbrace{b \cdot \nabla R_\lambda b \cdot \nabla \ldots R_\lambda}_{n\text{-times}} b \cdot \nabla R_\lambda f|$$

$$\leq \frac{c_3}{a_{\lambda,\alpha}} G(\underbrace{|b| \| \nabla R_\lambda b \cdot \nabla \ldots R_\lambda b \cdot \nabla R_\lambda f|}_{n\text{-times}})$$

$$\leq \frac{c_3}{a_{\lambda,\alpha}} \| \underbrace{\nabla R_\lambda b \cdot \nabla \ldots R_\lambda b \cdot \nabla R_\lambda f}_{n-1\text{-times}} \|_\infty G|b|$$

$$\leq \frac{c_3 c_2}{a_{\lambda,\alpha}} \| G| \underbrace{b \cdot \nabla \ldots R_\lambda b \cdot \nabla R_\lambda f}_{n-1\text{-times}} \|_\infty G|b|$$

$$\leq \frac{c_3 c_2}{a_{\lambda,\alpha}} \| \underbrace{\nabla R_\lambda b \cdot \nabla \ldots R_\lambda b \cdot \nabla R_\lambda f}_{n-1\text{-times}} \|_\infty \| G|b| \|_\infty^2$$

$$\leq \frac{c_3 c_2^2}{a_{\lambda,\alpha}} \| G(\underbrace{|b| \| \nabla R_\lambda b \cdot \nabla \ldots R_\lambda b \cdot \nabla R_\lambda f|}_{n-2\text{-times}}) \|_\infty \| G|b| \|_\infty^2$$

$$\leq \frac{c_3}{a_{\lambda,\alpha}} (c_2 \| G|b| \|_\infty)^n \| G|f| \|_\infty$$

$$\leq \frac{c_3}{a_{\lambda,\alpha}} \| G1 \|_\infty (c_2 \| G|b| \|_\infty)^n$$

so that

(6.8)

$$\|P_\lambda f\|_\infty \leq \|R_\lambda f\|_\infty + \frac{c_3}{a_{\lambda,\alpha}} \| G1 \|_\infty (1 - c_2 \| G|b| \|_\infty)^{-1} c_2 \| G|b| \|_\infty.$$

Since $\|G1\|_\infty \leq c_5/a_{\lambda,\alpha}$ and $1/a_{\lambda,\alpha} \leq c_6/\sqrt{|\lambda|}$, one has $\|P_\lambda f\|_\infty \leq \frac{c_7}{|\lambda|}$. Next, let $b_n = b1_{\{|b|\leq n\}}$, $n = 1, 2, \ldots$. Letting $P_\lambda(b) = P_\lambda$ and taking $b_n \cdot \alpha f$ with $\|f\|_\infty = 1$ instead of f in (6.8), one has

$$\|P_\lambda(b_n) b_n \cdot \alpha f\|_\infty$$

$$\leq \|R_\lambda b_n \cdot \alpha f\|_\infty + \frac{c_3}{a_{\lambda,\alpha}} \| G|b_n \cdot \alpha| \|_\infty$$

$$(1 - c_2 \| G|b_n| \|_\infty)^{-1} c_2 \| G|b_n| \|_\infty.$$

Since $\|G|b_n|\|_\infty \leq \|B|b|\|_\infty$, $\|G|b_n \cdot \alpha|\|_\infty \leq |\alpha|\|G|b|\|_\infty$, $\|R_\lambda b_n \cdot \alpha\|_\infty \leq \frac{|\alpha|c_3}{a_{\lambda,\alpha}}\|G|b_n|\|_\infty$, one has

(6.9)

$$\|P_\lambda(b_n)b_n \cdot \alpha f\|_\infty \leq \frac{|\alpha|c_3}{c_2 a_{\lambda,\alpha}}X\big(1+X(1-X)^{-1}\big) \leq c_8 X\big(1+X(1-X)^{-1}\big)$$

where $X = c_2\|G|b|\|_\infty$.

Let $C_\infty(\mathbb{R}^d) = \{f \in C_b(\mathbb{R}^d);\ \lim_{|x|\to\infty} f(x) = 0\}$ and $\|W\|$ denote the operator norm of $W\colon C_\infty(\mathbb{R}^d) \to C_\infty(\mathbb{R}^d)$. If λ_0 is large enough then, by (6.9), $\|P_\lambda(b_n)b_n \cdot \alpha\| \leq 1/2$ so that $\widetilde{R}_\lambda(b_n) = \big(1 - P_\lambda(b_n)b_n \cdot \alpha\big)^{-1} P_\lambda(b_n)$ (in the sense of bounded operators in $C_\infty(\mathbb{R}^d)$) and

(6.10) $$\|\widetilde{R}_\lambda(b_n)\| \leq 2\|P_\lambda(b_n)\| \leq \frac{2c_7}{|\lambda|}.$$

Thus we have proved that the semigroup $F^t(b_n)$ is holomorphic on $C_\infty(\mathbb{R}^d)$ and satisfies the bound

(6.11) $$\|F^t(b_n)f\|_\infty \leq c e^{ct}\|f\|_\infty, \qquad f \in L^1 \cap C_\infty,\ t \geq 0.$$

Since $\widetilde{R}_\lambda = R_\lambda(1 + \gamma_0 - \gamma_1)^{-1}$, one has $h := (b \cdot \nabla - b \cdot \alpha)\widetilde{R}_\lambda f \in L^1$ $(f \in L^1)$. By (6.10) $\|\widetilde{R}_\lambda(b_n)f - \widetilde{R}_\lambda(b)f\|_1 \leq \frac{2c_7}{|\lambda|}\|(1 - 1_{\{|b|\leq n\}})h\|_1$ so that $\widetilde{R}_\lambda(b_n) \to \widetilde{R}_\lambda$ strongly in L^1 as $n \to \infty$. Thus, due to (6.11), the proof of (6.7) is completed.

Next, due to $|b_n| \in L^\infty$, the semigroup $\exp[-tH_1(b_n)]$ admits a Gaussian upper bound. Using this fact, one can easily state the equality

$$S_\alpha \exp\big[-t\big(\delta\alpha^2 + H_1(b_n)\big)\big]S_{-\alpha} = \exp[-tH_1(b_n, \alpha)].$$

Since $\exp[-tH_1(b_n)] \rightarrow \exp[-tH_1(b)]$ and $\exp[-tH_1(b_n, \alpha)] \rightarrow \exp[-tH_1(b, \alpha)]$ strongly in L^1, one gets the equality

$$S_\alpha \exp\left[-t\left(\delta\alpha^2 + H_1(b)\right)\right]S_{-\alpha}f = F^t f, \qquad f \in L^1_{\text{com}}.$$

Thus, by (6.6) and (6.7),

$$\|S_\alpha e^{-tH_1(b)} S_{-\alpha} 1_\Omega\|_{1\to\infty} \leq e^{\delta\alpha^2 t} \|F^t\|_{1\to\infty} \leq ct^{-d/2} e^{t(c+\delta\alpha^2)}$$

where $\Omega = \{x; |x| \leq \rho\}$, for *any* finite ρ. This implies the Gaussian upper bound for the heat kernel of $e^{-tH_1(b)}$. The proof of Theorem 0.3 is completed.

Remark 6.2. It is an open questions whether or not $e^{-tH_1(b)}$ admits a Gaussian lower bound for *any* $b \in K_{d+1}$. Under the additional assumption "$b^2 \in K_d$" (LGB) can be easily proved using Theorem 0.3 and the Cameron-Martin-Kac formula. Finally, we note that the following conditions on b: $b^2 \in L^1 + L^\infty$, $b^2 \leq \beta(-\Delta) + c(\beta)$, for some $\beta < 1 \wedge 4/(d-2)^2$ guarantee (global) Hölder continuity of the solutions to the equation $(\omega - \Delta + b \cdot \nabla)u = f$, $\omega > c(\beta)$, $f \in L^2 \cap L^\infty$ (see [KS]). The latter indicates the unnaturalness of "$b^2 \in K_d$".

Appendix.   Form bounded perturbations

We will be assuming that the hypothesis $(\mathbf{H_1})$ is fulfilled and $b \cdot a^{-1} \cdot b \in PK_\beta(A)$ for some $\beta < 1$. As it has been mentioned in the Introduction the KLMN theorem allows one to construct the quasi m-sectorial operator $\Lambda = A \dotplus b \cdot \nabla$ in $L^2(\mathbb{R}^d)$ associated with the closure of the form $t(f,g) = E(f,g) + \langle \nabla \bar{f} \cdot bg \rangle$, $D(t) = D(E)$.

Theorem A.1. $(e^{-t\Lambda}, t \geq 0)$ is a positivity preserving L^∞-contraction C_0-semigroup. By means of $e^{-t\Lambda_p} :=$ $\left[e^{-t\Lambda}|[L^2 \cap L^p]\right]^{\sim}_{L^p \to L^p}$ a family of C_0-semigroups on L^p is defined for all $p \in [2/2 - \sqrt{\beta}, \infty[$, and

$$(A.1) \quad \|e^{-t\Lambda_p}\|_{p \to p} \leq e^{c(\beta)t/(p-1)}, \qquad 2/2 - \sqrt{\beta} \leq p < \infty, \ t \geq 0,$$

$$(A.2) \qquad \|e^{-t\Lambda_p}\|_{p \to q} \leq c e^{c(\beta)t/p\sqrt{\beta}} t^{-d(q-p)/2pq},$$

$$2/2 - \sqrt{\beta} < p < q \leq \infty, \ t > 0.$$

Before proving Theorem A.1, we recall the following well-known criterion for semigroups to be positivity preserving and L^∞-contraction.

Lemma A.2. (See e.g. [LS].) Let $(e^{-tT}, \ t \geq 0)$ be a C_0-semigroup in $L^2(M, \mu)$. Consider the following conditions:

(i) $e^{-tT}[L^2_+] \subset L^2_+$.

(ii) $(f \in L^2, \ |f| \leq 1) \Rightarrow (|e^{-tT}f| \leq 1 \ \mu \ a.e.)$.

(iii) $\langle Tf, \ f \vee 0 \rangle \geq 0, \ f \in D(T) \cap \mathrm{Re}\,L^2$.

(iv) $\mathrm{Re}\langle Tf, f - f_\wedge \rangle \geq 0, \ f \in D(T)$ where $f_\wedge := (|f| \wedge 1)\mathrm{sgn}\,f$.

Then (i)+(ii) $\Leftrightarrow$ (iii)+(iv).

The Proof of Theorem A.1. Put $T = \Lambda + r(\beta)$, $r(\beta) = \frac{1}{4}c(\beta)t(\beta)$, $t(\beta) = \frac{2}{1+\sqrt{1-\beta}}$. Let $f \in D(T)$. Then $\bar{f} \in D(T)$, $\overline{Tf} = T\bar{f}$ and hence $e^{-tT}[\mathrm{Re}\,L^2] \subset \mathrm{Re}\,L^2$. Moreover T satisfies (iii), (iv). Let us show for example how (iv) can be verified. Since $f \in D(T) = D(\Lambda) \subset D(A^{1/2})$, $f_\wedge \in D(A^{1/2})$ and since $D(A^{1/2}) \subset H^1(\mathbb{R}^d)$, $f_\wedge \in H^1(\mathbb{R}^d)$

and

$$\nabla(f - f_\wedge) = \left[\left(1 - \frac{1}{|f|}\right)\nabla f + \frac{f}{|f|^2}\nabla|f|\right]1_{|f|>1}.$$

Therefore

$$\langle Tf, f - f_\wedge\rangle = \left\langle 1_{|f|>1}\left(1 - \frac{1}{|f|}\right)\nabla\bar{f}\cdot a\cdot\nabla f\right\rangle$$

$$+ \left\langle 1_{|f|>1}\frac{f}{|f|}\nabla\bar{f}\cdot\frac{a}{|f|}\cdot\nabla|f|\right\rangle$$

$$+ \left\langle 1_{|f|>1}\frac{f}{|f|}\nabla\bar{f}\cdot b(|f| - 1)\right\rangle + r(\beta)\langle 1_{|f|>1}(|f| - 1)|f|\rangle$$

and

$$\mathrm{Re}\langle Tf, f - f_\wedge\rangle = \langle 1_{|f|>1}\nabla|f|\cdot a\cdot\nabla|f|\rangle + \langle 1_{|f|>1}\nabla|f|\cdot b(|f| - 1)\rangle$$

$$+ r(\beta)\langle 1_{|f|>1}(|f| - 1)\rangle.$$

Now,

$$|\langle 1_{|f|>1}\nabla|f|\cdot b(|f| - 1)\rangle| \le \frac{\varepsilon}{2}\langle\nabla\phi\cdot a\cdot\nabla\phi\rangle + \frac{1}{2\varepsilon}\langle\phi b\cdot a^{-1}\cdot b\phi\rangle, \qquad \varepsilon > 0$$

where $\phi = 1_{|f|>1}(|f| - 1) \in D(A^{1/2})$.

By the assumption $b\cdot a^{-1}\cdot b \in PK_\beta(A)$,

$$|\langle 1_{|f|>1}\nabla|f|\cdot b(|f| - 1)\rangle| = |\langle\nabla\phi\cdot b\phi\rangle|$$

$$\le \left(\frac{\varepsilon}{2} + \frac{\beta}{2\varepsilon}\right)\langle\nabla\phi\cdot a\cdot\nabla\phi\rangle + \frac{c(\beta)}{2\varepsilon}\|\phi\|_2^2$$

so that

$$\mathrm{Re}\langle Tf, f - f_\wedge\rangle \ge \left(1 - \frac{1}{2}\left(\varepsilon + \frac{\beta}{\varepsilon}\right)\right)\|A^{1/2}\phi\|_2^2$$

$$+ \frac{c(\beta)}{2}\left(\frac{t(\beta)}{2} - \frac{1}{\varepsilon}\right)\|\phi\|_2^2.$$

Putting $\varepsilon = \frac{2}{t(\beta)} = 1 + \sqrt{1 - \beta}$, we arrive at $\mathrm{Re}\langle Tf, f - f_\wedge\rangle \ge 0$.

Thus, due to Lemma A.2, $(e^{-t\Lambda}, t \ge 0)$ is positivity preserving and

$\|e^{-t\Lambda}f\|_\infty \le e^{r(\beta)t}\|f\|_\infty$, $f \in L^2 \cap L^\infty$. Let us prove the inequality $\|e^{-t\Lambda}f\|_\infty \le \|f\|_\infty$, $f \in L^2 \cap L^\infty$. Set $f_t = e^{-t\Lambda}f$, $f \in L^2 \cap L^\infty$. As shown above, $f_t \in D(A^{1/2}) \cap L^\infty$, $t > 0$, and using therefore the characterization of the Dirichlet forms [Fu, p. 5], we conclude that $f_t|f_t|^{p-2}$ and $v := f_t|f_t|^{p-2/2}$ belong to $D(A^{1/2}) \cap L^\infty$ for $p \ge 2$. Multiplying both sides of the equation $-\frac{d}{dt}f_t = \Lambda f_t$ by $f_t|f_t|^{p-2}$ and integrating over space, we obtain, after taking the real parts,

$$-\frac{1}{p}\frac{d}{dt}\|v\|_2^2 \ge 4\frac{p-1}{p^2}\langle \nabla|v| \cdot a \cdot \nabla|v|\rangle + \frac{2}{p}\langle \nabla|v| \cdot b|v|\rangle.$$

Applying the assumption $b \cdot a^{-1} \cdot b \in PK_\beta(A)$ to the term $\langle \nabla|v| \cdot b|v|\rangle$ yields

$$-\frac{d}{dt}\|v\|_2^2 \ge \left(4\frac{p-1}{p} - \varepsilon - \frac{\beta}{\varepsilon}\right)\langle \nabla|v| \cdot a \cdot \nabla|v|\rangle - \frac{c(\beta)}{\varepsilon}\|v\|_2^2$$

$$\ge -\frac{c(\beta)p}{2|p-1|}\|v\|_2^2 \left(\varepsilon = 2\frac{p-2}{p} + \sqrt{\left(2\frac{p-1}{p}\right)^2 - \beta}\right).$$

Thus

$$\|f_t\|_p \le e^{c(\beta)t/2(p-1)}\|f\|_p.$$

Since the mapping $p \to \|g\|_p$ ($g \in L^p \cap L^\infty$) is continuous, one has $\|f_t\|_\infty \le \|f\|_\infty$.

Let us now prove (A.1). It is clear that we need to prove (A.1) only for all $f \in L^1 \cap L^\infty$, $f \ge 0$, and $p = 2/2 - \sqrt{\beta}$. Set $v_\varepsilon = f_t(\varepsilon + f_t)^{p-2/2}$, $\varepsilon > 0$. One has

$$(A.3) \quad -\frac{d}{dt}\|v_\varepsilon\|_2^2 = p\langle \Lambda f_t, f_t(\varepsilon+f_t)^{p-2}\rangle + \varepsilon(2-p)\langle \Lambda f_t, f_t(\varepsilon+f_t)^{p-3}\rangle.$$

Since $f_t(\varepsilon + f_t)^{p-k} \in D(A^{1/2})$, $k = 2, 3, \ldots$, one has, letting $I_k = \langle (\varepsilon + f_t)^{p-k} \nabla f_t \cdot a \cdot \nabla f_t \rangle$,

$$p \langle \Lambda f_t, f_t(\varepsilon + f_t)^{p-2} \rangle \geq \varepsilon (2 - p) I_3 - \varepsilon^2 (2 - p)^2 \frac{p-1}{2p} I_4$$

$$- \frac{c(\beta)}{2} \frac{p}{p-1} \|v_\varepsilon\|_2^2 \,,$$

$$\varepsilon (2 - p) \langle \Lambda f_t, f_t(\varepsilon + f_t)^{p-3} \rangle \geq \varepsilon (2 - p) \left(p - 2 - v_1 - \frac{p^2}{16} \frac{\beta}{v_1} \right) I_3$$

$$+ \varepsilon^2 (2 - p)(3 - p) \left[1 - \frac{p-1}{8} \frac{\beta}{v_1} \right] I_4 - \varepsilon^3 (2 - p)(3 - p)^2 \frac{\beta}{16 v_1} I_5$$

$$- \frac{c(\beta)}{4 v_1} (2 - p) \|v_\varepsilon\|_2^2$$

where $v_1 = \frac{p-1}{2}$, $\frac{\beta}{v_1} = 8 \frac{p-1}{p^2}$.

Substituting these estimates in (A.3) yields the inequality

$$-\frac{d}{dt} \|v_\varepsilon\|_2^2 \geq -c(\beta) \frac{p}{p-1} \|v_\varepsilon\|_2^2 - \frac{d}{dt} \|v_\varepsilon\|_2^2 \geq -c(\beta) \frac{p}{p-1} \|v_\varepsilon\|_2^2$$

or

$$\|v_\varepsilon\|_2^2 \leq e^{c(\beta) \frac{p}{p-1} t} \|f^2 (\varepsilon + f)^{p-2}\|_1 \,.$$

Passing here to $\varepsilon \to 0$ we arrive at (A.1).

Define r by $\frac{1}{p} = \frac{1}{2} + \frac{1}{r}$ and use Hölder's inequality and (A.1) to obtain the estimate $\|(e^{-t\Lambda} - 1)f\|_p^2 \leq \|(e^{-t\Lambda} - 1)f\|_2 c(\beta, p) \|f\|_r$, which shows that $e^{-t\Lambda_p}$ is a C_0-semigroup for $p = 2/2 - \sqrt{\beta}$ and hence for all required p.

Since Λ^* satisfies an inequality of Nash, one has $\|f\|_\infty \leq c e^{\omega t} t^{-d/4} \|f\|_2$ by (A.1) and duality. (A.2) now follows by extrapolation

Theorem A.2. *Let* $\Lambda = \Lambda(b)$, $b_n = b1_n$, *where* 1_n *denotes the indica-*
tor of the set $\{x \in \mathbb{R}^d;\ b \cdot a^{-1} \cdot b(x) \le n\}$. *Then* $s - L^p - \lim_n e^{-t\Lambda_p(b_n)} =$
$e^{-t\Lambda_p}$ *uniformly on* $t \in [0,1]$ *for all* $p \in [2/2 - \sqrt{\beta}, \infty[$.

Proof. Due to Theorem A.1, it suffices to treat the case $p = 2$.
The latter follows from the estimate based upon the second resolvent
equation:

$$\left\|\left(\mu + \Lambda(b_n)\right)^{-1} f - (\mu + \Lambda)^{-1} f\right\|_2$$

$$= \left\|\left(\mu + \Lambda(b_n)\right)^{-1}(1 - 1_n)b \cdot \nabla(\mu + \Lambda)^{-1} f\right\|_2$$

$$\le \left\|\left(\mu + \Lambda(b_n)\right)^{-1}\sqrt{b \cdot a^{-1} \cdot b}\right\|_{2\to 2}\left\|(1 - 1_n)A^{1/2}(\mu + \Lambda)^{-1} f\right\|_2$$

$$\le c\|(1 - 1_n)h\|_2$$

where $\mu > c(\beta)$, $h = A^{1/2}(\mu + \Lambda)^{-1} f$, $f \in L^2$.

Remark A.3. *Although it may be of very little practical value, we*
note that the bound $r(t,x,y) \le ct^{-d/2}e^{\omega t}$ *of Theorem 0.5, 4) can be*
specified as follows.

Under the conditions of Theorem 0.5 assume that there is a $\phi \in$
$L^\infty_{\mathrm{loc}}(\mathbb{R}^d)$ with

$$\nabla\phi \cdot a \cdot \nabla\phi \le 1 + Z, \qquad Z \in L^q(\mathbb{R}^d),\ q > d/2,\ d \ge 3.$$

Fix any $\mu \in \left]0, \frac{1}{2}\left(1 - \frac{d}{2q}\right)\right[$. Then there are finite constants $\omega, c > 1$
such that

$$r(t,x,y) \le ct^{-d/2}e^{\omega t - Y(t,x,y)} \qquad (t > 0,\ \text{a.e. } x, y \in \mathbb{R}^d)$$

where $Y(t,x,y) = \left(|\phi(x) - \phi(y)|^{1/\mu}/ct\right)^{\mu/1-\mu}$.

Acknowledgement. The work on these notes began when the author was a guest at the Max-Planck Institute of Mathematics, Potsdam. It is a pleasure to express here gratitude to his host Michael Demuth.

References

[A]   D.G. Aronson, Non-negataive solutions of linear parabolic equations, *Ann. Sci. Norm. Sup. Pisa* **22** (1968) 607–694.

[D]   E.B. Davies, *Heat kernels and spectral theory,* Cambridge Tracts in Mathematics **92** (Cambridge University Press, 1989).

[Fa] E.B. Fabes, Gaussian upper bounds on fundamental solutions of parabolic equations; the method of Nash, in *Dirichlet Forms, Lectures Notes in Math.* **1563** (Springer-Verlag, Berlin, 1993) 1–20.

[FS] E.B. Fabes and D.W. Stroock, A new proof of Moser's parabolic Harnack inequality using the old ideas of Nash, *Arch. Rational Mech. Anal.* **96** (1986) 327–338.

[Fu] M. Fukushima, *Dirichlet Forms and Markov Processes,* North Holland, 1980.

[Ka] T. Kato, *Perturbation Theory for Linear Operators,* Springer-Verlag, 1980.

[KS] V.F. Kovalenko and Yu.A. Semenov, C_0-semigroups in $L^p(\mathbb{R}^d)$ and $C_\infty(\mathbb{R}^d)$ spaces generated by $\Delta + b \cdot \nabla$, *Theory Probab. Appl.* **35** (1990) 443–453.

[Kr] N.V. Krulov, *Nonlinear Second Order Elliptic and Parabolic Equations,* Moscow, Nauka, 1971.

[L]   E.M. Landis, *Second Order Equations of Elliptic and Parabolic Type,* Moscow, Nauka, 1971.

[LS] V.A. Liskevich and Yu.A. Semenov, Some problems on Markov semigroups, in *Advances in Partial Diff. Equations,* eds. B.-W. Schulze, M. Demuth, E. Schrohe (1996) **99**, 163–217.

[M]   I. Miyadera, On perturbation for semigroups of linear operators, *Tôhuku Math. J.* **18** (1966) 299–310, and *School of Education, Waseda Univ.* **21** (1971) 21–24.

[N]   J. Nash, Continuity of solutions of parabolic and elliptic equations *Amer. Math. J.* **80** (1958) 931–954.

[NS]   J.R. Norris and D.W. Stroock, Estimates on the fundamental solution to heat flows with uniformly elliptic coefficients, *Proc. London Math. Soc.* (3) **62** (1991) 373–402.

[S]   D.W. Stroock, Diffusion semigroups corresponding to uniformly elliptic divergence form operators, *Lecture Notes in Math.* **1321** (Springer-Verlag, Berlin, 1988) 316–347.

[VSC]   N.Th. Varopoulos, L. Saloff-Coste and T. Coulhon, *Analysis and Geometry on Groups,* Cambridge Tracts in Mathematics **100** (Cambridge Univ. Press, 1992).

[V]   J. Voigt, On the perturbation theory for strongly continuous semigroups, *Math. Ann.* **229** (1977) 163–171.

Department of Mathematics, University of Toronto, Toronto, Ontario
M5S 3G3, Canada

Selected Titles in This Series